NanoScience and Technology

NANOSCIENCE AND TECHNOLOGY

Series Editors:
P. Avouris   B. Bhushan   D. Bimberg   K. von Klitzing   H. Sakaki   R. Wiesendanger

The series NanoScience and Technology is focused on the fascinating nano-world, mesoscopic physics, analysis with atomic resolution, nano and quantum-effect devices, nanomechanics and atomic-scale processes. All the basic aspects and technology-oriented developments in this emerging discipline are covered by comprehensive and timely books. The series constitutes a survey of the relevant special topics, which are presented by leading experts in the field. These books will appeal to researchers, engineers, and advanced students.

Please view available titles in *NanoScience and Technology* on series homepage
http://www.springer.com/series/3705/

Peter Michler

Editor

Single Semiconductor Quantum Dots

With 191 Figures

 Springer

Professor Dr. Peter Michler
Universität Stuttgart, Institut für Halbleiteroptik und Funktionelle Grenzflächen
Allmandring 3, 70569 Stuttgart, Germany
E-mail: p.michler@ihfg.uni-stuttgart.de

Series Editors:

Professor Dr. Phaedon Avouris
IBM Research Division
Nanometer Scale Science & Technology
Thomas J. Watson Research Center
P.O. Box 218
Yorktown Heights, NY 10598, USA

Professor Dr. Bharat Bhushan
Ohio State University
Nanotribology Laboratory
for Information Storage
and MEMS/NEMS (NLIM)
Suite 255, Ackerman Road 650
Columbus, Ohio 43210, USA

Professor Dr. Dieter Bimberg
TU Berlin, Fakutät Mathematik/
Naturwissenchaften
Institut für Festkörperphyisk
Hardenbergstr. 36
10623 Berlin, Germany

Professor Dr., Dres. h.c. Klaus von Klitzing
Max-Planck-Institut
für Festkörperforschung
Heisenbergstr. 1
70569 Stuttgart, Germany

Professor Hiroyuki Sakaki
University of Tokyo
Institute of Industrial Science
4-6-1 Komaba, Meguro-ku
Tokyo 153-8505, Japan

Professor Dr. Roland Wiesendanger
Institut für Angewandte Physik
Universität Hamburg
Jungiusstr. 11
20355 Hamburg, Germany

NanoScience and Technology   ISSN 1434-4904

ISBN 978-3-540-87445-4   e-ISBN 978-3-540-87446-1

Library of Congress Control Number: 2008943972

© Springer-Verlag Berlin Heidelberg 2009

This work is subject to copyright. All rights are reserved, whether the whole or part of the material is concerned, specifically the rights of translation, reprinting, reuse of illustrations, recitation, broadcasting, reproduction on microfilm or in any other way, and storage in data banks. Duplication of this publication or parts thereof is permitted only under the provisions of the German Copyright Law of September 9, 1965, in its current version, and permission for use must always be obtained from Springer. Violations are liable to prosecution under the German Copyright Law.

The use of general descriptive names, registered names, trademarks, etc. in this publication does not imply, even in the absence of a specific statement, that such names are exempt from the relevant protective laws and regulations and therefore free for general use.

Typesetting: Data prepared by SPi using a Springer TEX macro package
Cover concept: eStudio Calamar Steinen

SPIN: 12188024   57/3180/SPi
Printed on acid-free paper

9 8 7 6 5 4 3 2 1

springer.com

To my wife Silke and my sons Jan, Dennis, and Tim

Preface

Worldwide, many researchers are fascinated from the rich physics of semiconductor quantum dots (QDs) and their high potential for applications in photonics and quantum information technology. QDs are nanometer-sized three-dimensional structures which confine electrons and holes in dimensions of their corresponding De Broglie wavelength. As a result, the energy levels are quantized and for that reason they are also often referred as artificial atoms. Epitaxially grown QDs which are the subject of this book are embedded in a solid state semiconductor matrix and their size, shape, composition, and location can be tailored to a large extent by modern growth techniques. In QDs, excitations can involve more than a single carrier and interaction among the carriers modify or even dominate the emission properties. Therefore, a simple two-level description is only appropriate under certain well defined experimental conditions. Tremendous progress has been obtained in understanding their electronic, optical and spin properties mainly by performing single dot spectroscopy and using appropriate theoretical models.

Spectacular achievements within the last years include the generation of triggered polarization-entangled photon pairs, reaching the strong-coupling regime of interaction between a single quantum dot and a photonic cavity, controlling cavity reflectivity with a single quantum dot, coherent optical manipulation of single electron spins in quantum dots, and controlling the quantum coupling in quantum dot molecules. The progress has been tightly linked to the improvements in growth and nanoprocessing which has allowed the fabrication of micro and nanocavities with high quality factors and ultralow mode volumes, and to produce quantum dots with radiative efficiencies close to unity. The full advantage of their superior properties can be utilized in the future if a scalable deterministic technology for the spatial and spectral matching of the quantum dot with respect to the device structure can be realized. This includes, e.g., the positioning of the QD with respect to a cavity structure and a macroscopic periphery. This task is still a challenge although recent progress has been made on growing quantum dots on predefined positions or by forming cavities around selected quantum dots.

Exciting applications are compact and robust single and entangled photon sources with ultra-high repetition rates, quantum storage devices, and basic building blocks for, e.g., spin-based quantum information implementations. In addition, high density quantum dot systems are appealing for classical optoelectronic applications such as low-threshold lasers, ultra-fast amplifiers and modulators, and sensitive detectors.

The book is organized as follows: Readers will find an introduction into a microscopic theory to describe luminescence and lasing from semiconductor quantum dots in the contribution from *Christopher Gies, Jan Wiersig, and Frank Jahnke* (Chap. 1). Especially, the first- and second-order correlation functions to characterize the coherence properties and the photon statistics in microcavity lasers with high spontaneous emission coupling into the laser mode are discussed in detail. Special emphasis is placed on the differences between quantum dots and atoms.

Armando Rastelli, Suwit Kiravittaya, and *Oliver Schmidt* give an overview on the different methods to fabricate optically active quantum dots (Chap. 2). Bottom-up methods based on self-assembled growth, top-down lithographic techniques, and a combination of them are discussed in detail in their contribution. They show a promising route to fabricate QDs with well defined spatial and spectral properties required for scalable devices.

The fascinating optical, electronic, and magnetic properties of wide bandgap semiconductor single quantum dots based on II–VI and group III-nitrides are presented by *Gerd Bacher* and *Tilmar Kümmell*. Coherent state control, stimulated biexciton emission in individual quantum dots, and room temperature electroluminescence is achieved and superradiance in quantum dot ensembles is also discussed. They further show that a single spin state of an individual magnetic atom in a solid state matrix can be addressed.

Chapters 4 and 5 deal with electron and nuclear spin effects in quantum dots. *Manfred Bayer, Alex Greilich,* and *Dmitri R. Yakovlev* discuss the coherent spin dynamics of electrons confined in semiconductor quantum dots. To study their spin dynamic circular polarized laser excitation is used to orient the spins and their subsequent coherent precession about an external magnetic field is detected by a Faraday rotation technique. They demonstrate that a spin ensemble can be synchronized by and with a periodic train of laser pulses. In this way a locking of several electron spin precession modes is achieved. *Patrick Maletinsky* and *Atac Imamoglu* review optical investigations of nuclear spin effects in individual, self-assembled QDs. The coupled electron–nuclear spin system is studied by optically induced dynamical nuclear spin polarization (DNSP) in detail. Especially, time-resolved measurements of DNSP, both in low and in high external magnetic fields are presented and the dominant nuclear spin relaxation mechanisms are identified.

The topic of Chaps. 6 and 7 is on nonclassical light generation. The contribution on quantum dot single-photon sources is written by *myself* where I briefly recall basic concepts of the quantum optical properties of QDs and review the recent progress in this rapidly evolving and fascinating field. New generations of electrically driven single-photon LEDs, new developments on coherent state preparation and single-photon emission in the strong coupling regime are reviewed. The remaining challenges for practical single-photon sources are also discussed. *Andrew Shields, R. Mark Stevenson, and Robert J. Young* describe recent progress in generating pairs of polarization-entangled photons from the biexciton–exciton cascade in single QDs. The biexciton emission is analyzed in the general case of finite fine-structure splitting in the intermediate exciton state of the cascade. A model to describe the factors limiting the fidelity is presented.

Cavity electrodynamics experiments with single dots in high quality micropillar cavities and photonic crystal cavities are discussed in Chaps. 8 and 9. Important aspects in the growth and patterning of quantum dot–micropillar cavities are addressed by *Stefan Reitzenstein and Alfred Forchel*. Especially, the demonstration of the quantum nature in a strongly coupled quantum dot–micropillar system as well as coherent photonic coupling of QDs mediated by the strong light field in high quality micropillar cavities are presented. *Dirk Englund, Andrei Faraon, Ilya Fushman, Bryan Ellis, and Jelena Vučković* discuss quantum dot-embedded photonic crystal devices for classical and quantum information processing. This includes high-speed, low-power lasing dynamics and carrier-induced switching in photonic crystals for classical applications. For quantum information applications, weak and strong coupling regimes are demonstrated and a newly developed technique for coherent dipole access in cavity (CODAC) is introduced.

Chapter 10 is dealing with the physics of coupled quantum dot structures. *Matthew F. Doty, Michael Scheibner, Allan S. Bracker, and Daniel Gammmon* review experimentally measured spectra of coupled quantum dots and explain the interactions that give rise to the spin fine structure. They discuss the formation of molecular states through tunnel coupling of electron and holes and explain resonant changes in the single-spin g factor for holes in the molecular states of coupled quantum dots.

The last chapter of this book reports on applications of single photon sources based on semiconductor QDs to quantum information processing. *Matthias Scholz, Thomas Aichele, and Oliver Benson* give a brief review of the quantum optical properties of quantum dots and introduce free space and fiber-based quantum cryptography experiments. In addition, a first demonstration along linear optics quantum computation is given in their contribution.

Finally, I would like to thank all my colleagues for writing the various chapters and for spending much of their free time for our common book. Very special thanks to my family for their unconditional support and the appreciation for my absence during numerous weekends.

Stuttgart *Peter Michler*
January 2009

Contents

1 Quantum Statistical Properties of the Light Emission from Quantum Dots in Microcavities
C. Gies, J. Wiersig, and F. Jahnke 1

2 Growth and Control of Optically Active Quantum Dots
Armando Rastelli, Suwit Kiravittaya, and Oliver G. Schmidt 31

3 Optical Properties of Epitaxially Grown Wide Bandgap Single Quantum Dots
Gerd Bacher and Tilmar Kümmell 71

4 Coherent Electron Spin Dynamics in Quantum Dots
Manfred Bayer, Alex Greilich, and Dmitri R. Yakovlev 121

5 Quantum Dot Nuclear Spin Polarization
Patrick Maletinsky and Atac Imamoglu 145

6 Quantum Dot Single-Photon Sources
Peter Michler 185

7 Entangled Photon Generation by Quantum Dots
Andrew J. Shields, R. Mark Stevenson, and Robert J. Young 227

8 Cavity QED in Quantum Dot–Micropillar Cavity Systems
S. Reitzenstein and A. Forchel 267

9 Physics and Applications of Quantum Dots in Photonic Crystals
Dirk Englund, Andrei Faraon, Ilya Fushman, Bryan Ellis, and Jelena Vučković 299

10 Optical Spectroscopy of Spins in Coupled Quantum Dots
*Matthew F. Doty, Michael Scheibner, Allan S. Bracker,
and Daniel Gammon* 331

11 Quantum Information with Quantum Dot Light Sources
M. Scholz, T. Aichele, and O. Benson 367

Index 385

List of Contributors

T. Aichele
Nanooptics, Department of Physics
Humboldt-University Berlin
Hausvogteiplatz 5-7
10117 Berlin, Germany

Gerd Bacher
Werkstoffe der Elektrotechnik and
CeNIDE, Universität Duisburg-Essen
Bismarckstrasse 81
47057 Duisburg, Germany
gerd.bacher@uni-due.de

Manfred Bayer
Experimentelle Physik 2, Technische
Universität Dortmund, 44221
Dortmund, Germany
manfred.bayer@tu-dortmund.de

O. Benson
Nanooptics, Department of Physics
Humboldt-University Berlin
Hausvogteiplatz 5-7
10117 Berlin, Germany
oliver.Benson@physik.
hu-berlin.de

Allan S. Bracker
Naval Research Laboratory
Washington, DC 20375, USA
bracker@bloch.nrl.navy.mil

Matthew F. Doty
University of Delaware, Newark
DE, USA
doty@udel.edu

Bryan Ellis
Ginzton Laboratory, Stanford
University, Stanford, CA 94305
USA

Dirk Englund
Ginzton Laboratory, Stanford
University, Stanford, CA 94305
USA

Andrei Faraon
Ginzton Laboratory, Stanford
University, Stanford, CA 94305
USA

A. Forchel
Technische Physik, Physikalisches
Institut, Julius-Maximilians-
Universität, Würzburg
Am Hubland, 97074 Würzburg
Germany

Ilya Fushman
Ginzton Laboratory, Stanford
University, Stanford, CA 94305
USA

Daniel Gammon
Naval Research Laboratory
Washington, DC 20375, USA
gammon@nrl.navy.mil

C. Gies
Institute for Theoretical Physics
University of Bremen
P.O. Box 330 440
28334 Bremen, Germany

Alex Greilich
Experimentelle Physik 2
Technische Universität Dortmund
44221 Dortmund
Germany
alex.greilich@tu-dortmund.de

Atac Imamoglu
Institute of Quantum Electronics
ETH-Zürich, 8093 Zürich
Switzerland
imamoglu@phys.ethz.ch

F. Jahnke
Institute for Theoretical Physics
University of Bremen
P.O. Box 330 440
28334 Bremen, Germany

Suwit Kiravittaya
Institute for Integrative
Nanosciences, IFW Dresden
Helmholtzstrasse 20
D-01069 Dresden, Germany

Tilmar Kümmell
Werkstoffe der Elektrotechnik
and CeNIDE, Universität
Duisburg-Essen
Bismarckstrasse 81,
47057 Duisburg, Germany

Patrick Maletinsky
Institute of Quantum Electronics
ETH-Zürich, 8093 Zürich
Switzerland
patrickm@phys.ethz.ch

Peter Michler
Institut für Halbleiteroptik und
Funktionelle Grenzflächen
Universität Stuttgart
peter.michler@ihfg.
uni-stuttgart.de

Armando Rastelli
Institute for Integrative
Nanosciences, IFW Dresden
Helmholtzstr. 20
D-01069 Dresden, Germany
and
Max-Planck-Institute
for Solid State Research
Heisenbergstr. 1, D-70569 Stuttgart
Germany
a.rastelli@ifw-dresden.de

S. Reitzenstein
Technische Physik
Physikalisches Institut
Julius-Maximilians-Universität
Würzburg, Am Hubland, 97074
Würzburg, Germany
stephan.reitzenstein@physik.
uni-wuerburg.de

Michael Scheibner
Naval Research Laboratory
Washington, DC 20375, USA
scheibner@bloch.nrl.navy.mil

Oliver G. Schmidt
Institute for Integrative
Nanosciences, IFW Dresden
Helmholtzstrasse 20
D-01069 Dresden, Germany

M. Scholz
Nanooptics, Department of Physics
Humboldt-University Berlin
Hausvogteiplatz 5-7, 10117 Berlin
Germany

Andrew J. Shields
Toshiba Research Europe Ltd., 208
Cambridge Science Park, Milton
Road, Cambridge
CB4 0GZ, UK
andrew.shields@crl.
toshiba.co.uk

R. Mark Stevenson
Toshiba Research Europe Ltd.
208 Cambridge Science Park
Milton Road, Cambridge
CB4 0GZ, UK

Jelena Vučković
Ginzton Laboratory
Stanford University
Stanford, CA 94305
USA

J. Wiersig
Institut für Theoretische Physik
Otto-von-Guericke
Universität Magdeburg
P.O. Box 4120
39016 Magdeburg
Germany

Dmitri R. Yakovlev
Experimentelle Physik 2
Technische Universität Dortmund
44221 Dortmund
Germany
dmitri.yakovlev@tu-dortmund.de

Robert J. Young
Toshiba Research Europe Ltd.
208 Cambridge Science Park
Milton Road, Cambridge
CB4 0GZ
UK

1

Quantum Statistical Properties of the Light Emission from Quantum Dots in Microcavities

C. Gies, J. Wiersig, and F. Jahnke

Summary. A microscopic theory is presented and applied to describe luminescence and lasing from semiconductor quantum dots. Special emphasis is placed on the differences between quantum dots and atoms. We calculate the first- and second-order correlation functions to characterize the coherence properties and the photon statistics in current state-of-the-art microcavity lasers with high spontaneous emission coupling into the laser mode. To gain a deeper understanding of the derived laser theory and of the differences to atomic descriptions, a discussion of quantum-optical models is performed and placed in context to the semiconductor theory.

1.1 Introduction

Due to advances in growth and processing methods, together with ground breaking experimental achievements, the field of semiconductor quantum-dots is more rapidly evolving than ever. Part of this evolution is the need for appropriate theoretical models that encompass the experimental developments described in this book. The three-dimensional confinement of the carrier wave functions on a nanometer scale leads to a discrete part of the single-particle density of states. Semiconductor quantum dots (QDs) are therefore often viewed as "artificial atoms," and in the description of the emission properties of QDs models and approximations are frequently used that are well suited for atomic systems, but which have to be reconsidered when applied to semiconductor systems. In QDs, excitations can involve more than a single carrier. The resulting description of the interaction with the quantized light field is different when atomic systems with single-electron excitations and QDs with multiple excited carriers are compared, even though the elementary interaction processes remain the same. A central goal of our investigations is to reveal the influence of multiple excited carriers on the emission properties of QDs. In addition to the modification of the light-matter interaction itself, it is also well-known from various semiconductor systems that the interaction among the excited carriers leads to effects – many of them discussed throughout this book – that can modify or even dominate the emission properties and, hence, should be included in semiconductor models. Pauli-blocking of states, the Coulomb interaction of excited carriers, their interaction with phonons, and a variety of resulting effects like energy renormalizations, contributions of new quasi-particles, or interaction-induced dephasing have been intensively studied also in QD systems. In atomic systems, interaction-induced effects are typically of minor importance and the single-particle excitations are subject to scattering and dephasing that is usually described via constant rates.

The placement of QDs in optical microcavities allows to tailor the coupling to the electromagnetic field, as is described in Chaps. 8 and 9. Very small mode volumes in combination with a high-quality mode enhance the spontaneous emission rate into that mode, a phenomenon known as Purcell effect [1–3]. While the effect can be employed for various applications like single-photon emitters (see, e.g., Chaps. 6 and 9) or LEDs with improved efficiency, we focus in the following on laser applications.

Due to the small size of the resonator, the high quality modes are usually spectrally well separated so that the QD ensemble can be considered to couple mainly to a single cavity mode. Still, a continuum of leaky modes provides dissipation channels via spontaneous emission. These modes, as well as the more weakly-coupled cavity modes, provide the non-lasing modes. A central parameter in lasers is the spontaneous emission coupling factor β that determines the fraction of the total spontaneous emission (SE) coupled into the laser mode:

$$\beta = \frac{\text{SE rate into laser mode}}{\text{total SE rate}}.$$

The Purcell effect can be used to enhance the spontaneous emission into the laser mode and to suppress spontaneous emission into non-lasing modes. For β approaching unity, the intensity jump in the input/output curve gradually disappears, which has lead to the discussion of a thresholdless laser [4,5]. Latest advances in the growth and design of semiconductor-QD microcavity lasers have now attained the regime of β-values close to unity experimentally [6–10]. To characterize these systems one needs to study the coherence properties of the emitted light and its statistical properties as function of the pump rate. Following Glauber, the quantum states of light can be characterized in terms of photon correlation functions [11]. Coherence properties are reflected by the correlation function of first order,

$$g^{(1)}(\tau) = \frac{\langle b^\dagger(t)b(t+\tau)\rangle}{\langle b^\dagger(t)b(t)\rangle}. \tag{1.1}$$

In the stationary regime this quantity depends only on the delay time τ but not on the time t. Its decay in τ is determined by the coherence time of the emitted light. Here $b^\dagger$ and b are the creation and annihilation operators for photons in the laser mode. Information about the statistical properties of the emitted light can be obtained from the correlation function of second order at zero delay time

$$g^{(2)}(\tau=0) = \frac{\langle n^2 \rangle - \langle n \rangle}{\langle n \rangle^2} = \frac{\langle b^\dagger b^\dagger bb \rangle}{\langle b^\dagger b \rangle^2}, \tag{1.2}$$

where $n = b^\dagger b$ is the photon number operator for the laser mode. The function $g^{(2)}(\tau=0)$ reflects the possibility of the correlated emission of two photons at the same time. For the characterization of the light field, this quantity is of central importance throughout this book. Note that in other chapters a and $a^\dagger$ are used for the photon operators, which is the common notation in quantum optics. Often discussed are the limiting cases of light emission from a thermal, coherent, and single-photon source. Thermal light is characterized by an enhanced probability that two photons are emitted at the same time (bunching), reflected in a value of $g^{(2)}(0) = 2$. For coherent light emission with Poisson statistics one finds $g^{(2)}(0) = 1$. An ideal single-photon emitter shows antibunching with $g^{(2)}(0) = 0$. The full photon statistics corresponding to these limiting cases is discussed in Chap. 6.

This first chapter is concerned with the emission properties of microcavity lasers with QDs as active material. The considered QD and atomic models are introduced in Sect. 1.2. In Sect. 1.3 we discuss several well-established quantum optical models, namely the rate equations, a master equation approach and the Liouville/von-Neumann equation. These models are placed in relation to each other and to the equation-of-motion approach that we use in Sects. 1.4 and 1.5 to develop a microscopic semiconductor model to describe the interaction of the QD emitters with the laser and non-lasing modes. This model allows us to access the statistical properties of the light emission, described in terms of the second-order photon correlation function (1.2). The calculation of first-order correlation function (1.1) and the coherence time is the topic of Sect. 1.6.

1.2 Quantum Dots and Atoms

In QDs a three-dimensional confinement of carriers leads to localized states both for conduction- and valence-band carriers, and discrete interband transition energies between these so-called shells. In the frequently used self-assembled QDs the discrete states appear energetically close to a quasi-continuum of delocalized states that corresponds to the two-dimensional motion of carriers in a wetting layer (WL). A sketch of the corresponding energy levels of conduction and valence band states in the vicinity of the optical band gap is shown in Fig. 1.1.

The finite height of the confinement potential restricts the number of confined shells. As discussed in Chap. 2 one can control the number of shells and their separation in energy by variation of the growth parameters. For self-assembled QDs one usually finds a strong confinement in growth direction and weaker confinement in the WL plane. Because of the strong confinement only the energetically lowest state is important for the motion in

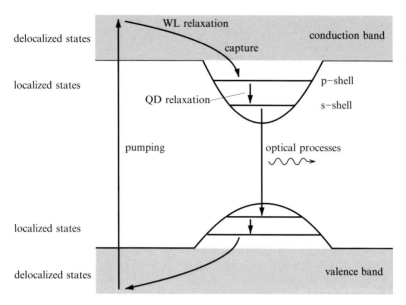

Fig. 1.1. Schematic representation of energy levels in a quantum dot (QD) with two shells for carriers in the conduction and valence band, respectively. The quasi-continuum of the wetting layer (WL) is shown as *shaded areas*

the growth direction, while for the in-plane problem one finds in general several bound states. Moreover, in the lens-shaped QDs the in-plane rotational symmetry leads to an angular-momentum degeneracy in addition to the spin degeneracy of the weakly confined states. In the following, we consider QDs where the confinement leads to two shells for conduction- and valence-band carriers. Then one s-shell and two p-shell states are available for each spin-subsystem. The unexcited state corresponds to filled valence-band states and empty conduction-band states.

To investigate the optical properties of QDs, the system may be off-resonantly excited by an optical pulse. The excitation creates carriers in the barrier-, WL-, or higher QD-states. Fast scattering (relaxation) into the lower QD states, as illustrated in Fig. 1.1, is facilitated by carrier–carrier and carrier–phonon interaction [12,13]. At low temperatures and at low to moderate carrier densities, the carriers populate solely the QD states. Then the WL states are mainly important for carrier-scattering processes if the excitation involves the quasi-continuum. In the following we are interested in the recombination dynamics due to carrier–photon interaction involving the localized QD states. We study a laser system based on QDs in optical microresonators, where we assume that the energetically lowest interband transition between the s-shells of the conduction and valence bands is dipole allowed and in resonance with the fundamental cavity mode.

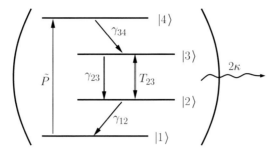

Fig. 1.2. Diagram of the atomic four-level laser system. With the pump rate $\tilde{P}$ an electron is excited from the ground state $|1\rangle$ into the pump level $|4\rangle$. The rates γ_{34} and γ_{12} determine the scattering into and from the laser levels $|2\rangle$ and $|3\rangle$, which are coupled to the laser mode via the transition amplitude T_{23}. Emission into non-lasing modes is described by the rate γ_{23}. The cavity mode is coupled to a reservoir of modes outside the cavity, giving rise to cavity losses 2κ

For the discussed QD system, it is tempting to exploit the similarities to an atomic four-level system displayed in Fig. 1.2.[1] The important difference to the semiconductor case is that the unexcited QD system contains filled valence-band states. Inevitably more than a single electron can be excited into the conduction-band states.

To obtain the closest possible analogy between the QD system of Fig. 1.1 and the atomic four-level system of Fig. 1.2, we assume that only two confined QD shells for electrons and holes, respectively, (denoted by s and p) are relevant and that the optical pump process is resonant with the p-shells.

The presence of more than one electron in the interacting discrete level system complicates the direct application of well-established models from quantum optics and requires their reformulation. At this point one might argue, that the elementary optical excitations are excitons, which consist of conduction-band electrons and the corresponding valence-band vacancies bound by the Coulomb interaction. However, the exciton operators obey bosonic commutators only in the absence of other carriers. While a formulation in a pure bosonic exciton picture clearly oversimplifies the problem, the inclusion of corrections is highly nontrivial and as such can be viewed as a reformulation of the fermionic electron–hole picture employed below.

---

[1] At this point we would like to add a general remark about the interpretation of the atomic model. The considered transitions take place between different configurations of the system, which may involve occupations of several states, again involving more than one carrier. However, in quantum-optical models transitions between these configurations are treated analogously to transitions between electronic states. Therefore, we speak of an atomic system with a single electron that can occupy any of the available levels, keeping in mind that while this is a simplified picture, formally the single-electron model is equivalent to the configuration picture.

To summarize this central point, the presence of multiple excited carriers and their mutual interaction is usually not an issue in atomic quantum optics, but it is of key importance in semiconductor optics. Other complications in the application of well established models in quantum optics to QD systems are of more practical nature and involve the applicability of approximations commonly applied to atomic systems. Examples are reabsorption of photons in connection with incomplete inversion of the laser transition, saturation effects, or interaction-induced dephasing.

In the following we start from standard models in quantum optics, that describe the interaction of the quantized light field with atomic systems characterized by "single-electron excitations" in the sense we have discussed above. These approaches are related in a second step to our semiconductor models that account for the excitation of multiple carriers and their interaction.

1.3 Light-Matter Interaction in Atomic Systems

1.3.1 Liouville/von-Neumann Equation

In quantum optics the most general approach to atom-photon interaction is to solve the Liouville/von-Neumann equation for the full density matrix ρ of the coupled carrier–photon system. From the density-matrix elements, arbitrary single-time expectation values can be obtained by calculating the trace. However, the full solution of the von-Neumann equation is only feasible for small systems (few atoms). Most familiar is the direct solution of the density-matrix equations for the Jaynes–Cummings Hamiltonian, that describes the interaction of a two-level system with the quantized light field. This model can be extended to a one-atom laser while still permitting a direct solution of the density-matrix equations, as discussed in [14]. For this purpose, the two-level system for the resonant optical interaction with the laser mode is augmented by two additional levels in order to facilitate the pump process and a rapid depletion of the lower laser level, as shown in Fig. 1.2. The model still contains a single electron that, if in the ground state $|1\rangle$, can be excited into the pump level $|4\rangle$. Transition processes between the levels and the corresponding dephasing of the off-diagonal density matrix elements are introduced by coupling the atom to reservoirs. The interaction with these reservoirs is treated in the so-called Born–Markov approximation [15] and is contained in the resulting transition rates $\tilde{P}$ and γ_{ij} indicated in Fig. 1.2. Of particular importance are fast transitions into the upper laser level $|3\rangle$ at the rate γ_{34} and rapid processes emptying the lower laser level $|2\rangle$ at the rate γ_{12}. For the transitions between the laser levels $|3\rangle$ and $|2\rangle$, the spontaneous emission into non-lasing modes, described by the rate γ_{23}, competes with the coupling to the laser mode via the transition amplitude T_{23}. By coupling also the laser mode to a reservoir as discussed above, cavity losses with a rate 2κ are introduced.

The time evolution of the full density matrix ρ can be obtained from its commutator with the Jaynes–Cummings Hamiltonian H_{JC} and the dissipative and pump processes are described by Lindblad terms $L\rho$ (that account for the above discussed coupling to the reservoirs [15]) according to

$$\frac{\mathrm{d}}{\mathrm{d}t}\rho = -\frac{i}{\hbar}[H_{\mathrm{JC}},\rho] + L\rho\,. \qquad (1.3)$$

The model can be extended to N atoms interacting with the quantized light field. Via the off-diagonal density-matrix elements coupling between different atoms is included. This coupling is connected to superradiance/superfluorescence. A direct numerical solution or the application of quantum Monte-Carlo techniques, however, is presently restricted to a small numbers of atoms.

1.3.2 Master Equations

A considerable simplification of the theory is possible if one can formulate closed equations for the diagonal density matrix elements, which represent probabilities. One can deduce such a treatment from the Liouville/von-Neumann equation by adiabatically eliminating the off-diagonal density matrix elements. In the adiabatic regime, the dephasing is sufficiently fast to dominate over the dynamics, so that the off-diagonal elements simply follow without delay their sources (incoherent regime). Furthermore, the contributions of the cavity losses to the dephasing of the off-diagonal density matrix elements need to be small [14].

An additional, commonly used approximation consists in neglecting the reabsorption of photons from the cavity mode. When the state $|2\rangle$ in Fig. 1.2 is rapidly depopulated via the scattering process γ_{12}, the reabsorption process is suppressed and the laser transition benefits from maximal inversion for a given pump rate $\tilde{P}$. In this case, the recombination rate is determined by the probability of finding the electron in the upper laser level together with a given number of photons in the cavity mode. A system containing many atoms can then be characterized by the probabilities $p_{n,N}$ of states with N excited atoms and n photons in the laser mode.

Deriving the master equation for $p_{n,N}$ from the Liouville/von-Neumann equation the way we have discussed it is a stringent approach. Often a phenomenological approach based on intuitive arguments is taken that leads to the same results. By considering all relevant processes that can act on a given state of the system, a birth/death model can be formulated where phenomenological transition rates are introduced. This is illustrated in Fig. 1.3 where λ is the cavity loss rate, P the pump rate and β is the spontaneous emission coupling factor. The arrows describe processes that can act on the state $p_{n,N}$, see the figure caption. As an example we consider the spontaneous emission into non-lasing modes, indicated as dashed lines in the vertical direction since

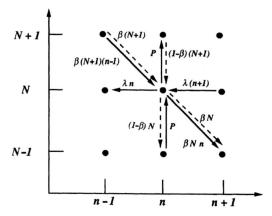

Fig. 1.3. Schematic representation of the relevant processes in the birth/death model. On the *vertical axis* the number of excited emitters N is shown, on the *horizontal axis* the number of photons n in the cavity. In the diagram, each *dot* stands for a matrix element of the diagonal density matrix $\rho_{n,N}$ and represents a state with the corresponding number of excited atoms and photons. Processes acting on a state with probability $\rho_{n,N}$ are: cavity losses, keeping the number of excited atoms unaltered and thus represented as *horizontal line*; pump process, keeping the number of photons unaltered and thus represented as *vertical line*; spontaneous and stimulated emission, where the number of excited emitters is reduced and the number of photons in increased, thus represented as *diagonal lines*. The rates related to each process are noted with the *arrows*. Figure reprinted with permission from P.R. Rice and H.J. Carmichael [5], copyright 1994 by the American Physical Society

the number of photons in the cavity remains unaltered by this process. The contribution to the equation of motion is read off the schematic as

$$\frac{d}{dt} p_{n,N}\bigg|_{nl} = -\frac{(1-\beta)}{\tau_{sp}} \left[N p_{n,N} - (N+1) p_{n,N+1} \right]. \tag{1.4}$$

The prefactor $(1-\beta)/\tau_{\rm sp}$ describes the rate of spontaneous emission into non-lasing modes, assuring that there is no emission at all into these modes in the case $\beta = 1$. There are two possibilities involving a state with N excited emitters, either a decay ("death") of the very same state (after emission a state with $N-1$ excited emitters is left), or its "birth" by the decay of a state with $N+1$ excited emitters. The full master equation with all contributions displayed in Fig. 1.3 can be found in [5].

Despite the involved approximations when going to the master equation from the Liouville/von-Neumann equation, the possibility remains to calculate the photon statistics $p_n = \sum_N p_{n,N}$, and hence $\langle n \rangle, \langle n^2 \rangle, \ldots$, as well as arbitrary photon correlation functions, where all operators have equal time arguments.

1.3.3 Rate Equations

From the Liouville/von-Neumann or the master equation, it is also possible to derive equations of motion for expectation values like the mean number of excited atoms and photons in the laser mode, $\langle N \rangle$ and $\langle n \rangle$, respectively. These equations couple to higher-order expectation values and as such form an infinite hierarchy of equations. The truncation of this hierarchy requires factorization approximations.

In the most simple form of such a truncation correlations between atoms and photons are completely neglected, i.e., $\langle nN \rangle = \langle n \rangle \langle N \rangle$. Since these terms appear only in the contributions representing stimulated emission, the treatment of these processes then corresponds to a semi-classical picture. Spontaneous emission, on the other hand, is not influenced by this approximation. Then one obtains from the equations of motion for $\langle n \rangle \equiv n$ and $\langle N \rangle \equiv N$ the well-established laser rate equations

$$\frac{d}{dt} n = -2\kappa n + \frac{\beta}{\tau_{\rm sp}} [1+n]\, N, \tag{1.5}$$

$$\frac{d}{dt} N = -\frac{\beta}{\tau_{\rm sp}} nN - \frac{1}{\tau_{\rm sp}} N + \tilde{P}. \tag{1.6}$$

The photon population is determined by the interplay of the cavity loss rate 2κ and the photon generation due to spontaneous processes $\propto N$ and stimulated processes $\propto nN$. The dynamics of the number of excited emitters N follows from the interplay of the carrier recombination and the pump rate $\tilde{P}$. The former comprises stimulated emission into the laser mode $\propto \beta/\tau_{\rm sp} = 1/\tau_{\rm l}$, and spontaneous emission $\propto 1/\tau_{\rm sp}$ into all available modes. For a detailed discussion of the laser rate equations, see, e.g., [5, 16].

In the following we discuss the stationary solution of the rate equations for a constant pump rate $\tilde{P}$. Results for the input/output curves and various values of the β-factor are shown in Fig. 1.4. We choose parameters that are typical for present microcavity lasers: $\tau_{\rm sp} = 50\,\rm ps$ and $\kappa = 20\,\rm \mu eV$. The corresponding cavity lifetime is about 17 ps, yielding a Q-factor of roughly 30,000. At the laser threshold that differentiates the regimes of dominant spontaneous and stimulated emission, the photon number exhibits a jump $\propto \beta^{-1}$. In the limit $\beta = 1$ this kink in the input/output curve disappears and the threshold is no longer well defined in this simple picture. Note that the photon statistics obtained from the master equation provides additional information about a possible transition from thermal to coherent light emission. The discussed behavior of the jump in Fig. 1.4 is commonly used to estimate the β-factor from measurements.

The customary form of the master equation and the rate equations have in common that saturation effects do not appear. The reason lies in negligence of reabsorption processes and in the fact that the total number of available

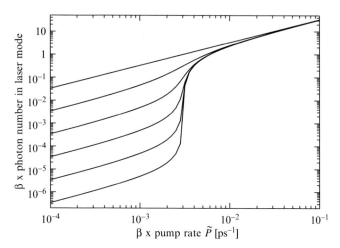

Fig. 1.4. Results from the laser rate equations (1.5) and (1.6) for $\beta = 1$ to 10^{-5} from *top* to *bottom*

atoms is not limited. This is important for the comparison to semiconductor-QD systems, where reabsorption and saturation effects can strongly modify the input/output curves.

It is by no means necessary to truncate the equations of motion at the simplest possible level that, as we have just shown, leads to the rate equations. The truncation can be performed on a higher level in the hierarchy. For the additional terms separate equations of motion have to be formulated. This way, for example, it is possible to obtain insight into the statistical properties of the light field, represented by the photon–photon correlations $\langle n^2 \rangle$, via the equation-of-motion approach in an analogous fashion to the rate equations. A critical issue remains with the inclusion of multiple excited carriers and their interaction. As discussed in Sect. 1.2, several carriers can be excited in semiconductor QDs.

Before the microscopic semiconductor theory is presented in Sect. 1.5, we give an illustrative explanation of how the emission properties are directly influenced by this difference to atomic systems.

1.4 Multiple Excited Carriers in Semiconductor QDs

1.4.1 Electron–Hole Picture

The QD level scheme of Fig. 1.1 with optical pumping into the p-states, fast relaxation of carriers from p to s-states, and resonant interaction with the laser mode via the s-states is translated into the electron–hole-picture in Fig. 1.5. Since for the unexcited system the valence-band states are filled (holes are absent) a richer set of possible states can be realized in comparison to the

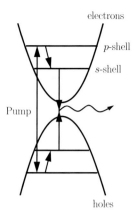

Fig. 1.5. QD laser model with carrier generation in the *p*-shells and the laser transition between the *s*-shells of electrons (*top*) and holes (*bottom*)

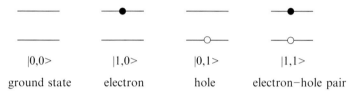

Fig. 1.6. Schematic level diagram of possible electron and hole occupations for the *s*-shell of a quantum dot. The photonic part of the states is of no relevance for the discussion in the text and, therefore, not explicitly shown

single-electron system in Fig. 1.2. For the *s*-states, the possible configurations are displayed in Fig. 1.6. In the second (third) configuration recombination is not possible since the hole (electron) is missing.

In the one-electron systems discussed in Sect. 1.3, the second configuration where electrons are present in both laser levels is impossible. In the third configuration no electrons are present in the *s*-shell at all. In the description based on the full density matrix in Sect. 1.3.1 this is possible if the carrier resides in a *p*-shell state (corresponding either to the state $|1\rangle$ or $|4\rangle$ in the scheme in Fig. 1.2). However, when the scattering processes γ_{12} and γ_{34} in Fig. 1.2 are sufficiently fast, this configuration is unlikely and, correspondingly, left out in the models discussed in Sects. 1.3.2 and 1.3.3.

1.4.2 Theory of QD Recombination

For the following discussion we consider the coupling of carriers in one spin subsystem to the corresponding circular polarization component of the light field. The four *s*-shell states are denoted by $|n_\mathrm{e}, n_\mathrm{h}\rangle$ with $n_\mathrm{e} = 0, 1$ and $n_\mathrm{h} = 0, 1$. For the calculation of the recombination dynamics we focus on the incoherent regime. Then the elements of the density matrix traced over the photonic and p-shell states are given by

$$\rho = P_{00}|0,0\rangle\langle 0,0| + P_{10}|1,0\rangle\langle 1,0| + P_{01}|0,1\rangle\langle 0,1| + P_{11}|1,1\rangle\langle 1,1| \tag{1.7}$$

with the probabilities $0 \leq P_{ij} \leq 1$ and the normalization condition

$$\text{tr}(\rho) = \sum_{i,j=0}^{1} P_{ij} = 1. \tag{1.8}$$

In the formalism of second quantization the states $|n_e, n_h\rangle$ are related to each other by creation and annihilation operators. The fermionic operator e ($e^\dagger$) annihilates (creates) an electron. The corresponding operators for holes are h and $h^\dagger$. For instance, applying the creation operator $e^\dagger$ to the ground state $|0,0\rangle$ yields the one-electron state $|1,0\rangle$. In the same way, the creation operator $h^\dagger$ turns the ground state into the one-hole state $|0,1\rangle$. The formalism of second quantization allows to describe many-particle effects in an elegant way. Relevant quantities can be expressed as expectation values of products of creation and annihilation operators. For example, the luminescence is determined by the dynamics of the photon number $\langle b^\dagger b \rangle$, where the bosonic operator b ($b^\dagger$) annihilates (creates) a photon in a given optical mode. Other relevant quantities are the occupations of electrons $f^e = \langle e^\dagger e \rangle$ and holes $f^h = \langle h^\dagger h \rangle$. The changes of the photon number $\langle b^\dagger b \rangle$ and of the s-shell occupation probabilities $\langle e^\dagger e \rangle$, $\langle h^\dagger h \rangle$ are determined by mixed expectation values $\langle b^\dagger he \rangle$ as discussed in detail in Sect. 1.5. The general form of these equations allows to describe not only the interplay of the different configurations shown in Fig. 1.6, but also the inclusion of many-body Coulomb effects on various levels of refinement.

In the following discussion we are only interested in the spontaneous recombination processes. For the purpose of uncovering essential differences between the single-electron atomic system and the many-electron QD system, we neglect only for the following discussion the effects of Coulomb interaction and use an adiabatic elimination of mixed expectation values $\langle b^\dagger he \rangle$ that represent interband transition amplitudes. With these approximations the decay of carrier population is determined by

$$\frac{d}{dt} f^{(e,h)}\bigg|_{\text{spont}} = -\frac{\langle e^\dagger e\, h^\dagger h \rangle}{\tau}, \tag{1.9}$$

where $1/\tau$ is the rate of spontaneous emission. Even though formula (1.9) is based on a considerable simplification in comparison to the general theory, it reveals the basic difference between QDs and atomic systems with single excited carriers. To see this, we express $\langle e^\dagger e\, h^\dagger h \rangle$ in terms of the four basis states illustrated in Fig. 1.6. We first note that $e|0, n_h\rangle = 0$ as one cannot remove an electron from a state that does not contain one. Equally, $h|n_\mathrm{e}, 0\rangle = 0$. From this follows immediately $e^\dagger e\, h^\dagger h |n_\mathrm{e}, n_\mathrm{h}\rangle \neq 0$ only if $n_\mathrm{e} = n_\mathrm{h} = 1$. Using this relation we find

$$\langle e^\dagger e\, h^\dagger h \rangle = \mathrm{tr}(\rho\, e^\dagger e\, h^\dagger h) = P_{11}. \tag{1.10}$$

The intuitive interpretation is, that the decay of carrier population described in (1.9) is proportional to the probability P_{11} of finding an electron–hole pair, since only in state $|1,1\rangle$ the electron and the hole can recombine via emission of a photon.

For the semiconductor system with multiple excited carriers, the probability of observing an electron–hole pair is in general different from the probability of finding an electron (or a hole). This can be seen by rewriting the occupation probability of electrons as

$$f^\mathrm{e} = \langle e^\dagger e \rangle = \mathrm{tr}(\rho\, e^\dagger e) = P_{10} + P_{11}. \tag{1.11}$$

Along the same lines one obtains for the holes

$$f^\mathrm{h} = P_{01} + P_{11}. \tag{1.12}$$

Comparing (1.11) or (1.12) with (1.10) reveals that the probability of finding an electron–hole pair is smaller than or equal to the probability of finding an electron or a hole, and equal to only if $P_{10} = P_{01} = 0$. In this particular case we can write (1.9) as

$$\left.\frac{\mathrm{d}}{\mathrm{d}t} f^{(e,h)}\right|_\mathrm{spont} = -\frac{f^{(e,h)}}{\tau}. \tag{1.13}$$

Assuming that no other mechanism contributes to the change of the population, it follows an exponential decay with rate $1/\tau$ for the population, which also carries over to the photoluminescence [17]. The conditions $P_{10} = P_{01} = 0$ reduce the four-state system for the electron and hole to a *two-state system* for the electron–hole pair with basis states $|0,0\rangle$ and $|1,1\rangle$. This situation corresponds to *fully correlated carriers*: the absence (presence) of an electron implies the absence (presence) of a hole and vice versa.

In the opposite limiting case of *uncorrelated carriers*, the joint probabilities P_{ij} factorize into the probabilities P_i^e for the electron and P_j^h for the hole, i.e., $P_{ij} = P_i^\mathrm{e} P_j^\mathrm{h}$. Accordingly the two-particle quantity $\langle e^\dagger e\, h^\dagger h \rangle$ factorizes into one-particle quantities

$$\langle e^\dagger e\, h^\dagger h \rangle_\mathrm{HF} = \langle e^\dagger e \rangle \langle h^\dagger h \rangle = f^\mathrm{e} f^\mathrm{h}. \tag{1.14}$$

This is the Hartree–Fock (HF) factorization. Note that polarization-like averages of the form $\langle e^\dagger h^\dagger \rangle$ vanish in the incoherent regime. The product $f^\mathrm{e} f^\mathrm{h}$ is the uncorrelated electron–hole population. Equations (1.10)–(1.12) show that one can interpret (1.14) as the factorization of the probability of finding an electron–hole pair into the product of the individual probabilities of finding an electron and a hole.

Replacing $\langle e^\dagger e\, h^\dagger h\rangle$ in (1.9) by its HF-factorization (1.14) yields

$$\frac{d}{dt} f^{(e,h)}\bigg|_{\text{spont}} = -\frac{f^e f^h}{\tau}. \tag{1.15}$$

From (1.15) it is obvious that the decay of the population f^e is non-exponential, unless f^h is held constant by some mechanism, like background doping. Furthermore, the decay rate depends on the carrier density and is higher for larger population.

In realistic situations the carriers in a semiconductor QD are neither fully correlated nor uncorrelated. To quantify the deviation from the discussed limiting cases, the correlation must explicitly be calculated.

The electron–hole correlation is defined as

$$\delta\langle e^\dagger e\, h^\dagger h\rangle = \langle e^\dagger e\, h^\dagger h\rangle - \langle e^\dagger e\, h^\dagger h\rangle_{\text{HF}} = \langle e^\dagger e\, h^\dagger h\rangle - \langle e^\dagger e\rangle\langle h^\dagger h\rangle. \tag{1.16}$$

The relation to classical correlation functions can be seen more clearly in the following representation

$$\delta\langle e^\dagger e\, h^\dagger h\rangle = \langle (e^\dagger e - \langle e^\dagger e\rangle)\,(h^\dagger h - \langle h^\dagger h\rangle)\rangle. \tag{1.17}$$

While a variance like $\langle (e^\dagger e - \langle e^\dagger e\rangle)^2\rangle$ quantifies the fluctuations around the expectation value of a single quantity, the correlation function (1.17) is a covariance that quantifies correlated fluctuations of two different quantities. It can have positive and negative contributions depending on the relative sign of the two brackets in (1.17). A positive correlation function here means that on average the fluctuations of the electron and hole number around their respective expectation values have the same sign. In the case of a negative correlation function, the fluctuations have mostly opposite signs. We can identify the sources of positive and negative correlations in terms of the basis states $|n_e, n_h\rangle$ using the normalization condition (1.8)

$$\delta\langle e^\dagger e\, h^\dagger h\rangle = \sum_{n_e,n_h=0}^{1} P_{n_e,n_h}(n_e - f^e)(n_h - f^h) = P_{00}P_{11} - P_{10}P_{01}. \tag{1.18}$$

The contribution from P_{00} and P_{11} enters with positive sign: if an electron is absent (present) the hole is absent (present). In contrast, the contribution from P_{10} and P_{01} enters with negative sign: if an electron is present (absent) the hole is absent (present). In other words, positive $\delta\langle e^\dagger e\, h^\dagger h\rangle$ implies that if we detect an electron then it is likely to find also a hole. Negative $\delta\langle e^\dagger e\, h^\dagger h\rangle$ implies that if we detect an electron then it is unlikely to find a hole. Vanishing correlation

$\delta\langle e^\dagger e\, h^\dagger h\rangle$ means that the two events of detecting an electron and detecting a hole are uncorrelated. In this particular case, the HF factorization (1.14) is exact.

1.4.3 QD Luminescence Dynamics

The fundamental behavior of carrier and luminescence dynamics can be visualized by considering a doped and an undoped system of semiconductor QDs. Details of this work can be found in [18]. In the doped case we choose the dopant density such that there is an average occupation of a single electron or hole per dot.

Figure 1.7 shows the time-resolved photoluminescence after excitation into the p-shell. The system is pumped with an equal electron and hole density of $N_\mathrm{e} = N_\mathrm{h} = 0.35N$ for a QD density of $N = 10^{10}\,\mathrm{cm}^{-2}$. In the doped case we assume on average one additional electron per QD, i.e., $N_\mathrm{e} = N_\mathrm{h} + N$ with again $N_\mathrm{h} = 0.35N$. Apart from this difference in the initial conditions both curves are calculated with exactly the same parameters. For the two different cases we see the following: In the situation of undoped QDs, we observe a non-exponential decay in agreement with our discussion. Again we would like to stress that the non-exponential decay cannot be predicted from an atomic theory. In contrast, the doped QDs show an exponential decay, which is faster by about a factor of two compared to the undoped case. This is due to the fact that the temporal change of the electron population relative to the doping level remains small. According to (1.15), a constant electron population f_ν^e leads to an exponential decay of the hole population f_ν^h, and, hence, of the photoluminescence intensity. This example nicely demonstrates the peculiarities of QD systems if thought of as "artificial atoms."

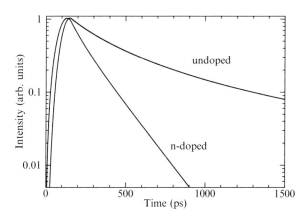

Fig. 1.7. Time-resolved photoluminescence for doped and undoped QDs excited into the p-shell

1.5 Quantum-Dot Microcavity Lasers

1.5.1 Semiconductor Theory

Our starting point is a Hamiltonian that contains the free carrier spectrum and the free electromagnetic field, as well as the light-matter interaction and Coulomb effects. For details we refer to [19]. The fermionic operators c_ν ($c_\nu^\dagger$) annihilate (create) electrons in the one-particle states $|\nu\rangle$ of energy ε_ν^c. The corresponding operators and single-particle energies for valence-band electrons are v_ν ($v_\nu^\dagger$) and ε_ν^v, respectively. The Bose operators b_ξ and $b_\xi^\dagger$ are the equivalents for photons in the mode ξ. We use the fact that polarization-like averages of the form $\langle v_\nu^\dagger c_\nu\rangle$ vanish due to the incoherent carrier generation, and so does the expectation value of the photon operators, $\langle b_\xi\rangle = 0$. Operator averages are obtained by means of the equation-of-motion technique and the cluster expansion method is used to truncate the inherent problem of an infinite number of coupled equations.

Cluster Expansion

The equation of motion of an average of N operators couples to $N+2$ operator averages due to the Coulomb and light-matter interactions. This hierarchy of equations must be truncated in an unambiguous way. One useful approach is the cluster expansion method [20]. As in (1.16), one schematically decomposes a four-operator average into

$$\langle a^\dagger a^\dagger aa\rangle = \langle a^\dagger a\rangle\langle a^\dagger a\rangle + \delta\langle a^\dagger a^\dagger aa\rangle, \tag{1.19}$$

denoting the fermionic carrier operators c, v by a. The first term on the right-hand side is called singlet contribution as it contains only single-particle quantities. For a four-operator average, the singlet factorization corresponds to the Hartree–Fock approximation. The second term is the correlation, which is a doublet contribution. It describes genuine two-particle effects.

The equation of motion for the four-operator (two-particle) correlation couples to averages of six operators, which are schematically factorized according to

$$\langle a^\dagger a^\dagger a^\dagger aaa\rangle = \langle a^\dagger a\rangle\langle a^\dagger a\rangle\langle a^\dagger a\rangle + \langle a^\dagger a\rangle\,\delta\langle a^\dagger a^\dagger aa\rangle + \delta\langle a^\dagger a^\dagger a^\dagger aaa\rangle \tag{1.20}$$

into singlet, singlet-doublet and triplet contributions. Note that all possible combinations of averages and correlations must be taken, meaning that each of the first two terms on the right-hand side may represent several terms of the same order. The triplet contribution, the last term in (1.20), contains only genuine three-particle effects. In this formulation, the truncation of the hierarchy can be consistently performed on all correlation functions of a certain order N, meaning that all correlations involving up to $N-1$ particles are included.

The same scheme can be applied to mixed averages of fermionic carrier and bosonic photon operators. Considering interband transitions, it must be borne in mind that the excitation of one electron is described as the destruction of a valence band carrier and the creation of a conduction band carrier. For the corresponding interaction processes, a photon operator is connected to *two* carrier operators [21, 22].

As an example for possible factorizations, we consider the factorization of the operator average $\langle b^\dagger b b^\dagger v^\dagger c \rangle$ into two doublets, which is obtained after normal order is established, i.e.,

$$\langle b^\dagger b b^\dagger v^\dagger c \rangle \bigg|_{\text{doublet}} = \{\langle b^\dagger v^\dagger c \rangle + \langle b^\dagger b^\dagger b v^\dagger c \rangle\}_{\text{doublet}} \qquad (1.21)$$
$$= \delta\langle b^\dagger v^\dagger c \rangle \left[1 + 2\delta\langle b^\dagger b \rangle\right].$$

In the equation-of-motion approach we pursue, the hierarchy of coupled equations must be extended to the so-called quadruplet-level in order to calculate the photon statistics. This can be seen by expressing (1.2) in terms of correlation functions, $\delta\langle b^\dagger b^\dagger b b \rangle = \langle b^\dagger b^\dagger b b \rangle - 2\langle b^\dagger b \rangle^2$, where the factor of two arises from the two possible realizations for this factorization. Since $\langle b \rangle = \langle b^\dagger \rangle = 0$ for a system without coherent excitation only a factorization into doublets is possible. Then the autocorrelation function can be written in terms of the quadruplet correlation function $\delta\langle b^\dagger b^\dagger b b \rangle$:

$$g^{(2)}(\tau = 0) = 2 + \frac{\delta\langle b^\dagger b^\dagger b b \rangle}{\langle b^\dagger b \rangle^2}. \qquad (1.22)$$

In experiments $g^{(2)}(\tau)$ can be determined, e.g., in a Hanbury–Brown Twiss setup, which is discussed in Chaps. 6 and 11.

Equations of Motion

For the dynamical evolution of the photon number $\langle b_\xi^\dagger b_\xi \rangle$ in the mode ξ and the carrier populations $f_\nu^{\text{e}} = \langle c_\nu^\dagger c_\nu \rangle$, $f_\nu^{\text{h}} = 1 - \langle v_\nu^\dagger v_\nu \rangle$, the contribution of the light matter (LM) interaction in the Heisenberg equations of motion leads to

$$\left(\hbar \frac{\text{d}}{\text{d}t} + 2\kappa_\xi\right) \langle b_\xi^\dagger b_\xi \rangle = 2\,\text{Re} \sum_{\nu'} |g_\xi|^2 \, \langle b_\xi^\dagger v_{\nu'}^\dagger c_{\nu'} \rangle, \qquad (1.23)$$

$$\hbar \frac{\text{d}}{\text{d}t} f_\nu^{\text{e,h}} \bigg|_{\text{LM}} = -2\,\text{Re} \sum_{\xi} |g_\xi|^2 \, \langle b_\xi^\dagger v_\nu^\dagger c_\nu \rangle. \qquad (1.24)$$

Note that we have scaled $\langle b_\xi^\dagger v_\nu^\dagger c_\nu \rangle \to g_\xi \langle b_\xi^\dagger v_\nu^\dagger c_\nu \rangle$ with the light-matter coupling strength g_ξ to have its modulus appear in the above equations. The notation is to be understood as follows: A new quantity $\widetilde{\langle b_\xi^\dagger v_\nu^\dagger c_\nu \rangle} = \langle b_\xi^\dagger v_\nu^\dagger c_\nu \rangle / g_\xi$ is used and the tilde is dropped in the following. In (1.23) we have introduced the loss rate $2\kappa_\xi/\hbar$. The mode index labels cavity as well as leaky modes. For the laser mode ξ_1, the loss rate is directly connected its Q-factor, $Q = \hbar\omega_{\xi_1}/2\kappa_{\xi_1}$. The dynamics of the photon number in a given mode is determined by the photon-assisted polarization $\langle b_\xi^\dagger v_\nu^\dagger c_\nu \rangle$ that describes the expectation value for a correlated event, where a photon in the mode ξ is created in connection with an interband transition of an electron from the conduction to the valence band. The sum over ν involves all possible interband transitions from various QDs, i.e., $\nu = \{\mu, \mathbf{R}\}$ with μ being the shell index and $\mathbf{R}$ the QD position. The dynamics of the carrier population in (1.24) is governed by contributions of photon-assisted polarizations from all possible modes (both lasing and non-lasing modes).

The time evolution of the photon-assisted polarization is given by

$$\left(\hbar\frac{d}{dt} + \kappa_\xi + \Gamma + i\left(\tilde{\varepsilon}_\nu^e + \tilde{\varepsilon}_\nu^h - \hbar\omega_\xi\right)\right) \langle b_\xi^\dagger v_\nu^\dagger c_\nu \rangle$$
$$= f_\nu^e f_\nu^h - \left(1 - f_\nu^e - f_\nu^h\right) \langle b_\xi^\dagger b_\xi \rangle$$
$$+ i\left(1 - f_\nu^e - f_\nu^h\right) \sum_\alpha V_{\nu\alpha\nu\alpha} \langle b_\xi^\dagger v_\alpha^\dagger c_\alpha \rangle$$
$$+ \sum_\alpha C^x_{\alpha\nu\nu\alpha} + \delta\langle b_\xi^\dagger b_\xi c_\nu^\dagger c_\nu \rangle - \delta\langle b_\xi^\dagger b_\xi v_\nu^\dagger v_\nu \rangle \ . \quad (1.25)$$

Here, the free evolution of $\langle b_\xi^\dagger v_\nu^\dagger c_\nu \rangle$ is determined by the detuning of the QD transitions from the optical modes. Hartree–Fock (singlet) contributions of the Coulomb interaction with the Coulomb matrix element $V_{\nu\alpha\nu\alpha}$ lead to the appearance of renormalized energies $\tilde{\varepsilon}_\nu^{e,h}$ and to the interband exchange contribution that couples the photon-assisted polarizations from different states α. The source term of spontaneous emission is described by an expectation value of four carrier operators $\langle c_\alpha^\dagger v_\alpha v_\nu^\dagger c_\nu \rangle$. We have discussed the implications arising from this source term in Sect. 1.4. For uncorrelated carriers, the Hartree–Fock factorization of this source term leads to $f_\nu^e f_\nu^h$, which appears as the first term on the right-hand side of (1.25). Correlations remaining after the factorization are provided by the Coulomb and light-matter interaction between the carriers and are included in $C^x_{\alpha\nu\nu'\alpha} = \delta\langle c_\alpha^\dagger v_{\nu'}^\dagger c_\nu v_\alpha \rangle$.

The stimulated emission/absorption term in the second line appears as the singlet-doublet factorization of the initial operator averages $\langle b_\xi^\dagger b_\xi c_\nu^\dagger c_\nu \rangle - \langle b_\xi^\dagger b_\xi v_\nu^\dagger v_\nu \rangle$. It is proportional to the photon number $\langle b_\xi^\dagger b_\xi \rangle$ in the mode ξ, thus providing feedback due to the photon population in the cavity. The correlations left after the factorization are given by the last two terms in (1.25). These are triplet-level carrier–photon correlations. These and higher order correlations are required for the calculation of the photon statistics in Sect. 1.5.2.

In (1.25) there are two terms acting as a dephasing: Photon dissipation κ_ξ and carrier–carrier and carrier–phonon interaction-induced dephasing. The latter is included via a phenomenological damping constant Γ in connection with transition amplitudes $\propto v_\nu^\dagger c_\nu$. A more rigorous treatment of the dephasing in connection with the scattering rates can be provided by reservoir interaction via Lindblad terms, as discussed in Sect. 1.3.1. The most stringent approach would be the inclusion of the relevant physical processes, such as interaction with phonons, on a microscopic level.

After this general introduction to the theoretical model, we now formulate the laser equations for a coupled QD-microcavity system. We consider QDs with two localized shells for electrons and holes. The dots are embedded in a microcavity that provides one long-lived mode that is in resonance with the QD s-shell emission. Higher cavity modes are assumed to be energetically well separated from this mode, and a continuum of leaky modes and other cavity modes constitute the background of non-lasing modes.

In the following scheme, several assumptions are included, which are justified by possible experimental conditions and which lead to a convenient formulation of the theory. They provide no principle limitations and their use can be circumvented at the cost of more complicated analytical and numerical formulations. We assume that optical processes involving the laser mode (stimulated and spontaneous emission as well as photon reabsorption) are exclusively connected to the s-shell transitions. The carrier generation is assumed to take place in the p-shell, from which down-scattering into the s-shell is treated in a relaxation-time approximation. At the considered low temperatures of around 10 K, up-scattering processes are negligible. Carrier excitation in the WL or the barrier material, in which the QDs are embedded is often used in experiments. While it has the same effect as p-shell excitation for the laser dynamics from the QD s-shells, the density of states of the excited states strongly influences the effects of Pauli blocking.

When using QDs as the active material in microcavities, the coupling strength is determined by the position of the dots and their resonance frequencies. Inhomogeneous broadening is accounted for in the estimation of the number of QDs in resonance with the laser mode, but is not explicitly included in the equations of motion. The number of strongly coupled QDs can be estimated from the dot area density and the overlap of the cavity resonance with the inhomogeneously broadened ensemble of QDs. For the considered micropillar cavities with diameters of a few microns, a small number of tens of QDs can be considered to be in perfect resonance with the cavity mode.

Regarding the influence of the Coulomb interaction, we distinguish between the single-particle renormalizations and carrier correlations on the two-particle level. The renormalizations of the single-particle properties are indirectly included by assuming QDs on resonance with the cavity mode and a certain radiative lifetime in this mode, which is adapted from experiments. From previous calculations for the photoluminescence of QDs in microcavities [17] and emission from QDs in unstructured samples [22] we expect that the influence of carrier correlations on the stationary properties of the laser light emission is small. In current calculations we have therefore neglected these contributions.

Under the discussed conditions we now derive the equations for the semi-conductor laser model. It is the greatest strength of our microscopic model that the accommodation of modifications is straightforward and follows a well-defined manner.

For the resonant s-shell/laser-mode transition, the equation of motion (1.25) for the photon-assisted polarization takes the form

$$\left(\hbar\frac{d}{dt} + \kappa + \Gamma\right)\langle b^\dagger v_s^\dagger c_s\rangle = f_s^e f_s^h - (1 - f_s^e - f_s^h)\langle b^\dagger b\rangle + \delta\langle b^\dagger b c_s^\dagger c_s\rangle - \delta\langle b^\dagger b v_s^\dagger v_s\rangle, \tag{1.26}$$

where, from now on, the index $\xi = \xi_1$ is omitted for the laser mode. In the equation of motion for the photon-assisted polarization of the non-lasing modes, the negligible photon population and the short lifetime of these modes allows for the omission of the feedback term and carrier–photon correlations,

$$\left(\hbar\frac{d}{dt} + \kappa_\xi + \Gamma + i\left(\varepsilon_s^e + \varepsilon_s^h - \hbar\omega_\xi\right)\right)\langle b_\xi^\dagger v_s^\dagger c_s\rangle\bigg|_{\xi\neq\xi_1} = f_s^e f_s^h. \tag{1.27}$$

As a result, (1.27) can be solved in the adiabatic limit and the part $\xi \neq \xi_1$ of the sum in (1.24) can be evaluated, yielding a time constant τ_{nl} for the spontaneous emission into non-lasing modes in a fashion similar to the Weißkopf–Wigner theory [23],

$$\frac{2}{\hbar}\operatorname{Re}\sum_{\xi\neq\xi_1}\frac{|g_\xi|^2}{\kappa_\xi + \Gamma + i\left(\varepsilon_s^e + \varepsilon_s^h - \hbar\omega_\xi\right)} = \frac{1}{\tau_{nl}}. \tag{1.28}$$

In a laser theory, one typically distinguishes between the rate of spontaneous emission into lasing and non-lasing modes, $1/\tau_l$ and $1/\tau_{nl}$, respectively. Both rates add up to the total spontaneous emission rate $1/\tau_{sp}$. Then the spontaneous emission factor is given by

$$\beta = \frac{\frac{1}{\tau_l}}{\frac{1}{\tau_{sp}}} = \frac{\frac{1}{\tau_l}}{\frac{1}{\tau_l} + \frac{1}{\tau_{nl}}} \tag{1.29}$$

and the rate of spontaneous emission into non-lasing modes can be expressed according to

$$\frac{1}{\tau_{nl}} = \frac{1-\beta}{\tau_{sp}}. \tag{1.30}$$

From (1.24) one can now determine the population dynamics in the s-shell. For the spontaneous emission into non-lasing modes, the adiabatic solution of (1.27) is used according to (1.28) and (1.30). Furthermore, we include a transition rate of carriers from the p- to the s-shell in an approximation where only downwards directed scattering is considered, $R_{p\to s}^{e,h} = (1 - f_s^{e,h})f_p^{e,h}/\tau_r^{e,h}$, and $g \equiv g_{\xi_1}$ to obtain

$$\frac{\mathrm{d}}{\mathrm{d}t} f_s^{\mathrm{e,h}} = -\frac{2|g|^2}{\hbar} \mathrm{Re}\, \langle b^\dagger v_s^\dagger c_s \rangle - (1-\beta)\frac{f_s^{\mathrm{e}} f_s^{\mathrm{h}}}{\tau_{\mathrm{sp}}} + R_{p\to s}^{\mathrm{e,h}}. \tag{1.31}$$

Here the first term describes the carrier dynamics due to the interaction with the laser mode, while the second term represents the loss of carriers into non-lasing modes. The blocking factor $1 - f_s^{\mathrm{e,h}}$ in $R_{p\to s}^{\mathrm{e,h}}$ ensures that the populations cannot exceed unity.

The carrier dynamics for the p-shell can be written as

$$\frac{\mathrm{d}}{\mathrm{d}t} f_p^{\mathrm{e,h}} = P(1 - f_p^{\mathrm{e}} - f_p^{\mathrm{h}}) - \frac{f_p^{\mathrm{e}} f_p^{\mathrm{h}}}{\tau_{\mathrm{sp}}^p} - R_{p\to s}^{\mathrm{e,h}}, \tag{1.32}$$

where a carrier generation rate P is included together with the Pauli-blocking factor $(1 - f_p^{\mathrm{e}} - f_p^{\mathrm{h}})$. As the carrier generation takes place in the p-shell of each QD, P is to be understood as a "pump rate per emitter," in contrast to the pump rate $\tilde{P} = N_{\mathrm{emitter}} P$ appearing in (1.6). The second term describes spontaneous recombination of p-shell carriers and the third contribution is the above-discussed carrier relaxation.

Without the carrier–photon correlations $\delta\langle b_\xi^\dagger b_\xi c_\nu^\dagger c_\nu \rangle$ and $\delta\langle b_\xi^\dagger b_\xi v_\nu^\dagger v_\nu \rangle$, the resulting set of equations (1.31) and (1.32), together with (1.23) and (1.26) constitute the basic equations of the semiconductor laser model on the doublet level. They describe the coupled dynamics for the photon number and the carrier population.

1.5.2 Photon Statistics

To access intensity correlations, we must go beyond the doublet level and include all terms consistently up to the quadruplet level in the cluster expansion. Regarding these higher-order correlations, it is helpful to assume that only photons from the laser mode linger long enough to build up correlations.

The photon statistics follows from (1.22). The time evolution of the intensity correlation function $\delta\langle b^\dagger b^\dagger bb \rangle$ is given by

$$\left(\hbar \frac{\mathrm{d}}{\mathrm{d}t} + 4\kappa\right) \delta\langle b^\dagger b^\dagger bb \rangle = 4|g|^2 \,\mathrm{Re} \sum_{\nu'} \delta\langle b^\dagger b^\dagger b v_{\nu'}^\dagger c_{\nu'} \rangle, \tag{1.33}$$

where the sum involves all resonant laser transitions from various QDs. In this equation enters another quadruplet function, which represents a correlation between the photon-assisted polarization and the photon number. For it and the two correlation functions in the last line of (1.26) more equations of motion must be solved, which we do not spell out here. Within reasonable approximations, the system of coupled equations now describing both the photon number and the second-order photon autocorrelation function closes on the quadruplet level.

From the derived equations, it is possible to obtain approximate analytic expressions for the second-order correlation function. Before the numerical results of the semiconductor model are presented, it is instructive to study the two limiting cases of strong and weak pumping. Above the threshold the photon number becomes large, so that the limit $\langle b^\dagger b\rangle/N \gg 1$ can be fulfilled. In this case we obtain $g^{(2)}(0) = 1$, i.e., well above threshold the light is coherent. For the limiting case of weak pumping, stimulated emission and the higher-order correlations $\delta\langle b^\dagger bc_\nu^\dagger c_\nu\rangle$, $\delta\langle b^\dagger bv_\nu^\dagger v_\nu\rangle$ in (1.26) can be neglected due to the lack of a photon population. In the "bad cavity limit" [5], where the cavity loss rate is much larger than the total rate of spontaneous emission into the laser mode, we obtain

$$g^{(2)}(0) = 2 - \frac{2}{N}. \tag{1.34}$$

This is an important result, as it provides the statistics of thermal light in the limit of many QDs, $g^{(2)}(0) = 2$, and in the opposite limit of a single QD it gives the statistics of a single-photon emitter, $g^{(2)}(0) = 0$. The correlation properties and possible realizations of single-photon sources are the topic of Chap. 6.

1.5.3 Numerical Results

We now present numerical solutions of the extended semiconductor laser theory including carrier–photon correlations based on (1.23), (1.26), (1.31)–(1.33) and equations for higher-order correlations not explicitly given here. We consider a typical parameter set: The number of emitters is $\tilde{N} = 20$. The number used in the calculations is increased with decreasing β in order to have the thresholds occur at the same pump rate, i.e., $N = \tilde{N}/\beta$. For the spontaneous emission time enhanced by the Purcell effect we use $\tau_{\rm sp} = 50\,{\rm ps}$, and the cavity damping is $\kappa = 20\,\mu{\rm eV}$. The corresponding cavity lifetime is about 17 ps, yielding a Q-factor of roughly 30,000. The effective relaxation times for electrons and holes are taken to be $\tau_r^e = 1\,{\rm ps}$ and $\tau_r^h = 500\,{\rm fs}$, respectively.

In Fig. 1.8 the autocorrelation function $g^{(2)}(\tau = 0)$ is shown atop the input/output curve for various values of the spontaneous emission coupling β. There are several striking features:

1. The jump of the intensity curve from below to above threshold is no longer determined by $1/\beta$, like in Fig. 1.4 and many examples found in the literature [5, 24, 25] that are obtained from an atomic laser theory.
   The origin of the jump lies in the transition from dominantly spontaneous to stimulated emission in the system. As the non-lasing modes are only fed by spontaneous emission, their effect is that a fraction $1 - \beta$ of the total spontaneous emission is lost for the laser mode. Above the intermediate threshold region, stimulated emission into the laser mode dominates, so that losses into non-lasing modes are irrelevant. In the regime where spontaneous emission dominates, i.e., below the threshold, the output-power is reduced accordingly, visible as a jump in the input/output curve. In atomic systems operating at full inversion, like conventional four-level gas

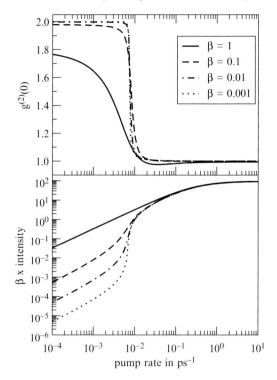

Fig. 1.8. Calculated input/output curve (*lower panel*) and autocorrelation function $g^{(2)}(\tau = 0)$ (*upper panel*) for $\beta = 1, 0.1, 0.01$, and 0.001. The system is excited at a constant pump rate, corresponding to continuous wave excitation

lasers, the height of the resulting jump is truly given by $1/\beta$. The fact that QD-based laser shows different behavior lies in the reabsorption present in the system, i.e., the laser does not operate at full inversion.

This is of particular importance since measurements of the input/output characteristics are often used to experimentally deduce the β-factor according to the predictions of the two-level models. If the atomic $1/\beta$-behavior would be used to extract the β-factors from the curves in Fig. 1.8, one would obtain 0.017 instead of 0.1, 0.0017 instead of 0.01, and 0.00017 instead of 0.001.

2. The usual laser threshold in conventional lasers with low spontaneous emission coupling ($\beta \sim 10^{-6}$) is very abrupt. We see that for larger β values ($\beta > 0.1$), the s-shaped jump in the input/output curve becomes smeared out, and also the drop of the autocorrelation function becomes softer.

3. For small β values, the intensity jump is accompanied by a decrease of the second-order coherence from the Poisson value $g^{(2)}(0) = 2$ for thermal light to $g^{(2)}(0) = 1$ for coherent laser light. For higher spontaneous emission coupling β, $g^{(2)}$ remains smaller than two below the

threshold. Corresponding to the discussion above, this behavior can arise either in the bad cavity regime if the number of emitters is small (compare (1.34)), or if the loss rate from the cavity becomes smaller than the rate of spontaneous emission into the cavity. In the latter case, and this is the reason for the decrease of sub-threshold value of $g^{(2)}(0)$ in Fig. 1.8, a substantial population of photons can build up in the cavity mainly due to spontaneous emission processes, and exhibit a deviation from the signature of incoherent thermal light. Furthermore, for the largest β value an antibunching signature is observed just above the threshold region, where $g^{(2)}(0)$ is smaller than 1.
4. At high pump intensities saturation effects due to Pauli blocking become visible in the input/output curve, effectively limiting the maximum output that can be achieved. The strength of the Pauli blocking depends on the number of available states in the pump levels and the number of QDs in resonance with the laser mode. When pumping into the barrier, where the density of states is larger than for the localized states, saturation effects will be less influential compared to the situation where higher localized states are pumped.

1.6 First-Order Coherence and Two-Time Quantities

Coherence is usually associated with the occurrence of fringes in an interference experiment. Chapter 6 gives details on experiments using a Michelson interferometer, where a quasi-monochromatic beam is divided into two by a beam splitter. By means of a moving mirror, a time delay τ is introduced to one of the beams before they are reunited. Only if this time delay is shorter than the coherence time τ_c, interference fringes can be observed between the two beams. The visibility of the interference fringes is directly described by the first-order correlation function, which we have written in terms of photon operators for the laser mode in (1.1). The loss of coherence carries over to the second-order correlation function $g^{(2)}(\tau)$. No matter if the light is thermal, coherent or exhibits an antibunching signature – the correlation function converges to a value of unity on a timescale of the coherence time. The coherence time can be calculated from the first-order correlation function:

$$\tau_c = \int_{-\infty}^{\infty} |g^{(1)}(\tau)|^2 \mathrm{d}\tau. \tag{1.35}$$

In quantum optics two-time operator averages, like (1.1), are accessed by invoking the quantum regression theorem [15, 26]. The quantum regression theorem in its standard formulation in quantum optics, however, requires the equations of motion to be linear and, therefore, applies to systems where only a single excitation is possible. Due to the unavoidable factorization of the equations of motion in the semiconductor model, the initial linearity of the model is spoiled. With the source-term of spontaneous emission, our equations are nonlinear already on the singlet level, so that the validity of approaches that can be traced back to the quantum regression theorem rely on additional assumptions.

A straightforward approach for the calculation of $g^{(1)}(\tau)$ lies in the equation-of-motion-technique itself, where the time derivative is now taken with respect to the delay time τ [27]. In order to obtain the dynamics of a quantity $F(t, t+\tau)$ with respect to the time difference τ, in a first step the single-time problem is solved for $\tau = 0$. In a second step, the τ-evolution is evaluated according to its own equation of motion. The initial value is given by the $\tau = 0$ result obtained in the first step.

One has to be aware that for operators with different time arguments the commutation relations do not apply, i.e., $\langle b^\dagger(t)b^\dagger(t+\tau)b(t+\tau)b(t)\rangle \neq \langle b^\dagger(t)b^\dagger(t+\tau)b(t)b(t+\tau)\rangle$. Therefore, the number of equations in the resulting hierarchy that needs to be solved quickly scales with the order of the initial two-time expectation value. This method is now demonstrated in the calculation of the first-order coherence function. We use the Hamiltonian and the methods introduced in Sect. 1.5. In order to obtain non-rotating dynamical equations, we introduce

$$G(\tau) = e^{i\omega\tau}\langle b^\dagger(t)b(t+\tau)\rangle. \tag{1.36}$$

The quantity defined in (1.36) obeys the following equation of motion

$$\left(\hbar\frac{d}{d\tau} + \kappa\right)G(\tau) = \sum_\nu g_\nu^* P_\nu(\tau), \tag{1.37}$$

where we have introduced the two-time photon-assisted polarization

$$P_\nu(\tau) = e^{i\omega\tau}\langle b^\dagger(t)v_\nu^\dagger(t+\tau)c_\nu(t+\tau)\rangle. \tag{1.38}$$

We now invoke the same assumptions that we have used in the definition of the laser system in Sect. 1.5.1. In the following we consider N identical QDs that are on resonance with the laser mode. Furthermore, terms coupling different modes are neglected. With $g^{(1)}(\tau)$ being a doublet quantity (two-particle average) in the cluster expansion scheme, we truncate the hierarchy consistently at the same level. With these approximations, we finally arrive at the closed set of equations

$$\left(\hbar\frac{d}{d\tau} + \kappa\right)G(\tau) = g^*P(\tau), \tag{1.39}$$

$$\left(\hbar\frac{d}{d\tau} + \Gamma\right)P(\tau) = gN(f^c - f^v)G(\tau), \tag{1.40}$$

where we have introduced $P(\tau) = \sum_\nu P_\nu(\tau)$. The solution of these equations yields the normalized first-order correlation function

$$|g^{(1)}(\tau)| = \frac{-\gamma_-}{\gamma_+ - \gamma_-}e^{-\gamma_+|\tau|} + \frac{\gamma_+}{\gamma_+ - \gamma_-}e^{-\gamma_-|\tau|} \tag{1.41}$$

with

$$\hbar\gamma_\pm = \frac{\kappa + \Gamma}{2} \pm \sqrt{|g|^2 N(f^c - f^v) + \frac{(\kappa - \Gamma)^2}{4}}\ . \tag{1.42}$$

Thus, on the doublet level the first-order coherence properties are determined by the carrier populations in the lowest confined QD states, which are known from the stationary solutions of the dynamic laser equations. From (1.35) and the equations above, we find for the coherence time

$$\tau_c = \frac{1}{\gamma_+} + \frac{1}{\gamma_-} + \frac{\hbar}{\kappa + \Gamma}\ . \tag{1.43}$$

In Fig. 1.9 the coherence times obtained from (1.42) and (1.43) are shown together with the input/output curves for three different values of the spontaneous emission coupling factor β. The parameters for the $\beta = 0.01$ curve are chosen to meet the characteristics of a real existing micropillar: $N = 500$ QDs, total spontaneous emission time $\tau_{\rm sp} = 80$ ps, homogeneous QD broadening

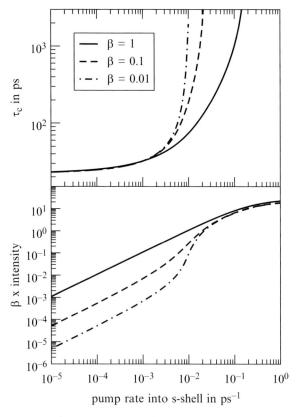

Fig. 1.9. Theoretical i/o curves (*bottom*) and coherence times (*top*) for various values of the β parameter. For $\beta = 0.01$, $\beta = 0.1$ and $\beta = 1$, 500, 50 and 5 QDs have been used, respectively. All other parameters remain unaltered

$\Gamma \approx 200\,\mu\text{eV}$, and cavity losses $2\kappa = 30\,\mu\text{eV}$. For the $\beta = 0.1$ ($\beta = 1$) curve, $N = 50$ (5) QDs was used. As the threshold region is approached, a strong increase in the coherence time is observed. While below threshold the value lies between 20 and 30 ps for all three curves, the coherence time is found to increase slower with increasing pump power in cavities with larger spontaneous emission coupling: At comparable points on the input/output curves, we find that devices with a larger β factor display shorter coherence times. An illustrative explanation is that fluctuations introduced by spontaneous emission processes decrease the coherence in the system, and at higher β values more spontaneous emission is coupled into the laser mode. At the same time, even in the "thresholdless" case of $\beta = 1$, the slower, but nevertheless distinct rise in the coherence time indicates the beginning of the threshold region.

The decay of the first-order correlation function is presented in Fig. 1.10. A clear qualitative change of the decay from a Gaussian-like profile to a more exponential behavior can be observed within the transition regime. This qualitative change of the lineshape can be seen in the analytical solutions of the coherence function. Expanding (1.41) into a Taylor series reveals the Gaussian-like characteristic in the decay of $|g^{(1)}(\tau)|$, as the term linear in τ drops out. Considering the solutions $\gamma_\pm$ at transparency, $f^c - f^v = 0$, we get $\gamma_+ = \Gamma \gg \gamma_- = \kappa$, yielding a decay that is close to exponential $|g^{(1)}(\tau)| = e^{-\gamma_- \tau}$.

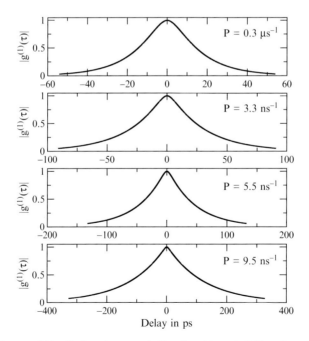

Fig. 1.10. Decay of the first-order correlation function at different excitation powers. A gradual change in the profile from Gaussian-like to a more exponential behavior becomes evident with increasing excitation power P

1.7 Concluding Remarks

We have reviewed the common methods in quantum optics for the computation of laser properties and placed them in relation to each other and to the equation-of-motion approach that our semiconductor theory is based on. As the most fundamental method we have discussed the Liouville/von-Neumann equation that yields the time-evolution of the full density matrix. This model can be reduced to a master equation that describes probabilities, or to the well-known rate equations. Two points are important to notice: Firstly, in both approaches where the density matrix elements are calculated, information about the statistical properties of the light are preserved, whereas in the rate equation formulation this information is lost. And secondly, we have discussed that the rate equations are the most simple possible factorization in a reformulation of the first two approaches in a hierarchy of coupled equations of motion.

In this context we place our semiconductor theory: It is based on an equation-of-motion approach. While rate equations typically describe the carrier system in terms of the expectation value of the number of excited two-level systems, the semiconductor approach is based on population expectation values and transition amplitudes in a semiconductor basis. Also the hierarchy of equations is not truncated on the semiclassical level, but higher-order correlation functions are kept in a consistent manner. This enables us to calculate the coherence properties of the emitted light in terms of the second-order correlation function. As the equation-of-motion method does not require the equations to be linear, Coulomb correlations and other effects specifically important in semiconductor systems can be included in a straightforward way. To give an illustrative example for a deviation from the atomic behavior, we have discussed the source term of the spontaneous emission, which is modified due to the possibility to have more than single excitation in a QD. As a consequence, the carrier decay and the photoluminescence exhibits a non-exponential signature with a density-dependent decay rate. This effect is naturally included in our semiconductor approach. While it is possible to modify existing quantum-optical models to include more than a single excitation per emitter, it has, to the best of our knowledge, not been done so far. In this sense, our approach can be understood as an extension to these models. However, being based on a microscopic Hamiltonian, the full spectrum of semiconductor effects, including Coulomb effects and interaction with phonons, can be considered on a microscopic level.

In order to calculate the first-order coherence properties we have shown how the described model can be used to calculate two-time operator averages. With this and the second-order correlation function, we obtain a consistent overall picture of the laser transition in QD-based microcavity laser devices. At the large spontaneous emission coupling factors β that are currently obtained in state-of-the-art devices, the transition in the input/output characteristics is washed-out and cannot be clearly identified anymore. At the same time, around the threshold region we observe a distinct rise in the coherence time by about two orders of magnitude, a change in the lineshape of the first-order correlation function from Gaussian-like to exponential, and a change in the second-order coherence properties from (close to) the signature of thermal light ($g^{(2)}(0) = 2$) to that of coherent ($g^{(2)}(0) = 1$) light. These results characterize clearly the physics in the threshold region even if a threshold is no longer directly visible in the intensity of the emitted light.

From atomic models it is expected that the output intensity shows a jump by a factor of $1/\beta$ at the threshold. For several reasons this does not apply to semiconductor systems. The reabsorption present in the system modifies the height of the jump. Saturation effects of the pump levels and due to the small numbers of QDs typically present in microcavity lasers, can have an effect at higher excitation powers. In this case the upper branch of the input/output curve can be masked to an extent where even the threshold is not fully developed. If in this case atomic models are used to extract parameters from measured data, the smaller jump in the input/output characteristics may be mistaken to be caused by a spontaneous emission coupling factor β larger than it truly is. This effect is even more pronounced for pulsed excitation [28].

The close relation to the discussed atomic models allows for a direct comparison and a verification of the truncation scheme if the semiconductor model is considered in the limit of a single possible excitation. We have performed such tests for an ensemble of emitters and find that the truncation of the hierarchy introduced by the light-matter coupling on the quadruplet level delivers an accurate description of the laser properties including the second-order correlation function for realistic parameters [19].

Acknowledgments

In the end we would like to thank our colleagues in the experimental groups in Stuttgart (S. Ates, S.M. Ulrich, P. Michler) and Dortmund (T. Auer, T. Berstermann, M. Aßmann, M. Bayer), as well as Paul Gartner, Michael Lorke and Sandra Ritter in our own group for stimulating discussions and fruitful collaboration. Financial support from the DFG research group "Quantum optics in semiconductor nanostructures" and a grant for CPU time at the Forschungszentrum Jülich (Germany) is gratefully acknowledged.

References

1. E.M. Purcell, H.C. Torrey, R.V. Pound, Phys. Rev. **69**, 37 (1946)
2. P. Lodahl, A. Floris van Driel, I.S. Nikolaev, A. Irman, K. Overgaag, D. Vanmaekelbergh, W.L. Vos, Nature **430**, 654 (2004)
3. K.J. Vahala, Nature **424**, 839 (2003)
4. F. DeMartini, G.R. Jacobovitz, Phys. Rev. Lett. **60**, 1711 (1988)
5. P.R. Rice, H.J. Carmichael, Phys. Rev. A **50**, 4318 (1994). http://prola.aps.org/abstract/PRA/v50/p4318

6. G.S. Solomon, M. Pelton, Y. Yamamoto, Phys. Rev. Lett. **86**, 3903 (2001)
7. H.G. Park, S.H. Kim, S.H. Kwon, Y.G. Ju, J.K. Yang, J.H. Baek, S.B. Kim, Y.H. Lee, Science **305**, 1444 (2004)
8. S. Strauf, K. Hennessy, M.T. Rakher, Y.S. Choi, A. Badolato, L.C. Andreani, E.L. Hu, P.M. Petroff, D. Bouwmeester, Phys. Rev. Lett. **96**, 127404 (2006)
9. Y.S. Choi, M.T. Rakher, K. Hennessy, S. Strauf, A. Badolato, P.M. Petroff, D. Bouwmeester, E.L. Hu, Appl. Phys. Lett. **91**, 031108 (2007)
10. S.M. Ulrich, C. Gies, S. Ates, J. Wiersig, S. Reitzenstein, C. Hofmann, A. Löffler, A. Forchel, F. Jahnke, P. Michler, Phys. Rev. Lett. **98**, 043906 (2007)
11. R.J. Glauber, Phys. Rev. **130**, 2529 (1963)
12. T.R. Nielsen, P. Gartner, F. Jahnke, Phys. Rev. B **69**, 235314 (2004)
13. J. Seebeck, T.R. Nielsen, P. Gartner, F. Jahnke, Phys. Rev. B **71**, 125327 (2005)
14. Y. Mu, C.M. Savage, Phys. Rev. A **46**, 5944 (1992)
15. H.J. Carmichael, *Statistical Methods in Quantum Optics 1* (Springer, Berlin Heidelberg New York, 1999)
16. H. Yokoyama, S.D. Brorson, J. Appl. Phys. **66**, 4801 (1989)
17. M. Schwab, H. Kurtze, T. Auer, T. Berstermann, M. Bayer, J. Wiersig, N. Baer, C. Gies, F. Jahnke, J.P. Reithmaier, A. Forchel, M. Benyoucef, P. Michler, Phys. Rev. B **74**, 045323 (2006)
18. T. Berstermann, T. Auer, H. Kurtze, M. Schwab, M. Bayer, J. Wiersig, C. Gies, F. Jahnke, D. Reuter, A. Wieck, Phys. Rev. B **76**, 165318 (2007)
19. C. Gies, J. Wiersig, M. Lorke, F. Jahnke, Phys. Rev. A **75**, 013803 (2007)
20. J. Fricke, Ann. Phys. **252**, 479 (1996)
21. G. Khitrova, H.M. Gibbs, F. Jahnke, M. Kira, S.W. Koch, Rev. Mod. Phys. **71**, 1591 (1999)
22. N. Baer, C. Gies, J. Wiersig, F. Jahnke, Eur. Phys. J. B **50**, 411 (2006)
23. P. Meystre, M. Sargent III, *Elements of Quantum Optics* (Springer, Berlin, 1999)
24. G. Björk, Y. Yamamoto, IEEE Journal of Quant. Electr. **27**, 2386 (1991)
25. Y. Yamamoto, S. Machida, G. Björk, Phys. Rev. A **44**, 657 (1991)
26. Y. Yamamoto, A. Imamoğlu, *Mesoscopic Quantum Optics* (John Wiley & Sons, Inc., 1999)
27. S. Ates, C. Gies, S.M. Ulrich, J. Wiersig, S. Reitzenstein, A. Löffler, A. Forchel, F. Jahnke, P. Michler, Phys. Rev. B **78**, 155319 (2008)
28. C. Gies, J. Wiersig, F. Jahnke, Phys. Rev. Lett. **101**, 067401 (2008)

2

Growth and Control of Optically Active Quantum Dots

Armando Rastelli, Suwit Kiravittaya, and Oliver G. Schmidt

Summary. We provide a general overview on the different methods employed to fabricate optically active quantum dots, i.e., quantum dots which confine both electrons and holes in three dimensions. All the techniques are based on epitaxial growth of semiconductor heterostructures. We first discuss bottom-up methods based on self-assembled growth, since they are the most used for single dot investigations. We then mention top-down lithographic techniques, which allow the fabrication of dots at controlled positions. Finally we show how a clever combination of top-down and bottom-up approaches may open the route to the fabrication of quantum dots with well defined spatial and spectral properties, required for scalable device integration.

2.1 Introduction

Semiconductor optically active quantum dots (QDs) are structures confining the motion of charge carriers (electrons and holes) in regions of space with nanometric sizes. They are typically based on the use of direct-bandgap semiconductors, so that transitions are direct both in real and in reciprocal space. In order to achieve such confinement, both conduction band and valence band profiles should be modulated, as schematically illustrated in Fig. 2.1.

As a result of the confinement, the energy levels of the charge carriers are quantized. Quantization effects become apparent when the dimensions of the confining region are comparable to the De Broglie wavelength of the charge carriers: $\lambda_{\rm DB} \sim h/\sqrt{2m^* k_{\rm B} T}$, where h and $k_{\rm B}$ are the Planck's and Boltzmann's constants, T the absolute temperature and m^* the effective mass of electrons/holes in the semiconductor crystal. Since m^* is usually much smaller than the free-electron mass, $\lambda_{\rm DB}$ is of the order of 10–100 nm at low temperatures. For such sizes the spacing between energy levels can be larger than the thermal energy. Moreover, when $k_{\rm B}T$ is smaller than the electron–hole binding energy due to Coulomb attraction, electrons and holes bind to form excitons, which can decay radiatively generating light with spectra consisting

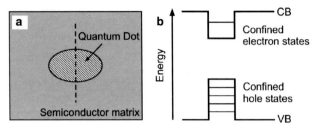

Fig. 2.1. (a) Schematic illustration of a QD defined in a crystalline semiconductor matrix. (b) Schematics of the conduction band (CB) and valence band (VB) edges, which are modulated on nanometric length scales

of sharp peaks, similar to atoms. The spatial extent of an exciton in a dielectric material is given by its Bohr radius $a = \frac{4\pi \epsilon_0 \epsilon_r \hbar^2}{\mu^* e^2}$, where ϵ_r is the relative permittivity of the medium and μ^* the electron–hole reduced effective mass. Because of the small value of μ^* compared to the free electron mass and the large values of ϵ_r for typical semiconductors, a is substantially larger than the Bohr radius for an hydrogen atom. Instead of considering λ_{DB} for electrons and holes separately, it is usually more convenient to describe a QD as a region of space with sizes comparable to the exciton Bohr radius a.

With respect to real atoms, QDs (also dubbed "artificial atoms") present the advantage that their position is fixed in space since they can be embedded in a solid state matrix. Moreover the material surrounding the QDs can be processed and structured using approaches which are well established in semiconductor industry. This allows, e.g., the QDs to be embedded in the active region of a light-emitting-diode (LED) [1,2] (see Chap. 6), a field-effect device [3] (see Chaps. 5 and 10) or an optical microcavity [2,4,5] (see Chaps. 1, 8, and 9).

The understanding of the electronic properties of QDs has seen tremendous progress in recent years. In turn, this has opened the route to the demonstration of concepts in which single-QDs are the building blocks of optical and optoelectronic devices with novel functionalities. For instance QDs can be used as triggered sources of single photons [1, 4], where a classical pulse of light or of electric current is deterministically converted into a light quantum (see Chaps. 6, 7, and 11). This application is interesting, e.g., in the field of secure data transmission based on quantum cryptography. Excitons in QDs can be used as two-level systems, i.e., quantum bits, whose state can be set by optical pulses [6]. Weak and strong-coupling of excitons and optical modes in semiconductor microcavities [4, 7–11] are allowing exciting solid-state-based quantum electrodynamics (QED) experiments (see Chaps. 8 and 9).

The progress has been tightly linked to the availability of QDs with excellent optical properties, meaning exciton dephasing times comparable with the lifetime [12] and reduced probability of nonradiative decay. A number of different fabrication approaches has been explored starting from the end of the 1980s, culminating with the observation of the first single-QD optical spectra characterized by sharp lines at the beginning of the 1990s [13–16]. Nowadays the most commonly used methods to fabricate high quality QDs are those based on self-assembled growth (see Sect. 2.2). In particular, the Stranski–Krastanow (SK) growth mode occurring during lattice-mismatched heteroepitaxy is an elegant and efficient mechanism to produce defect-free nanostructures of a narrow bandgap material embedded in a higher-bandgap matrix. Since most of the results presented in this book are obtained with SK dots, this growth mode will be discussed in some detail in Sect. 2.2.2 with reference to the most studied material system, i.e., In(Ga)As on GaAs(001) substrates. Another successful method to confine the motion of charge carriers is to exploit thickness fluctuations which naturally occur during the epitaxial growth of thin quantum wells (QWs).

The combination of SK growth with selective in situ etching opens up the opportunity to fabricate unstrained QDs with large confinement, which is not possible with the SK mode. The method is also useful to create advanced QD structures, such as quantum rings (QRs) and groups of closely spaced QDs (QD molecules) and will be discussed in Sect. 2.2.4. Other approaches based on so-called "droplet epitaxy" and its variants, strain modulation of a QW and QDs in nanowires will be mentioned in Sect. 2.2.5.

Because of the statistical nature of the self-assembled growth, novel concepts must be developed to create useful devices based on single-QDs, where the spatial (see Sect. 2.3) and optical properties (see Sect. 2.4) should be controlled with tolerances depending on the application. Historically, the first attempts to fabricate QDs were actually based on top-down methods, where shape, size, and position of the QDs on the substrate were nominally fixed by electron-beam lithography (EBL) and etching of a buried QW (see Sect. 2.3.2). In spite of these advantages, defects created during the etching process limited the performances of such QDs for optical investigations and applications.

A clever combination of bottom-up and top-down approaches represents nowadays the most promising route to fabricate QDs with well defined properties. We will therefore discuss how QDs with well defined position can be fabricated by growth of heterostructures on different substrate orientations or on substrates patterned with nano or micro-holes in Sect. 2.3.3 and by combining SK growth with patterned arrays in Sect. 2.3.4. Since positioned QDs are still affected by size/shape fluctuations we will conclude this chapter by discussing an approach to permanently tune the emission of single dots in a broad spectral range and resolution limited accuracy by in situ laser processing (ILP).

2.2 Self-Assembled III–V Quantum Dots

2.2.1 Basic Concepts in Epitaxial Growth

Epitaxial deposition takes place when a single-crystalline material A is deposited (or grown) on a clean surface of a single crystal B (substrate) under proper conditions. When A and B are the same (different), the process is referred to as homoepitaxy (heteroepitaxy). In most of the cases discussed below B is a GaAs substrate with (001) crystal orientation. The material A can be provided, e.g., by thermal evaporation of material from hot crucibles (in molecular beam epitaxy (MBE [17])) or through a molecular gas, which decomposes at the substrate surface (in metal organic vapour phase epitaxy (MOVPE [18])). At the substrate temperatures required for epitaxial growth (typically 500°C or higher for epitaxy on GaAs), both atoms from the A and B species diffuse on the surface as adsorbed atoms (*adatoms*), can attach or detach from steps or other inhomogeneities, can aggregate to form two-dimensional islands, or can be desorbed. All these processes can occur with a certain probability per unit time ν, depending on a characteristic activation energy E_a: $\nu \propto \exp(-E_a/k_B T_s)$, where T_s is the substrate temperature. In thermodynamic equilibrium all processes proceed in two opposite directions at equal rates, as required by the principle of "detailed balance." Thus, surface processes such as condensation (*adsorption*) and evaporation (*desorption*) must counterbalance. Hence, crystal growth must clearly be a nonequilibrium process. The final macroscopic state generally depends on the path taken through various single events and it is not necessarily the most stable one. In spite of these considerations, thermodynamic models succeed in describing some growth aspects, as we will see below.

According to Bauer [19], the film growth may be classified in either of the following three modes: layer-by-layer or Frank–van der Merwe (FM), island growth or Volmer–Weber (VW), or layer-plus-island or Stranki–Krastanow (SK) (see Fig. 2.2).

A simple distinction between the FM and the VW modes can be done on the basis of surface and interface tension, assuming that growth occurs in vacuum (like in MBE). In the FM mode the interaction between neighboring A atoms in the film is weaker than that with the B atoms of the substrate:

$$\gamma_B \geq \gamma_A + \gamma_{IF},$$

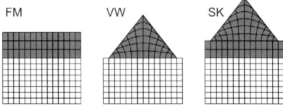

Fig. 2.2. Schematic representation of the three modes for heteroepitaxial growth of a material A on a substrate B: layer-by-layer or FM, island growth or VW, or layer-plus-island or SK. In the drawing it is assumed that the lattice constant of material A is larger than the lattice constant of B (strained heteroepitaxy)

where γ_A and γ_B are the film/vacuum and substrate/vacuum surface tensions and γ_{IF} is the film/substrate surface tension. The opposite occurs for the VW growth mode:

$$\gamma_B < \gamma_A + \gamma_{IF}.$$

Many factors may account for the mixed SK growth mode, but a certain lattice mismatch between the film and the substrate is the most relevant for the creation of QD.

2.2.2 Stranski–Krastanow QDs

Three-Dimensional Island Formation

The deformation (tetragonal for InAs on GaAs(001)) undergone by the flat film to match the substrate lattice, costs elastic energy and, when this deformation energy becomes too high, it must be relieved in some way. In the case of InAs on GaAs the lattice mismatch is about 7%. A possibility is to introduce misfit dislocations, when the thickness of the deposited film exceeds a certain critical value. Such crystal defects generally create deep traps for charge carriers.

Another possible way to relieve the energy due to strain is the nucleation of three-dimensional (3D) islands on top of the flat film (the wetting layer (WL)). Qualitatively, the absence of lateral constraints allow the atomic planes to laterally relax (see rightmost panel of Fig. 2.2). Only in 1990 [20–22] it became clear that such islands can be free of crystal defects opening the way to their exploitation as QDs. In fact, if the energy bandgap of the material composing the 3D islands is smaller than the bandgap of the substrate, the islands can effectively confine the motion of charge carriers and act as QDs. This is the case for In(Ga)As islands on GaAs, where the notation "(Ga)" means that the same argument applies for pure InAs or for an alloy containing some Ga.

A simple approach to describe the formation of 3D islands is the *capillary model*. This name comes from the fact that it is based only on the thermodynamically defined surface and interface tensions, γ_A, γ_B and γ_{IF}. The free energy $\Delta F(n)$ for an island with n atoms with respect to a flat film is

$$\Delta F(n) = -n\Delta\nu + n^{2/3} X, \quad (2.1)$$

where $\Delta\nu$ is the difference between the bulk energies of the atoms contained in the two phases, 3D island and film. X is the surface contribution to the free energy and can be evaluated as:

$$X = \sum_k C_k \gamma_k + C_{IF}(\gamma_{IF} - \gamma_B).$$

The first term on the right-hand side is the contribution of the various island facets, and the second is the energy variation due to the interface, replacing some portion of the substrate free surface. C's represent geometric constants. The first term on the right-hand side of (2.1) represents the relaxation energy, i.e., the difference (negative) between the elastic energy of a partially relaxed island and that of a flat film containing the same number of atoms n. The second term is the energy increase due to the additional surface. ΔF is positive up to a certain $n = N$, suggesting that islands containing less than N atoms have a free energy higher than a flat film. However, larger islands can effectively release energy, since the elastic relaxation energy overcomes the increase of surface energy. N is obviously given by

$$\Delta F(N) = 0, \qquad N = \left(\frac{X}{\Delta \nu}\right)^3.$$

In this picture, 3D islands are the result of local fluctuations of the density of adatoms. During growth, *nuclei* composed of groups of adatoms form and may either (1) dissolve, if they contain less than n^* atoms, where n^* is the position of the maximum of the ΔF curve (in this case they are called *subcritical*), or (2) expand, if their size is larger than that of the *critical nucleus* containing n^* atoms. The process involves an activation energy $\Delta F^* = \Delta F(n^*)$ given by

$$\frac{\partial \Delta F(n^*)}{\partial n} = 0, \qquad n^* = \left(\frac{2X}{3\Delta \nu}\right)^3, \qquad \Delta F^* = \frac{4X^3}{27 \Delta \nu^2}.$$

Instead of using the total free energy of the island ΔF, the nucleation concept can be better understood by using the chemical potential $\Delta \mu = \partial \Delta F / \partial n$, i.e., the free energy "gained" by each adatom attaching to the nucleus. When $n < n^*$, $\Delta \mu$ is positive an thus it is not favorable for an adatom to attach to the nucleus. However, if the nucleus is supercritical, $\Delta \mu$ will be negative and the island will tend to increase its size.

Several other models have been developed to describe island formation and growth. In particular the contribution of edges to the island energy, discussed in [23,24], is not considered in (2.1). Moreover islands are usually composed of an alloy of the components A and B and entropy of mixing should be included. For very small islands, composed of just a few atoms, it is questionable whether they can be treated as continuous thermodynamic systems. Atomistic models have been developed to tackle this problem. Finally the nucleation theory only applies if one assumes the nuclei to have a well defined shape and a finite slope. Formation of islands without nucleation, i.e., with no activation energy nor critical nucleus size can occur if the film continuously evolves from a planar to a rippled morphology [25].

Morphological Transitions

Morphological transitions are one of the most intriguing aspects of island growth and have been observed for different SK systems, such as InAs/GaAs(001) or Ge/Si(001) discussed below. To illustrate their origin, we can rewrite (2.1) in a different form, by replacing the number of atoms n with the island volume V and expanding $\Delta \nu$ to include the contribution of island shape and of misfit strain ϵ between substrate and island material. With the simple hypothesis that the transition involves just two pyramidal islands with facet inclination given by α and α' ($\alpha < \alpha'$), $\Delta F(V, \alpha)$ can be parameterized as [26, 27]

$$\Delta F(V, \alpha) = -K \epsilon^2 V \alpha + \Gamma V^{2/3} \alpha^{4/3}, \qquad (2.2)$$

where Γ contains the surface tensions of island facets and WL and K includes the elastic properties of the material.

Figure 2.3 illustrates qualitatively ΔF and the chemical potential per unit volume $\Delta \mu = \partial \Delta F / \partial V$ as a function of size for the two shapes. The two shapes with facet inclination α and α' (labeled as Pyramid and "Dome," respectively) are degenerate in energy at the critical size V^* where the transition occurs. V^* is simply given by imposing $\Delta F(V^*, \alpha) = \Delta F(V^*, \alpha')$:

$$V^* = \left(\frac{\Gamma' \alpha'^{4/3} - \Gamma \alpha^{4/3}}{\alpha' - \alpha} \right)^3 \frac{1}{K^3 \epsilon^6}, \qquad (2.3)$$

where Γ' contains the surface tensions of the facets of the steep shape and is in general different from Γ. Thus, the transition between the shallow and the steep morphology occurs at a critical volume proportional to ϵ^{-6}. The

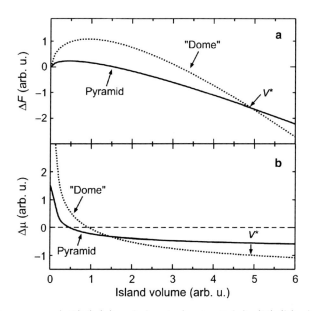

Fig. 2.3. Free energy $\Delta F(V)$ (**a**) and chemical potential $\Delta \mu(V)$ (**b**) of islands with two types of facets with inclinations $\alpha = 11°$ (pyramids) and $\alpha' = 25°$ ("domes"), calculated from (2.2) with fixed values of the other parameters. The shape transition occurs at volume V^*, where the energy curves cross and $\Delta \mu$ has a discontinuity

chemical potential $\Delta\mu$ drops discontinuously at V^*, indicating that the transition is of the first order. The drop in chemical potential also implies that islands with steep morphology act as a sink of material and their appearance leads to rapid coarsening of the ensemble and consequent disappearance of shallower and smaller islands. Such phenomena are well documented in the case of SiGe/Si(001) islands [26, 28].

Properties of InAs/GaAs(001) Islands

In(Ga)As/GaAs(001) QDs represent nowadays the most studied example of optically active QDs, both for single-QD (see Chaps. 5–10) or ensemble-QD based studies (Chap. 4) and applications [29]. For this reason we present the results of a detailed investigation on the morphology of such dots. To underline the similarities with other QDs obtained via the SK mode, we provide a direct comparison to (Si)Ge islands grown on Si(001) substrates. Such material system is in fact considered as a prototype for understanding the SK growth mode.

Figure 2.4a shows a room temperature (RT) scanning tunneling microscopy (STM) image of the surface of a sample obtained by MBE-growth of 1.8 monolayers (ML) of InAs on top of a clean GaAs substrate at a temperature T_s of 500°C and a growth rate of 0.01 ML s^{-1}. Arsenic was provided in the form of As$_4$. To enhance the different facets composing the island surface the

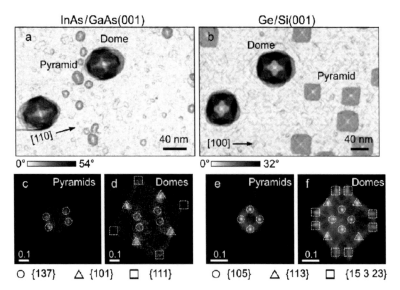

Fig. 2.4. RT-STM images of (**a**) InAs/GaAs(001) and (**b**) Ge/Si(001) 3D islands (SK-QDs). (**c**)–(**f**) Facet plots for pyramids (small islands in (**a**) and (**b**)) and domes (large and dark islands in (**a**) and (**b**)). The facets corresponding to each spot in such histograms are labeled. Adapted from [34, 35]

topograph is displayed using a grayscale with tones which depend on the local surface slope $|\vec{n}|$, where $\vec{n}$ is the local surface gradient. Dark and light areas correspond to large and small angle orientations with respect to the substrate orientation. Based on the size and shape one can immediately distinguish two kinds of islands: small islands bound by shallow facets, named pyramids and large islands bound by steeper facets with different orientations, named domes. In order to identify the various facets, we construct a "facet plot" [30], i.e., a two-dimensional histogram of the values of $\vec{n}$ separately for pyramids (Fig. 2.4c) and domes (Fig. 2.4d). In such plots each spot corresponds to a different facet. The distance of the spot from the center of the plot gives the slope of the facet with respect to the (001) plane while the position in the plot gives the orientation with respect to the in-plane crystal direction marked in Fig. 2.4a. The facets bounding the surface of pyramids are four {137} facets, as previously reported in [31]. Domes contain such facets at their top and foot and steeper {101} and {111} at their body. According to the discussion in the previous section it is reasonable to assume that domes are the result of the growth of pyramids beyond a critical size, at which a morphological transition occurs. Shapes intermediate between pyramid and dome have been in fact reported in [32]. In the same work Kratzer et al. calculated the elastic strain relief for different island shapes with continuum elasticity theory and the surface energies and stresses by density-functional theory. The shape transition was found to be of the first order, as for the case of SiGe islands [26, 33].

For comparison, Fig. 2.4b shows a RT-STM image of the surface of a sample obtained by growing 7 ML Ge on top of a clean Si(001) surface at $T_s = 550°C$ at a growth rate of $0.3\,\mathrm{ML\,s^{-1}}$. As in the case of InAs growth on GaAs(001) we observe two different island shapes: small and shallow pyramids and large multifaceted domes. A facet plot analysis allows us to identify the facets composing the surface of the different islands. As for InAs pyramids, Ge pyramids are bound by four shallow facets (in this case {105} planes), while domes have a more complex surface with {105} facets at their top and foot and steeper {113} and {15 3 23} facets at their body. The morphological transition from pyramid to dome was first reported by Medeiros-Ribeiro et al. [24]. Intermediate shapes between pyramids and domes were observed in [36, 37] and the transition modeled in detail in [37]. Recently it has been observed that at some critical size domes transform into even steeper morphologies [38] and that pyramids are the result of the transition of unfaceted "prepyramids" into partially faceted truncated pyramids [25].

The similarities between the two material systems extend beyond those presented here and are discussed in [34, 35]. In particular, there are indications that when an alloy is deposited (i.e., InGaAs or SiGe instead of pure InAs or Ge) the observed shapes are similar to those observed for pure components. On the other hand the typical size of the islands and the critical size for the morphological transitions increase because of the reduced misfit between deposited material and substrate (see (2.3)).

An important point which deserves to be mentioned is that even when a pure material is deposited on the substrate alloying between the two materials takes place. Alloying is favored by entropy and allows for a reduction of elastic energy in the layer and substrate. Intermixing generally occurs through surface diffusion and is clearly documented in the literature both for InAs QDs on GaAs and for Ge QDs on Si (see, e.g., [39–41]). It is therefore more appropriate to describe such systems as In(Ga)As QDs and (Si)Ge QDs. Because of space limitations we do not discuss such results here. It is enough to keep in mind that intermixing becomes more pronounced at high T_s and low growth rates and the resulting composition profiles of QDs is rather complex and critically depends on the growth parameters. Since the exact composition profiles affect the strain in the QD and surrounding matrix and hence the band-edge profiles, predicting the optical properties of QDs requires a detailed knowledge of such profiles. A recent attempt to determine the full 3D composition profiles of SiGe islands is presented in [42].

The situation gets even more involved when islands are "capped," i.e., overgrown with material (usually the same as the substrate) to create a well defined confinement potential, as described in the following section.

Besides the island shape, size and composition, for the single-QD investigations discussed in this book it is useful to be able to tune the density of QDs on the substrate. In the case of InAs QDs grown on GaAs(001), this can be easily done by properly tuning the amount of deposited InAs, as illustrated in Fig. 2.5, where InAs islands were obtained by InAs deposition at a low rate of $0.01\,\mathrm{ML\,s^{-1}}$ and relatively high substrate temperature ($T_s = 500°C$). The InAs deposition is followed by 30 s annealing at T_s. In MBE growth the appearance of islands on the surface is usually monitored in situ by means of reflection high energy electron diffraction (RHEED). When islands appear the RHEED pattern changes from streaky to spotty. The island density can be semi-quantitatively tuned by visual inspection of the intensity of the RHEED spots. Saturated spots correspond (in our case) to about 1.8 ML of InAs and

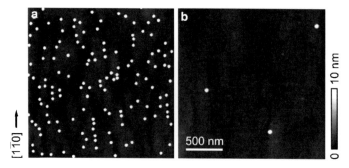

Fig. 2.5. AFM images of InAs/GaAs QDs with surface density ranging from $\sim 4 \times 10^9$ (**a**) down to $<10^8\,\mathrm{cm^{-2}}$. QDs with low surface density (**b**) are particularly useful for single QD studies. Adapted from Ref. [43]

an island density of the order of $4 \times 10^9 \, \mathrm{cm}^{-2}$ (see Fig. 2.5a). On the other hand, if the InAs deposition is interrupted right after the appearance of spots in the RHEED pattern, densities of the order of less than $10^8 \, \mathrm{cm}^{-2}$ can be obtained (see Fig. 2.5b). This density is suitable for single-QD optical investigations in which a laser focused to a spot with diameter of the order of a micrometer is usually employed. Another commonly used method to obtain QDs with low surface density consists in exploiting the gradient in the amount of deposited InAs across a substrate (wafer) [43].

Overgrowth of InAs/GaAs QDs

In order to provide a well-defined confinement potential, islands need to be overgrown. Even if the morphology is well characterized before capping, strong changes can occur during the subsequent capping [44–46]. Therefore, one must pay attention in not using the shape of uncapped QDs to argue the morphology of buried QDs. Vice versa, data on buried QDs, which can be obtained by cross-sectional transmission electron microscopy (TEM) or STM generally cannot be used to argue the morphology of the QDs prior to capping. Attribution of facets from such analyses is generally erroneous. Typically buried QDs are shallower than uncapped QDs. To understand why, it is useful to study the initial stages of the capping process. An example is shown in Fig. 2.6, where InAs QDs were grown at $T_s=500°\mathrm{C}$ and overgrown at $T_s=460°\mathrm{C}$. In

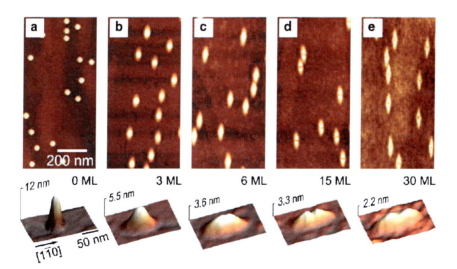

Fig. 2.6. Sequence of AFM images illustrating the evolution of the surface morphology after overgrowth of InAs QDs with the indicated amount of GaAs at $T_s = 460°\mathrm{C}$. The *bottom panels* are magnified 3D views of representative surface structures. Adapted from [46, 49]

spite of the low temperature used for the overgrowth, which limits the surface diffusion, the deposition of a few ML of GaAs produces strong changes in the surface morphology and of the height of the mounds observed on the surface.

Higher resolution images collected by STM for capping thicknesses ranging from 1 to 3 ML show that the collapse in height is accompanied by substantial morphological changes: the shallow {137} facets at the island top rapidly expand so that domes transforms into pyramid-like islands [34, 47]. The same phenomenon is observed in the case of Si capping of Ge islands indicating that during capping the islands undergo a morphological change which is the reverse of that occurring during growth. Qualitatively this can be explained as due to surface-mediated alloying and consequent strain decrease, which destabilizes the steep dome morphology [48] (see (2.3)). A detailed account of the changes undergone by InAs QDs during capping was given in [46, 47].

Depending on the conditions used for capping the islands, a more or less pronounced smearing of the QD profile can be achieved. This is important because it allows the QD emission properties to be finely tuned. If for instance the QD emission needs to be blue-shifted, a partial capping and annealing step (PCA) may be used [44]. This approach, employed by several groups to tune the emission of InAs QDs in the sensitive range of silicon charge-coupled-device (CCD) cameras, is illustrated in Fig. 2.7 where InAs QDs are grown, overgrown, and annealed at a substrate temperature of $T_s=500°C$. The effect of the PCA treatment is to reduce the height of the QDs, because In migrates away from the QD apex during annealing and mixes with the surrounding GaAs, as illustrated in the insets of Fig. 2.7a,b. On the other hand the lateral extension of the QDs is practically not affected by the process [50]. As a consequence of the height decrease, the QD emission energy gradually blue-shifts with increasing annealing time, as shown in Fig. 2.7c.

A further improvement of the method consist in an "In-flush" approach [51], where the partial capping is followed by annealing at higher T_s, which allow residual indium to be desorbed from the surface and the height of the QDs to be finely tuned (see also Chap. 10).

As a final remark, we should point out that while the RT AFM images shown in Figs. 2.6 and 2.7 provide information on the changes undergone by the QDs during overgrowth, their interpretation is not straightforward. In fact scanning probe microscopy alone is not able to distinguish between cap and QD material. Furthermore, since the sample has to be cooled to RT for the measurements, further changes may occur during such a step. A standard approach to obtain the shape/composition of buried QDs is represented by cross-sectional TEM and STM. A more recent approach to extract the 3D shape of buried QDs consists in TEM-based tomography [52]. A simpler method, briefly discussed in Sect. 2.2.4, is to remove the cap layer by selective wet chemical etching.

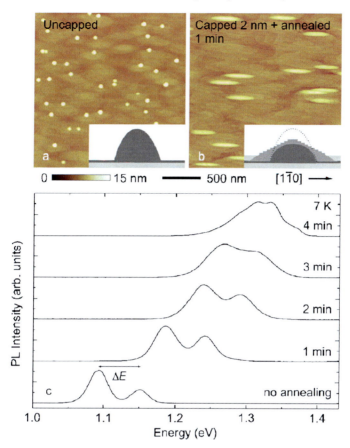

Fig. 2.7. Illustration of the PCA process. AFM images of InAs QDs (**a**) overgrown with a 2-nm thick layer of GaAs and annealed for 2 min at $T_s = 500°C$ (**b**). *Insets* illustrate the effect of PCA. Evolution of the photoluminescence (PL) spectrum for increasing annealing time (**c**), showing how this capping procedure can be used to blue-shift the emission of QDs. Adapted from [50]

Vertical Stacks of InAs/GaAs QDs

Once InAs QDs are overgrown, the GaAs surface above them can be used as "substrate" for the deposition of a new layer of QDs. This is useful to increase the volume density of QDs in a sample (which is particularly important for applications such as lasers based on QDs), or to create coupled QD systems (vertical QD molecules, see Chap. 10). If the spacer between two subsequent layers is thin enough, QDs in the upper layer tend to form right on top of the buried QDs because of the tensile strain present on the GaAs regions above buried QDs [53–55]. In general, each QD in a QD molecule have different structural and electronic properties due to the influence of the strain from the buried QDs on the growth of second QD layer. AFM, cross-sectional TEM or STM studies indicate that the QDs in the second layer are generally larger in size and are more intermixed [56–58]. By changing the growth or overgrowth conditions for the second layer, the emission energy of the two layers can be tuned to be close [59]. On the other hand intentional asymmetries in the top and bottom QDs can be useful to study in detail the electronic coupling in single QDMs (see Chap. 10). The fine tuning of the energy levels in a QD molecule can be done by applying electrical, magnetic, or stress fields (Chap. 10).

2.2.3 Thickness Fluctuations in Quantum Wells: "Natural QDs"

Semiconductor QWs are easily fabricated by sandwiching a layer of low bandgap material between two barriers with larger bandgap. Ideally, in QWs the motion of charge carriers is confined along the growth direction and is free in the QW plane. However, unavoidable interface roughness, i.e., fluctuations in the QW thickness, may provide additional lateral confinement as illustrated in Fig. 2.8. A typical example consists of GaAs/AlGaAs QWs. The almost perfect match of lattice parameters between GaAs and AlGaAs lets the growth proceed according to the FM mode. Depending on the growth parameters the interfaces between AlGaAs and GaAs can be more or less smooth. Typically the bottom surface is rougher than the top one because of the smaller diffusivity of Al compared to Ga. Such roughness may effectively generate traps for the motion of excitons in regions with sizes comparable with the Bohr radius, giving rise to sharp emission lines, as first reported in [14, 15].

Compared to SK dots, such "natural QDs" are characterized by smaller confinement energies E_c, which can be defined as the energy difference between continuum (QW emission) and QD emission. In a simplified picture in which the QW nominal thickness consists of n ML and thickness fluctuations are one ML deep, $E_c \simeq E_n - E_{n+1}$, where E_n is the recombination energy. Because E_n scales approximately with n^{-2}, thin QWs offer the possibility of having larger E_c. For GaAs/AlGaAs E_c can be of the order of 10 meV, which is much smaller than the corresponding value for SK QDs, where E_c of a few hundreds meV can be easily obtained. Therefore, while for SK QDs the confinement

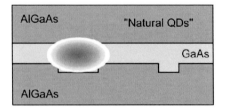

Fig. 2.8. Schematic illustration of excitons confined at thickness fluctuations of a thin QW

is sufficient to study their emission even at RT, natural QDs provide a 3D confinement only at low temperatures. In fact, at temperatures such that $k_B T \simeq E_c$, carriers escape from the shallow traps because of their thermal energy.

In reality not only monolayer thickness fluctuations, but also local fluctuations in the Al content can generate traps for excitons, so that the fabrication of "natural QDs" simply requires a thin QW and does not need any special procedure. It is worth to notice that even the In(Ga)As WL present between InAs/GaAs(001) SK QDs can be considered as a thin QW and shows sharp lines indicating the presence of excitons whose motion is laterally confined by local thickness and/or composition fluctuations [60].

Natural QDs have a noteworthy advantage with respect to SK QDs for applications in solid-state QED: their oscillator strength (related to the dipole moment of the optical transitions) can be much larger [61]. This allowed the observation of strong coupling with an optical mode confined in a microdisk microcavity with large Rabi splitting [9]. The large dipole moment also means that the lifetime is substantially shorter than for typical SK QDs and that natural QDs can be used as triggered single photon sources operating at higher frequencies than SK QDs (see also Chaps. 6, 7, and 11).

2.2.4 QD Structures by Combination of SK Growth and In Situ Etching

In this section we present a special class of QDs and QD structures which can be fabricated by combining standard solid-source III–V MBE growth with an in situ etching system. The employed gas is $AsBr_3$, which can be used to remove the deposited material (Ga/In/Al–As) in a layer-by-layer fashion [62–64]. Such an etching gas shows moderate selectivity both to composition and to local strain, allowing the creation of novel QD structures with excellent optical properties.

The first step for the fabrication of the structures discussed here consists in the growth of standard InAs QDs on GaAs(001) substrates at $T_s = 500°C$ followed by a 30 s growth interruption, which produces a narrowing of the size distribution of domes [65] and the almost complete disappearance of small pyramids (see the part concerning the properties of InAs QDs in Sect. 2.2.2). The amount of deposited InAs is varied according to the desired QD density. T_s is subsequently decreased to 470°C and a 10-nm thick GaAs layer is deposited while ramping T_s back to 500°C. The cap layer deposition is followed by an in-situ etching step at $T_s = 500°C$, where the etching depth depends only on the amount of supplied $AsBr_3$ (supply rate-limited regime). The gas flow is fixed at 80 msccm, corresponding to an etching rate of $0.23\,\mathrm{ML\,s^{-1}}$ for planar GaAs(001). The nominal etching depth H_e is defined as the etching depth of an unstrained GaAs(001) surface.

Nanoholes and Nanorings

The surface morphology above buried InAs QDs consists of rhombus-shaped structures with a tiny hole at the center, as shown in Fig. 2.9a. In order to disclose the buried QDs we removed the GaAs capping layer by dipping the sample in a wet chemical etching solution consisting of $NH_4OH{:}H_2O_2{:}H_2O$, which selectively removes GaAs leaving the InAs QDs [66]. The result is shown in Fig. 2.9e. This method is particularly useful to study the evolution of the InAs QDs during subsequent etching and was validated by comparison with cross-sectional TEM data [66]. Compared to the as grown QDs, the height of the buried QDs is reduced from about ~11 to ~8 nm.

The surface morphology of the GaAs cap layer changes when the etching gas is supplied. By increasing the nominal etching depth, the holes at the center of the rhombus structures show an increase both in depth and diameter as shown in Fig. 2.9b–d for $H_e = 1, 2, 3$ nm, respectively. By continuing the etching the holes assume a "bow-tie" shape which is qualitatively attributed to the anisotropic shape of the initial rhombus structures and to anisotropic surface diffusion of Ga and possibly of the etching rate [67]. By analyzing in detail the etching process it was concluded that material selectivity alone (InAs is etched faster than GaAs [63]) is not able to explain the relatively fast removal of the InAs QDs with respect to the surrounding GaAs. Thus, strain selectivity of $AsBr_3$ is probably the main responsible for the observed behavior [68].

Figure 2.9f–h show what happens to the buried InAs QDs during etching: their top is gradually etched away so that they assume a "volcano"-like shape. By solving the Schödinger equation for the envelope function of electrons and

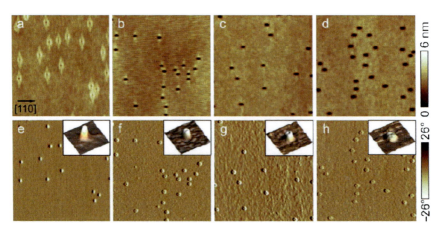

Fig. 2.9. $1 \times 1 \mu m^2$ AFM images illustrating the development of the surface above buried InAs QDs (**a–d**) and of the underlying InAs QDs (**e–h**) after in situ etching with $AsBr_3$ for 1 (**b**, **f**), 2 (**c**, **g**) and 3 nm (**d**, **h**). The *insets* of (**e–h**) show 3D views of representative QDs before (**e**) and after (**f–h**) in situ etching. Adapted from [66]

holes for the observed structures we found that such "nano-volcanos" may behave as QRs, as the wave function maxima for electrons and holes move from the QD center to their periphery [66]. This suggests a simple root to the fabrication of QRs. Other methods proposed by other groups rely on PCA under As_2 flux [44,69] or on droplet epitaxy described below.

For $H_e \geq 5$ nm, the buried InAs QDs/QRs are completely removed, as suggested by PL measurements of samples with different H_e [67]. In particular, for $H_e = 5$ nm the nanoholes observed on the GaAs surface have an optimum size homogeneity and can be used as a template for the fabrication of novel QD structures discussed below.

We should point out that the size/shape of the nanoholes can be tuned to a certain extent by varying the conditions of etching or by in situ annealing under an As_4 flux [67]. Moreover the hole depth can be substantially increased by etching a stack of vertically aligned InAs QDs [70].

Unstrained GaAs/AlGaAs QDs

As mentioned above, the SK growth mode provides a simple route to fabricate QDs with large confinement energy E_c. Since the main driving force behind the SK process is the lattice mismatch, the SK mode cannot be used to produce unstrained QDs in systems such as GaAs/AlGaAs. The GaAs surface with nanoholes described in the previous section allows us to obtain QDs with large E_c also for this material system [71], as illustrated in Fig. 2.10a–d. We first "transfer" the nanoholes to an AlGaAs surface by overgrowing them with a layer of $Al_{0.45}Ga_{0.55}As$ with thickness D at a rate of 1.1 ML s^{-1} and at $T_s = 500°C$. Such temperature is low enough to prevent the recovery of a flat surface, provided that D is not too large. The AlGaAs nanoholes are then filled with GaAs by overgrowing them with a thin layer of GaAs with thickness d and annealing the surface for 2 min at 500°C. The structure is finally completed by growing a 100-nm thick $Al_{0.35}Ga_{0.65}As$ barrier followed by a 20-nm thick $Al_{0.45}Ga_{0.55}As$ cladding layer and a 10-nm thick GaAs cap layer, while ramping T_s to 610°C to improve the material quality. (The presence of cladding layers allows us to observe luminescence from single QDs up to RT [72].) Depending on d we obtain "inverted" GaAs QDs below a thin GaAs QW or, if d is sufficiently small (~ 0.5 nm) we can completely suppress the QW.

Since GaAs/AlGaAs intermixing is negligible at 500°C, the QD morphology is determined by that of the AlGaAs nanohole, which we measured by AFM and STM on samples with growth interrupted right after the $Al_{0.45}Ga_{0.55}As$ barrier. A RT STM image of an AlGaAs nanohole is shown in Fig. 2.10e, which was used in [71,73] for predicting the optical properties of QDs with no adjustable structural parameters. This is different from strained In(Ga)As QDs where assumptions on the composition profiles must be made in order to reproduce the experimental data. The good optical properties of such GaAs/AlGaAs QDs are witnessed by micro-PL spectroscopy, as shown

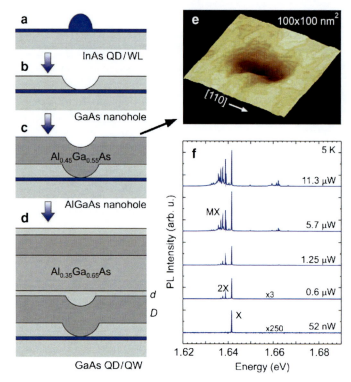

Fig. 2.10. Schematic illustration of the steps used to fabricate unstrained GaAs/AlGaAs QDs starting from InAs QDs (**a–d**). RT-STM image of an AlGaAs nanohole ($D = 7$ nm) with depth of about 4 nm (**e**). Micro-PL spectra of a single QD ($D = 7$ nm, $d = 2$ nm) at increasing excitation intensity (**f**). Partially adapted from [71, 72]

in the example of Fig. 2.10f, where the spectrum of a single QD is shown as a function of excitation power. (The excitation wavelength was 532 nm and the laser was focused down to a spot with 1 μm diameter through a microscope objective.)

Lateral QD Molecules

The GaAs surface with nanoholes can also be used as a template for the fabrication of groups of closely spaced QDs, which may act as lateral QD molecules [67]. The growth process is illustrated in Fig. 2.11a–c. Nanoholes are obtained as described at the beginning of this section, using a nominal etching depth of 5 nm. Bow-tie shaped nanoholes, with a depth of 5–6 nm are then overgrown with InAs at 500°C. At an InAs coverage of 0.8–1.4 ML the surface is planarized, as illustrated schematically in Fig. 2.11b. X-ray diffraction data indicate that holes are filled by a diluted InGaAs alloy [74]. By increasing

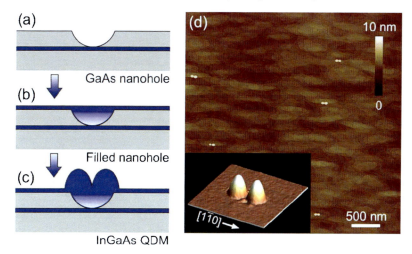

Fig. 2.11. Schematic illustration of the growth process leading to the formation of In(Ga)As/GaAs(001) QDMs (**a**–**c**). (**d**) AFM image of the surface containing QDMs with low density. The *inset* shows a 3D view of one of such QDMs. Adapted from [67, 76, 77]

the amount of deposited InAs, pairs of small QDs start forming on top of the original nanoholes and are invariably aligned along the $[1\bar{1}0]$ crystal direction. With further growth such QDMs grow in size and then single QDs start appearing. The mean center-to-center distance of the two QDs composing a QDM is about 45 nm. If one considers that each QD has a base diameter of about 40 nm, electronic coupling between the two QDs can be expected [75] and was demonstrated in [76].

By properly choosing the amount of deposited InAs, which depends on the initial density of nanoholes, occurrence of single QDs can be strongly suppressed [77]. Further improvement of the ensemble homogeneity is provided by a surface decorated with steps aligned along the $[1\bar{1}0]$ direction [77]. Figure 2.11d shows an AFM image of QDMs with density suitable for single QDM optical spectroscopy.

By overgrowing the holes at lower substrate temperatures and higher growth rate, the number of QDs composing each QDM can be increased [67].

2.2.5 Other Fabrication Approaches

QD Structures by Droplet Epitaxy

Nanocrystals acting as QDs with large confinement energy can be created with methods not relying on the SK growth mode. An approach for GaAs-based QDs is represented by the so-called "low temperature droplet epitaxy" [78, 79].

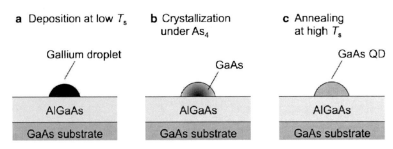

Fig. 2.12. Basic illustration of the droplet epitaxy method. (**a**) Formation of droplets by deposition of Ga on top of an AlGaAs layer; (**b**) Recrystallization by exposure to an As flux; (**c**) Annealing for improvement of crystal quality. Similar approaches are used to create InGaAs QDs

Basically, the method consists in the deposition of Ga, In or a mixture of the two components in absence of an arsenic flux at low substrate temperatures ($T_s \leq 300°C$) on a layer suitable for creating carrier confinement. Because Ga and In are metals which are in the liquid state at such temperatures, the deposited material forms droplets on the surface (Fig. 2.12a), which can be monitored during MBE growth by RHEED. In order to use such droplets as QDs, an As flux must be provided under suitable conditions to allow the droplets to recrystallize (Fig. 2.12b). Because the driving force for droplet formation is surface tension, QDs without strain can be fabricated, similar to the GaAs/AlGaAs QDs discussed in Sect. 2.2.4. Moreover, differently from SK QDs, QDs with or without a surrounding WL can be obtained with the method [80]. The recrystallization step is critical in this approach, since crystal defects generally form during this process, thus affecting the electronic and optical properties of the structures. An annealing step at $T_s \simeq 350$–$400°C$ is therefore often employed to improve the crystal quality (Fig. 2.12c) prior to QD overgrowth. Alternatively, a thermal treatment after capping can be used [81] to limit significant morphological changes of the QD structures. The formation of GaAs/AlGaAs QDs by droplet epitaxy was recently investigated and modeled in detail in [82].

Besides single QDs a number of QD structures ranging from QRs [83] to QD pairs (QDP [81,84]) have been fabricated by droplet epitaxy by tuning the growth parameters during droplet formation or subsequent recrystallization. In particular, Wang et al. have shown that droplet epitaxy can be performed also at high T_s ($\sim 500°C$) and that the so obtained QDPs have excellent optical properties [85].

Similar to the nanoholes discussed in Sect. 2.2.4, nanostructures created by droplet epitaxy can also be used as a template to guide the formation of closely spaced QD groups, which may act as QD molecules (see [86] and references therein).

Strain-Induced QDs

We have seen above that 3D confinement of excitons can occur in QWs because of local thickness or composition fluctuations. However, the confinement energy E_c is generally small for such QDs. A self-assembled approach to provide a deeper lateral confinement in a QW was presented in [87,88]. It is based on the modulation of the energy bandgap in a near-surface QW produced by a stressor located in vicinity of the QW. The stressors are SK dots grown on the surface of the upper barrier of the QW, as illustrated in Fig. 2.13 for InAs islands grown on the surface of an InGaAs/InP QW. The regions right under InAs islands are characterized by tensile strain, which produces a lowering of the energy bandgap and hence lateral confinement for both electrons and holes. Figure 2.13b illustrates the profile of the bottom of the conduction band in the QW. The areas under the island edges are compressed and the CB profile is expected to have a local increase. The approach is general and can yield ensembles of strain-induced QDs (SIQDs) with small inhomogeneous broadening (i.e., narrow emission) for several material combinations (see [89] and references therein).

Compared to SK dots, which provide vertical and lateral confinement, in SIQDs the QW provides vertical confinement and the stressor the lateral confinement. The latter can be well approximated with a harmonic potential $V_i = \frac{1}{2}m_i^*\omega_i^2 r^2$, where m_i^* is the electron/hole effective mass, ω_i the angular frequency of the oscillator and r the lateral size of the confining potential. With this assumption one expects equally spaced radiative transitions with energy separation given by $\hbar(\omega_e + \omega_h)$, in good agreement with the experimental observation of equally spaced excited states in power-dependent PL spectroscopy [87].

QDs in Nanowires

In SIQDs the starting point is a QW, which provides confinement in one direction. The confinement in the other two directions (in plane) is then provided

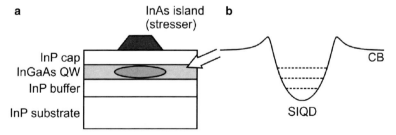

Fig. 2.13. (a) Sample structure used for creating strain-induced-QDs in a near surface QW (*right panel*) and (b) schematic illustration of the conduction band (CB) profile modulated by the presence of a stressor (a SK island) above the QW. Adapted from [89]

by a stressor. An alternative is to start with a structure providing confinement in two directions (quantum wire) and insert a QW in it to obtain 3D confinement.

A common method to produce nanowires consists in using MOVPE and gold droplets which act as catalysts for vapor–liquid–solid growth (see [90] and references therein). By properly tuning the precursors provided during the wire growth, vertical heterostructures can be fabricated. Bright single photon emission from QDs in wires was reported in [91] in the case of GaPAs/GaP QDs and attributed to the geometry of the structures. In fact, because of total internal reflection, the light extraction efficiency for QDs in planar structures is generally very small, while in wires light can readily escape from the walls of the wire. Recently a light-emitting diode based on a QD in nanowire was also fabricated [92].

2.3 Control of QD Position

In the methods discussed above, QDs are generally arranged randomly on the substrate surface. While this does not represent a problem for fundamental studies on single QDs and basic demonstrations, this problem will need to be addressed if the goal is to realize scalable devices with single QDs at their core.

Interestingly, the first attempts to fabricate QDs were based on top-down approaches, where the QD position on the substrate was determined by lithography. Defects associated with the processing and the concurrent emergence of high quality QDs obtained by self-assembled growth let the research on top-down methods shrink. In recent years, the need of QDs with controlled position encouraged new approaches in which the advantages of lithographic positioning is combined with the high quality of self-assembled growth.

2.3.1 Cleaved Edge Overgrowth

The capability of growing planar layers and QWs by epitaxy can be used to create more complex structures by cleaving the sample in vacuum and growing heterostructures on different substrate orientations. This approach, called "cleaved edge overgrowth" (CEO) has been successfully employed to create QDs at well defined positions and to create linear chains of SK dots.

The fabrication of QDs by CEO proceeds as follows. First an AlGaAs/GaAs/AlGaAs QW is grown by MBE on a conventional GaAs(001) substrate (see Fig. 2.14a). After in-situ cleavage, the sample is flipped and a GaAs/AlGaAs layer sequence is grown on the (110) and ($\bar{1}$10) surfaces. Three QWs lying on planes perpendicular to each other are obtained. The intersection of these QWs act as a "T-shaped" QD, as demonstrated by μ-PL spectroscopy [93]. In spite of the rather weak confinement (a few meV), the technique has been applied to fabricate coupled QDs (QD molecule) with well-defined separation [94].

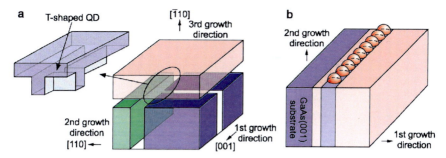

Fig. 2.14. (a) Schematic illustration of a T-shaped QD structure. First a GaAs/AlGaAs QW is grown on the (001) surface. The sample is then cleaved in situ. The (110) surface is overgrown by GaAs and AlGaAs. Finally, the sample is cleaved and another GaAs/AlGaAs layer is grown on the ($1\bar{1}0$) plane. The intersection of three QWs constitutes a QD. (b) Schematic illustration of an 1D array of InAs QDs on the ($1\bar{1}0$) plane. First an AlAs/GaAs multilayer is grown on the (001) surface. The sample is then cleaved and the ($1\bar{1}0$) surface is overgrown with InAs. Due to the preferential nucleation of InAs on the AlAs surface, SK QDs form there. Adapted from [93,96]

Chains of InAs QDs can be obtained by depositing InAs on the cleaved edge of a heterostructure. A AlAs/GaAs multilayer is first grown on a GaAs(001) substrate. Then the sample is cleaved and flipped. By depositing InAs on the cleaved ($1\bar{1}0$) surface, the InAs QDs preferentially form on the AlAs surface (see Fig. 2.14b). They arrange in a QD chain and their size can be controlled by changing the initial AlAs layer width [95, 96].

2.3.2 Top-down Fabrication Methods

As in the case of SIQDs mentioned in the last section, QDs can be fabricated by starting from a QW and adding a lateral confinement potential. Much work on this was done at the end of the 1980s and beginning of the 1990s. Here we just mention the basic ideas with reference to Fig. 2.15. A more detailed overview can be found in the introductory chapter of [97]. Lateral confinement can be achieved by defining mesas with diameters less than some 100 nm through EBL and etching. If the QW and surrounding barriers are etched (Fig. 2.15a) a deep confining potential (exceeding 1 eV) can be obtained. In this case carriers confined in the QDs are in "touch" with the processed surface and strong reduction of the luminescence intensity is often observed. The effective size of the QDs can be actually smaller than the physical size of the processed mesas so that it becomes comparable to the exciton Bohr radius. A softer approach consists in the selective etching of the top QW barrier (Fig. 2.15b). In this case the recombination energy of the regions below the unetched areas is smaller than the corresponding energies under the etched areas because of the higher potential barrier provided by the vacuum interface.

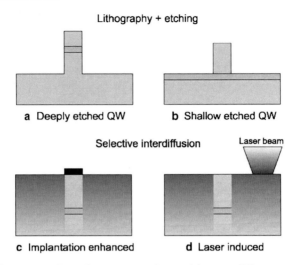

Fig. 2.15. Schematics of top-down approaches to fabricate QD structure. (**a**) Deeply etched QWs; (**b**) Modulated barrier QWs by shallow etching; (**c**) Selective interdiffusion by ion-implantation; (**d**) Local interdiffusion based on a focused laser

A lateral modulation of the QW barrier is therefore achieved and improvement of the optical properties is observed.

An alternative to etching is represented by selective interdiffusion. In this case the lateral barrier is obtained by blue-shifting the QW emission in a large area of the sample with exclusion of a region with size comparable with the exciton Bohr radius. Such selective interdiffusion can be obtained by creating defects in the areas out of the QD followed by rapid thermal processing (RTP, Fig. 2.15c). The defects, generated by ion implantation, favor interdiffusion and are healed out during the RTP. Another approach (Fig. 2.15d) consists in using a focused laser to locally intermix the QW and create barriers for the lateral motion of excitons [13].

Besides the benefit of positioning, top-down approaches offer a wide flexibility of design. For instance QRs were fabricated with the "modulated barrier" approach and their optical quality was sufficient to observe the Aharonov–Bohm effect for a charged exciton in magnetic field [98].

We should point out that, although lithographic approaches allow an accurate positioning of the QDs to be achieved, different QDs usually show different emission energies. This means that PL spectra of QD ensembles are affected by inhomogeneous broadening, similar to self-assembled QDs. With the present technology it appears therefore that QDs with well defined spatial and electronic properties cannot be obtained. A further top-down step to control the emission energies of QDs will be presented at the end of this chapter.

2.3.3 Growth on Prepatterned Substrates

QDs in Inverted Pyramids

A successful method to fabricate positioned QDs, based on MOVPE, is illustrated in Fig. 2.16. A GaAs substrate with (111)B orientation is first patterned

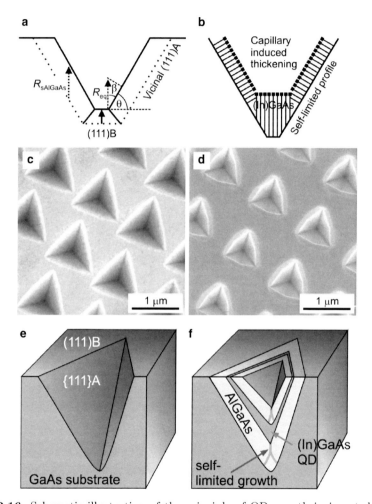

Fig. 2.16. Schematic illustration of the principle of QD growth in inverted pyramids. (**a**) Growth by MOVPE of an AlGaAs layer, governed by strong growth rate anisotropies. (**b**) When the system has reached a self-limited profile, the deposition of a (In)GaAs layer results in a strong capillarity contribution which widens the bottom profile, without contributing significantly to the sidewall growth. (**c**) SEM image of the patterned substrate with arrays of inverted pyramids. (**d**) SEM image of ordered (In)GaAs QD surface. (**e**) and (**f**) are cross-section schematics of the structures shown in (**c**) and (**d**), respectively. Adapted from [99–101]

by photolithography or EBL to obtain apertures on resist. Wet chemical etching is then used to create holes with shape of inverted tetrahedral pyramids, mainly bounded by {111}A crystallographic planes, which are slowly etched (Fig. 2.16a,c). When a first AlGaAs barrier layer is grown on this patterned substrate, the morphology is controlled by strong growth rate anisotropies. The sidewalls with (111)A facets tend to grow with a higher growth rate (R_sAlGaAs) than the bottom (111)B surface if the system has not reached yet his self-limited profile. An inverted pyramid with a flat bottom thus evolves with a constant reduction of the lateral dimensions of the bottom (111)B facet, until it reaches a self-limited profile of ($\sim$20–30 nm). This happens when the (111)B growth rate (R_eq) equals the sidewalls growth rate due to capillarity contributions. Once the system has reached a self-limited profile as shown in Fig. 2.16b, the growth of an (In)GaAs layer takes place with a thickening of the bottom profile with little growth of the sidewalls because of capillarity. Figure 2.16c,d shows scanning electron microscopy (SEM) images of the surface before and after MOVPE growth.

Subsequent growth of an appropriate multilayer yields a QD within each inverted pyramid. Figure 2.16f shows a schematic of a (In)GaAs QD surrounding by AlGaAs barrier. This QD has a lens shape and locates at the bottom of the pyramid due to the combination of growth rate anisotropy and capillarity. Such QDs are characterized by excellent optical properties. Moreover, because of the self-limited growth, they have the smallest inhomogeneous broadening reported so far. More details on the growth and the optical investigation of this type of QDs can be found in [99, 102–105].

Ordered GaAs/AlGaAs QDs on Patterned GaAs(001) Substrates

Ordered GaAs/AlGaAs QDs with excellent optical properties can also be obtained by growth on patterned GaAs substrates with conventional (001) orientation [106, 107]. The growth protocol for these ordered GaAs QDs is analogue to the one for GaAs/AlGaAs QDs on nanohole templates discussed in Sect. 2.2.4. For ordered QDs, a nanohole array is created by conventional EBL instead of in situ etching. Dilute arrays (5 μm period) of nanoholes over an area of $100 \times 100\,\mu m^2$ are defined on the center of a square mesa and then transferred to the GaAs surface by $SiCl_4$ reactive ion etching. After etching, the holes have an average diameter of $\sim$80 nm and a depth of 20–30 nm. Details on the patterned substrate processing conditions are reported elsewhere [108, 109]. After the final chemical cleaning of the patterned sample with HF, the sample is introduced into the MBE system, where the surface is irradiated by atomic hydrogen to remove the residual oxide.

Figure 2.17a shows an integrated PL intensity map of a $15 \times 15\,\mu m^2$ area at the center of the patterned region. The PL signal is integrated over the energy range between 1.57 and 1.69 eV, which is the emission wavelength of these GaAs QDs. Within this area, an array of 3×3 ordered GaAs QDs is clearly observed. The relative intensity difference between the ordered QDs

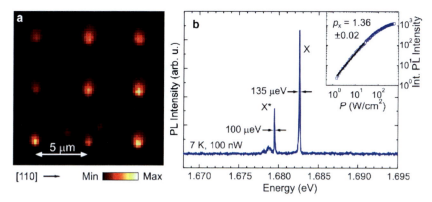

Fig. 2.17. (a) $15 \times 15\,\mu m^2$ integrated PL intensity map of site-controlled GaAs/AlGaAs QDs on GaAs(001). (b) PL spectrum of a single site-controlled GaAs QDs. The dominant peaks are assigned to neutral exciton and charged exciton recombinations. The *inset* of (b) shows the integrated PL intensity as a function of excitation power for the neutral exciton line. Adapted from [106, 107]

is due to the fact that no illumination/collection optimization is performed during the scan. Apart from the signal from the ordered GaAs QDs, some small spots are also observed on the unpatterned area. We attribute these spots to naturally formed QDs originating from QW thickness fluctuations (see Fig. 2.10).

Figure 2.17b shows the PL spectrum of a representative site-controlled QD collected at the excitation power of 100 nW (12.7 W cm^{-2}). The spectrum is dominated by a neutral exciton peak (at 1.6826 eV for this QD) labeled as X. A second sharp peak is observed at lower energy and is labeled as X*. This line becomes more pronounced at higher excitation (not shown) and is attributed to a singly charged exciton (trion) transitions. The broad peak at lower energies ($\sim$1.6787 eV) probably results from the transition of charged and neutral multiexcitonic states. The neutral exciton peak exhibits a narrow line with a linewidth of 135 µeV while the charged exciton line has a linewidth of 100 µeV. The slightly larger linewidth of X line compared to the X* line is due to the fact that this neutral exciton line consists of two lines, which are separated by a fine structure splitting (see Chap. 7). Polarization-dependent PL reveals the intrinsic linewidth of $\sim$100 µeV for such lines. Analyzing the integrated PL intensity of the X peak as a function of the excitation power (see upper right inset of Fig. 2.17b), we extract an exponent p_X of 1.36 ± 0.02 for the X peak. The deviation of the exponent from the ideal value expected for the excitonic line ($p_X = 1$) might be due to the effect of nonradiative recombination centers near the QDs, which affect the carrier distribution at different carrier densities. However, negligible background signal is observed around the exciton peak (X) even at high excitation power, indicating a possible use of the exciton transition from ordered QDs as a source of triggered single photons at high repetition rate (see Chaps. 6, 7, and 11).

2.3.4 Guided Self-Assembly by Growth on Hole-Patterned GaAs(001)

The nucleation sites of self-assembled SK QDs can be controlled by performing the growth on substrates patterned with nanoholes [106, 108–110]. This approach has been applied to different material systems and is here illustrated for In(Ga)As/GaAs(001) QDs. Figure 2.18a shows a 3D AFM image of a GaAs surface with holes regularly arranged along [100] and [010] directions. After atomic hydrogen cleaning and deposition of an 18 ML GaAs buffer, an enlargement of the hole diameter and a reduction of the hole depth are observed (Fig. 2.18b). When the deposition proceeds further the holes show clear facets (Fig. 2.18c) [109]. These facetted holes elongate along the [110] direction. Comparing the observed sidewall slope ($\sim 20°$) with a previous study of the growth on patterned substrates [111], the facet in [1$\bar{1}$0] direction is assigned to a (1$\bar{1}$4)B plane. This regularly faceted hole array can be used as a template for the InAs growth. With increasing GaAs deposition we observe two closely spaced holes forming at each patterned site (not shown). Finally the holes disappear and a flat GaAs surface is recovered. Therefore a careful control of the growth of the GaAs buffer layer is important for archiving ordered QD arrays.

For the growth of ordered InAs QDs, 2 ML InAs are deposited on top of a 18 ML thick GaAs buffer. The AFM morphology of this surface is shown in Fig. 2.18d. We observe that the InAs material preferentially attach onto

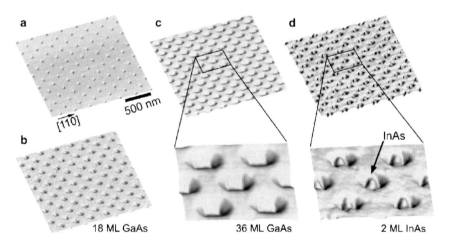

Fig. 2.18. Demonstration of the guided self-assembled growth on a substrate patterned with nanoholes. AFM images of the initial GaAs(001) surface (**a**) and after deposition of 18 ML (**b**) and 36 ML GaAs (**c**). AFM image of the surface after 18 ML GaAs and 2.0 ML InAs deposition (**d**). Self-assembled QDs are observed in the patterned holes. *Insets* of (**c**) and (**d**) are zooms. Adapted from [109]

the B-related facets and forms a ridge-like structure. Once the thickness of deposited InAs is beyond the critical thickness, self-assembled QDs form. They generally form on the ridge structure at the center of the hole. In some case, QDs separately form on each side of the B-related facets, thus producing a QD pair aligned along the $[1\bar{1}0]$ direction. This observation is consistent with the study of Kohmoto et al. [112], who showed that InAs QDs preferentially form on B-related facets. Further stacking of GaAs/InAs layers on this surface results in vertically aligned QDs. Such three-dimensionally ordered QD arrays have several interesting predicted properties [113–116].

2.4 Control of Quantum Dot Emission Energy

Postgrowth thermal treatments (also referred to as "rapid thermal annealing" or "rapid thermal processing" (RTP)) have successfully been employed to controllably blue-shift the emission of QD ensembles by promoting intermixing between the QD material and the surrounding matrix [43, 117, 118]. This approach is however not local and can only reduce the inhomogeneous broadening of QD ensembles. A local enhancement of the blue-shift can be obtained by combining RTP with selective ion implantation [119] or by using a laser as a local heat source [13]. The latter approach is particularly appealing, since a laser can also be employed as an excitation source to probe the optical properties of QDs. Here we first discuss the RTP process and then the so-called "ILP," which allows the emission of single QDs to be adjusted after fabrication.

2.4.1 Ensemble QD Emission Tuning by Rapid Thermal Processing

To illustrate the RTP process we consider a sample containing InAs QDs grown by MBE at a substrate temperature $T_s = 500°C$ and overgrown at $T_s = 470°C$. A 200-nm SiO_x layer was deposited on top of the structure to limit As desorption during the heat treatment. The RTP was performed by heating small pieces of the same sample for 30 s at different temperatures T_p in forming gas atmosphere. Figure 2.19 illustrates the effect of RTP on the low-temperature luminescence of QDs. At low excitation intensity (Fig. 2.19a) the ground state recombination is observed. With increasing T_p the PL peak gradually blue-shifts because of intermixing between the QD material and the surrounding GaAs matrix, as summarized in Fig. 2.19c. The full-width at half-maximum (FWHM) of the peak, reflecting inhomogeneities in QD size/shape/composition first increases and then drops monotonically (Fig. 2.19d). This effect is well reproduced by a simple model which approximates the QD ensemble with an ensemble of QWs with different thickness [43]. For a given diffusion length, the emission of thin QWs (or small QDs) is expected to shift faster than the emission of thicker QWs (or larger QDs),

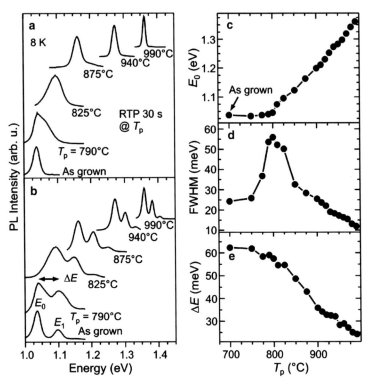

Fig. 2.19. Low temperature PL spectra of a sample with InAs/GaAs QDs RTP-treated for 30 s at the indicated temperature T_p collected at low (**a**) and high (**b**) excitation intensity. E_0 indicates the ground state QD transition and E_1 the transition from the first excited state. The effect of the RTP treatment on E_0, its FWHM and the level splitting are summarized in (**c**–**e**). Adapted from [43]

so that the high energy tail of the emission blue-shifts faster than the low energy tail. For high T_p (larger diffusion lengths) the bandgap of the material composing the QDs gets very close to the GaAs bandgap, so that variations in size produce small effects on the emission energy. For this reason the FWHM drops.

At higher excitation intensity, also the recombination from the first QD excited state becomes visible (labeled as E_1 in Fig. 2.19b). The energy separation between E_1 and the ground state emission E_0 drops monotonically with increasing T_p (Fig. 2.19e), which is attributed to a reduction of confinement with increasing diffusion lengths. The presence of excited states, together with single-QD PL spectroscopy investigations [43] clearly demonstrate that the QDs preserve their nature even at the highest T_p considered here.

2.4.2 Single QD Emission Tuning by In Situ Laser Processing

The controllable and smooth variation of emission energy produced by RTP (Fig. 2.19c) suggests a root to solve the long standing problem of inhomogeneously broadened emission of QD ensembles.

In spite of a substantial reduction of the inhomogeneous broadening of the QD ensemble, RTP produces a simultaneous shift of all the QDs in the processed sample. It is therefore evident that the only way to obtain QDs with identical emission energies (in the same sample) consists in processing each single QD independently. This can be done by means of a local heat source. Besides the lack of locality, RTP has also the disadvantage of being an ex situ method, in the sense that the processing is performed with an equipment independent from that used for the characterization. In practice the samples have to be cooled to cryogenic temperature to check the result of the RTP. This would render it extremely difficult to fine tune the emission energy of a single QD to a given value. Both issues can be solved by using the laser of a micro-PL setup as local heat source [120]. This process is referred to as "in situ laser processing" (ILP).

In order to be able to heat a QD structure from cryogenic to elevated temperatures with moderate laser powers it is necessary to reduce the thermal contact with the substrate. This is easily achieved by introducing an AlGaAs layer below the active layer and by applying an underetching step with diluted HF, as shown in Fig. 2.20a for a microdisk structure standing on a thin post. The protective SiO_x layer deposited after fabrication is also indicated. Figure 2.20b shows an example of the evolution of the PL from a microdisk

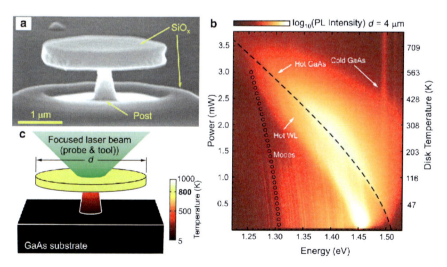

Fig. 2.20. (a) SEM image of a SiO_x-coated GaAs microdisk containing InAs QDs. (b) PL intensity map as a function of energy and laser power. (c) Calculated temperature profile for a disk similar to that used for the measurement shown in (b)

with 4 µm diameter while the laser power is ramped continuously from 150 nW up to 3.8 mW in 100 s. (The intensity map is obtained by combining spectra acquired at increasing powers.) At low power the microdisk temperature is the same as the temperature of the substrate (5 K). With increasing power the emission of the WL and GaAs matrix are observed to red-shift (see dashed lines). The same occurs for the disk optical modes (sharp lines), although the amount of the shift is much smaller. This behavior indicates a progressive heating of the microdisk. The GaAs substrate below the disk remains cold, as indicated by the vertical straight features in the PL intensity map. The red-shift of the GaAs and WL emission and that of the modes are due mainly to band-gap shrinkage and increase of the effective refractive index of the material with increasing temperature, respectively. By knowing the temperature dependence of the optical properties of GaAs we can estimate the temperature reached by the structure at a certain laser power. Since the GaAs and WL peak broaden with increasing temperature, we extract the temperature values shown on the right axis of Fig. 2.20b from the mode positions, which we use as "local thermometers." The displayed values indicate that it is possible to heat a disk from cryogenic to elevated temperatures with a few mW of laser power.

We modeled the structure as shown in Fig. 2.20c and we assumed the laser to have a Gaussian beam profile with FWHM of 1.5 µm and to be centered above the microdisk center. The power absorbed in the disk represents the heat source while the substrate is the thermal bath kept at cryogenic temperatures (5 K). The temperature profile calculated by solving the heat conduction equation by means of a finite element method is color-coded in Fig. 2.20c for a laser power of 4 mW. With this geometry the disk temperature is rather homogeneous across the disk area, with a slight drop on top of the post. The calculation also suggests that the temperature reaches a stable value within a few µs after the heat source is "switched on." For the experiments considered here, where the power was ramped to the desired value in about 1 s, the heating and cooling can thus be considered instantaneous. The post is the channel through which the heat flows towards the substrate. Thus, its size critically affects the value of temperature reachable for a given laser power. This is confirmed by the experimental observation that disks with relatively large posts require several tens of mW to be appreciably heated. Since the structure is heated inside the cryostat, it is sufficient to reduce the laser power back to a few nW (typical power used for single QD spectroscopy) to measure the effect of the thermal treatment. This allows us to controllably blue-shift the emission of selected QDs by applying different in situ heating steps followed by the measurement of the resulting spectrum.

Figure 2.21a demonstrates that ILP can indeed be used to bring into resonance spatially separated QDs. For this experiment we considered single InAs QDs grown at $T_s = 500°C$ and treated by partial capping and annealing (see Sect. 2.2.2). We took the positively charged trion X_T^+ of a quantum dot, labeled as QDT, as our target. By laser heating for 7 s at increasingly high

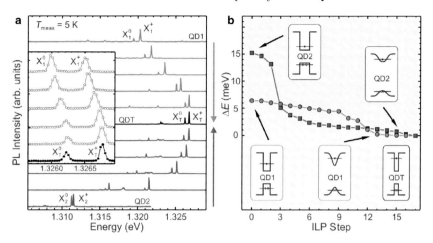

Fig. 2.21. (a) PL spectra illustrating the ILP-tuning of the positive trion emission of three spatially separated QDs into resonance. *Inset:* PL spectra showing the fine-tuning of the QD1 emission. (b) Shifts of the QDs shown in (a) as a function of heating step. *Insets:* schematics of the band structure of the QDs prior to and after the ILP treatment. Adapted from [120]

powers (up to a few mW) we gradually blue-shifted the emission of two QDs, QD1 and QD2, until their X^+ transitions, X_1^+ and X_2^+, reached the same energy as X_T^+. At the beginning of the experiment we performed relatively large steps, while at the end, when X_1^+/X_2^+ approached X_T^+ we performed smaller steps. Figure 2.21b shows a summary of the X_1^+ and X_2^+ positions as a function of ILP step. We can illustrate the mechanism leading to the blue-shift by assuming that the different QDs have the same homogeneous composition, but different sizes prior to processing. The interdiffusion occurring during the heat treatment smoothes the interfaces between QD and surrounding barrier, leading to shallower confinement potentials for electrons and holes, as depicted in the insets of Fig. 2.21b. In reality different QDs are characterized by different atomic arrangements, since all the processes taking place during InAs deposition and subsequent GaAs-overgrowth can be described by statistics. This renders the structure, and hence the optical spectra of each QD unique (note, e.g., the X^0–X^+ separation in the three QDs shown in Fig. 2.21a prior to ILP). From Fig. 2.21 we observe that ILP allows the QD emission to be shifted in a broad range (about 15 meV for QD2), comparable with typical inhomogeneous broadening values. The emission lines remain resolution-limited throughout the process, suggesting that no significant damage is produced by the ILP. The processing is at the same time at least as accurate as the resolution of our spectrometer (70 μeV). This is illustrated more in detail in the inset of Fig. 2.21a, which displays a series of spectra corresponding to the final ILP steps for QD1. Since the heating steps can be made arbitrarily small, we are confident that lines of different QDs can be brought into perfect resonance, i.e., to coincide within their intrinsic linewidths.

2.5 Summary and Outlook

In this chapter we have provided an overview on different approaches to fabricate optically active QDs. At present, the most successful methods to create QDs with excellent optical properties rely on self-assembled growth. In particular, Stranski–Krastanow (SK) In(Ga)As QDs in GaAs matrix have become a paradigm for studying the properties of single QDs and test their potential applications. We have therefore illustrated the basic principles behind SK growth and the morphological properties of InAs/GaAs(001) QDs, also with reference to another well-known SK system, namely SiGe on Si(001). Techniques which are commonly employed to tune the QD density and the emission energy based were also discussed.

Other self-assembled growth strategies were presented, such as thickness fluctuations in thin QWs, QDs obtained by combination of SK growth and in situ etching, QDs grown by droplet epitaxy, strain induced QDs and QDs in nanowires.

Thanks to the availability of high quality QDs, enormous progress has been reached in the understanding of their properties. With little doubt future progress will be tightly linked to technological advances in the fabrication of QDs with deterministically controlled spatial and electronic properties.

For this reason we have discussed approaches to control the position of QDs, either by pure top-down methods or by a combination of top-down lithography and bottom-up self-assembled growth. The latter ("hybrid") approach is promising because it can provide QDs which do not suffer from defects created during processing.

The ultimate level of control will be the tuning of the electronic structure of optically active QDs. Since the creation of QDs with structures which are known a priori seems at present out of reach, we have presented a recent top-down method which allows the emission of single QDs to be finely tuned after growth.

By combining bottom-up epitaxial growth with top-down control on spatial and spectral properties, we envision the creation of QD structures with known emission and position which will be required for their use in QD-based devices.

Acknowledgments

The authors thank E. Pelucchi, M. Sopanen and J. Riikonen for providing figures. L. Wang, F. Ding, M. Benyoucef, C. Manzano, G. Costantini and R. Songmuang are gratefully acknowledged for their contribution. This work was supported by the DFG FOR 730, the SFB/TR21, the BMBF (no. 01BM459 and 03N8711) and by the EU (no. 012150).

References

1. Z. Yuan, B.E. Kardynal, R.M. Stevenson, A.J. Shields, C.J. Lobo, K. Cooper, N.S. Beattie, D.A. Ritchie, M. Pepper, Science **295**, 102 (2002)
2. C. Böckler, S. Reitzenstein, C. Kistner, R. Debusmann, A. Löffler, T. Kida, S. Höfling, A. Forchel, L. Grenouillet, J. Claudon, J.M. Gérard, Appl. Phys. Lett. **92**, 1107 (2008)
3. H. Drexler, D. Leonard, W. Hansen, J.P. Kotthaus, P.M. Petroff, Phys. Rev. Lett. **73**, 2252 (1994)
4. P. Michler, A. Kiraz, C. Becher, W.V. Schoenfeld, P.M. Petroff, L.D. Zhang, E. Hu, A. Imamoğlu, Science **290**, 2282 (2000)
5. A. Badolato, K. Hennessy, M. Atatüre, J. Dreiser, E. Hu, P.M. Petroff, A. Imamoğlu, Science **308**, 1158 (2005)
6. H.J. Krenner, S. Stufler, M. Sabathil, E.C. Clark, P. Ester, M. Bichler, G. Abstreiter, J.J. Finley, A. Zrenner, New J. Phys. **7**, 184 (2005)
7. J.P. Reithmaier, G. Sek, A. Löffler, C. Hofmann, S. Kuhn, S. Reitzenstein, L.V. Keldysh, V.D. Kulakovskii, T.L. Reinecke, A. Forchel, Nature **432**, 197 (2004)
8. T. Yoshie, A. Scherer, J. Hendrickson, G. Khitrova, H.M. Gibbs, G. Rupper, C. Ell, O.B. Shchekin, D.G. Deppe, Nature **432**, 200 (2004)
9. E. Peter, P. Senellart, D. Martrou, A. Lemaître, J. Hours, J.M. Gérard, J. Bloch, Phys. Rev. Lett. **95**, 067401 (2005)
10. K. Hennessy, A. Badolato, M. Winger, D. Gerace, M. Atatüre, S. Gulde, S. Fält, E.L. Hu, A. Imamoğlu, Nature **445**, 896 (2007)
11. K. Srinivasan, O. Painter, Nature **450**, 862 (2007)
12. P. Borri, W. Langbein, S. Schneider, U. Woggon, R.L. Sellin, D. Ouyang, D. Bimberg, Phys. Rev. Lett. **87**, 157401 (2001)
13. K. Brunner, U. Bockelmann, G. Abstreiter, M. Walther, G. Böhm, G. Tränkle, G. Weimann, Phys. Rev. Lett. **69**, 3216 (1992)
14. K. Brunner, G. Abstreiter, G. Böhm, G. Tränkle, G. Weimann, Phys. Rev. Lett. **73**, 1138 (1994)
15. A. Zrenner, L.V. Butov, M. Hagn, G. Abstreiter, G. Böhm, G. Weimann, Phys. Rev. Lett. **72**, 3382 (1994)
16. J.Y. Marzin, J.M. Gérard, A. Izraël, D. Barrier, G. Bastard, Phys. Rev. Lett. **73**, 716 (1994)
17. A. Cho (ed.), *Molecular Beam Epitaxy*, 1st edn. (Springer, Berlin, 1997)
18. G.B. Stringfellow, *Organometallic Vapor-Phase Epitaxy: Theory and Practice*, 2nd edn. (Academic, London, 1999)
19. E. Bauer, Z. Kristtallogr. **110**, 372 (1958)
20. D.J. Eaglesham, M. Cerullo, Phys. Rev. Lett. **64**, 1943 (1990)
21. Y.W. Mo, D.E. Savage, B.S. Swartzentruber, M.G. Lagally, Phys. Rev. Lett. **65**, 1020 (1990)
22. D. Leonard, M. Krishnamurthy, C.M. Reaves, S.P. Denbaars, P.M. Petroff, Appl. Phys. Lett. **63**, 3203 (1993)
23. V.A. Shchukin, N.N. Ledentsov, P.S. Kop'ev, D. Bimberg, Phys. Rev. Lett. **75**, 2968 (1995)
24. G. Medeiros-Ribeiro, A.M. Bratkovski, T.I. Kamins, D.A.A. Ohlberg, R.S. Williams, Science **279**, 353 (1998)
25. J. Tersoff, B. Spencer, A. Rastelli, H. von Känel, Phys. Rev. Lett. **89**, 196104 (2002)

26. F.M. Ross, J. Tersoff, R.M. Tromp, Phys. Rev. Lett. **80**, 984 (1998)
27. J. Tersoff, F.K. LeGoues, Phys. Rev. Lett. **72**, 3570 (1994)
28. A. Rastelli, M. Stoffel, U. Denker, T. Merdzhanova, O.G. Schmidt, Phys. Stat. Sol. (a) **203**, 3506 (2006)
29. D. Bimberg, M. Grundmann, N.N. Ledenstov, *Quantum Dot Heterostructures*, 1st edn. (Wiley, Chichester, 1999)
30. A. Rastelli, H. von Känel, Surf. Sci. Lett. **515**, L493 (2002)
31. J. Márquez, L. Geelhaar, K. Jacobi, Appl. Phys. Lett. **78**, 2309 (2001)
32. P. Kratzer, Q.K.K. Liu, P. Acosta-Diaz, C. Manzano, G. Costantini, R. Songmuang, A. Rastelli, O.G. Schmidt, K. Kern, Phys. Rev. B **73**, 205347 (2006)
33. I. Daruka, J. Tersoff, A. Barabási, Phys. Rev. Lett. **82**, 2753 (1999)
34. G. Costantini, A. Rastelli, C. Manzano, R. Songmuang, O.G. Schmidt, K. Kern, H.V. Känel, Appl. Phys. Lett. **85**, 5673 (2004)
35. G. Costantini, A. Rastelli, C. Manzano, P. Acosta-Diaz, G. Katsaros, R. Songmuang, O.G. Schmidt, H. von Känel, K. Kern, J. Cryst. Growth **278**, 38 (2005)
36. F. Ross, R. Tromp, M. Reuter, Science **286**, 1931 (1999)
37. F. Montalenti, P. Raiteri, D.B. Migas, H. von Känel, A. Rastelli, C. Manzano, G. Costantini, U. Denker, O.G. Schmidt, K. Kern, L. Miglio, Phys. Rev. Lett. **93**, 216102 (2004)
38. M. Stoffel, A. Rastelli, J. Tersoff, T. Merdzhanova, O.G. Schmidt, Phys. Rev. B **74**, 155326 (2006)
39. P.B. Joyce, T.J. Krzyzewski, G.R. Bell, B.A. Joyce, T.S. Jones, Phys. Rev. B **58**, 15981 (1998)
40. I. Kegel, T.H. Metzger, A. Lorke, J. Peisl, J. Stangl, G. Bauer, J.M. García, P.M. Petroff, Phys. Rev. Lett. **85**, 1694 (2000)
41. T.I. Kamins, G. Medeiros-Ribeiro, D.A.A. Ohlberg, R. Stanley Williams, J. Appl. Phys. **85**, 1159 (1999)
42. A. Rastelli, M. Stoffel, A. Malachias, T. Merdzhanova, G. Katsaros, K. Kern, T.H. Metzger, O.G. Schmidt, Nano Lett. **8**, 1404 (2008)
43. A. Rastelli, S.M. Ulrich, E.M. Pavelescu, T. Leinonen, M. Pessa, P. Michler, O.G. Schmidt, Superlattice Microst. **36**, 181 (2004)
44. J.M. García, G. Medeiros-Ribeiro, K. Schmidt, T. Ngo, J.L. Feng, A. Lorke, J. Kotthaus, P.M. Petroff, Appl. Phys. Lett. **71**, 2014 (1997)
45. P.B. Joyce, T.J. Krzyzewski, P.H. Steans, G.R. Bell, J.H. Neave, T.S. Jones, Surf. Sci. **492**, 345 (2001)
46. R. Songmuang, S. Kiravittaya, O.G. Schmidt, J. Cryst. Growth **249**, 416 (2003)
47. G. Costantini, A. Rastelli, C. Manzano, P. Acosta-Diaz, R. Songmuang, G. Katsaros, O.G. Schmidt, K. Kern, Phys. Rev. Lett. **96**, 226106 (2006)
48. A. Rastelli, M. Kummer, H. von Känel, Phys. Rev. Lett. **87**, 256101 (2001)
49. S. Kiravittaya, R. Songmuang, A. Rastelli, H. Heidemeyer, O.G. Schmidt, Nanoscale Res. Lett. **1**, 1 (2006)
50. L. Wang, A. Rastelli, O.G. Schmidt, J. Appl. Phys. **100**, 4313 (2006)
51. S. Fafard, Z.R. Wasilewski, C.N. Allen, D. Picard, M. Spanner, J.P. McCaffrey, P.G. Piva, Phys. Rev. B **59**, 15368 (1999)
52. T. Inoue, T. Kita, O. Wada, M. Konno, T. Yaguchi, T. Kamino, Appl. Phys. Lett. **92**, 1902 (2008)
53. Q. Xie, A. Madhukar, P. Chen, N.P. Kobayashi, Phys. Rev. Lett. **75**, 2542 (1995)

54. G.S. Solomon, J.A. Trezza, A.F. Marshall, J.S. Harris, Jr., Phys. Rev. Lett. **76**, 952 (1996)
55. J. Tersoff, C. Teichert, M.G. Lagally, Phys. Rev. Lett. **76**, 1675 (1996)
56. M.O. Lipinski, H. Schuler, O.G. Schmidt, K. Eberl, Appl. Phys. Lett. **77**, 1789 (2000)
57. P.B. Joyce, T.J. Krzyzewski, P.H. Steans, G.R. Bell, J.H. Neave, T.S. Jones, J. Cryst. Growth **244**, 39 (2002)
58. D.M. Bruls, P.M. Koenraad, H.W.M. Salemink, J.H. Wolter, M. Hopkinson, M.S. Skolnick, Appl. Phys. Lett. **82**, 3758 (2003)
59. P. Howe, E.C. Le Ru, E. Clarke, B. Abbey, R. Murray, T.S. Jones, J. Appl. Phys. **95**, 2998 (2004)
60. A. Babinski, M. Potemski, S. Raymond, Z.R. Wasilewski, Physica E **40**, 2078 (2008)
61. L.C. Andreani, G. Panzarini, J. Gérard, Phys. Rev. B **60**, 13276 (1999)
62. T. Kaneko, P. Šmilauer, B.A. Joyce, T. Kawamura, D.D. Vvedensky, Phys. Rev. Lett. **74**, 3289 (1995)
63. H. Schuler, T. Kaneko, M. Lipinski, K. Eberl, Semicond. Sci. Technol. **15**, 169 (2000)
64. H. Schuler, N.Y. Jin-Phillipp, F. Phillipp, K. Eberl, Semicond. Sci. Technol. **13**, 1341 (1998)
65. S. Kiravittaya, Y. Nakamura, O.G. Schmidt, Physica E **13**, 224 (2002)
66. F. Ding, L. Wang, S. Kiravittaya, E. Müller, A. Rastelli, O.G. Schmidt, Appl. Phys. Lett. **90**, 173104 (2007)
67. R. Songmuang, S. Kiravittaya, O.G. Schmidt, Appl. Phys. Lett. **82**, 2892 (2003)
68. S. Kiravittaya, R. Songmuang, N.Y. Jin-Phillipp, S. Panyakeow, O.G. Schmidt, J. Cryst. Growth **251**, 258 (2003)
69. N.A.J.M. Kleemans, I.M.A. Bominaar-Silkens, V.M. Fomin, V.N. Gladilin, D. Granados, A.G. Taboada, J.M. García, P. Offermans, U. Zeitler, P.C.M. Christianen, J.C. Maan, J.T. Devreese, P.M. Koenraad, Phys. Rev. Lett. **99**, 146808 (2007)
70. A. Rastelli, R. Songmuang, O.G. Schmidt, Physica E **23**, 384 (2004)
71. A. Rastelli, S. Stufler, A. Schliwa, R. Songmuang, C. Manzano, G. Costantini, K. Kern, A. Zrenner, D. Bimberg, O.G. Schmidt, Phys. Rev. Lett. **92**(16), 166104 (2004)
72. M. Benyoucef, A. Rastelli, O.G. Schmidt, S.M. Ulrich, P. Michler, Nanoscale Res. Lett. **1**, 172 (2006)
73. Y. Sidor, B. Partoens, F.M. Peeters, N. Schildermans, M. Hayne, V.V. Moshchalkov, A. Rastelli, O.G. Schmidt, Phys. Rev. B **73**, 155334 (2006)
74. B. Krause, T.H. Metzger, A. Rastelli, R. Songmuang, S. Kiravittaya, O.G. Schmidt, Phys. Rev. B **72**, 085339 (2005)
75. L. Wang, A. Rastelli, S. Kiravittaya, P. Atkinson, F. Ding, C.C. Bof Bufon, C. Hermannstädter, M. Witzany, G.J. Beirne, P. Michler, O.G. Schmidt, New J. Phys. **10**, 5010 (2008)
76. G.J. Beirne, C. Hermannstädter, L. Wang, A. Rastelli, O.G. Schmidt, P. Michler, Phys. Rev. Lett. **96**, 137401 (2006)
77. L. Wang, A. Rastelli, S. Kiravittaya, R. Songmuang, O.G. Schmidt, B. Krause, T.H. Metzger, Nanoscale Res. Lett. **1**, 74 (2006)
78. C.D. Lee, C. Park, H. Lee, K.S. Lee, S.J. Park, C. Park, S. Noh, N. Koguchi, Jpn. J. Appl. Phys. **37**, 7158 (1998)

79. K. Watanabe, N. Koguchi, Y. Gotoh, Jpn. J. Appl. Phys. **39**, 79 (2000)
80. T. Mano, K. Watanabe, S. Tsukamoto, N. Koguchi, H. Fujioka, M. Oshima, C.D. Lee, J.Y. Leem, H.J. Lee, S.K. Noh, Appl. Phys. Lett. **76**, 3543 (2000)
81. M. Yamagiwa, T. Mano, T. Kuroda, T. Tateno, K. Sakoda, G. Kido, N. Koguchi, F. Minami, Appl. Phys. Lett. **89**, 113115 (2006)
82. C. Heyn, A. Stemmann, A. Schramm, H. Welsch, W. Hansen, Á. Nemcsics, Phys. Rev. B **76**, 075317 (2007)
83. T. Kuroda, T. Mano, T. Ochiai, S. Sanguinetti, K. Sakoda, G. Kido, N. Koguchi, Phys. Rev. B **72**, 205301 (2005)
84. Z.M. Wang, K. Holmes, Y.I. Mazur, K.A. Ramsey, G.J. Salamo, Nanoscale Res. Lett. **1**, 57 (2006)
85. R. Pomraenke, C. Lienau, Y.I. Mazur, Z.M. Wang, B. Liang, G.G. Tarasov, G.J. Salamo, Phys. Rev. B **77**, 075314 (2008)
86. J.H. Lee, K. Sablon, Z.M. Wang, G.J. Salamo, J. Appl. Phys. **103**, 4301 (2008)
87. H. Lipsanen, M. Sopanen, J. Ahopelto, Phys. Rev. B **51**, 13868 (1995)
88. M. Sopanen, H. Lipsanen, J. Ahopelto, Appl. Phys. Lett. **66**, 2364 (1995)
89. J. Riikonen, J. Sormunen, M. Mattila, M. Sopanen, H. Lipsanen, Jpn. J. Appl. Phys. **44**, L518 (2005)
90. L. Samuelson, C. Thelander, M.T. Björk, M. Borgström, K. Deppert, K.A. Dick, A.E. Hansen, T. Mårtensson, N. Panev, A.I. Persson, W. Seifert, N. Sköld, M.W. Larsson, L.R. Wallenberg, Physica E **25**, 313 (2004)
91. M. Borgstrom, V. Zwiller, E. Muller, A. Imamoglu, Nano Lett. **5**, 1439 (2005)
92. E. Minot, F. Kelkensberg, M. vanKouwen, J. vanDam, L. Kouwenhoven, V. Zwiller, M. Borgstrom, O. Wunnicke, M. Verheijen, E. Bakkers, Nano Lett. **7**, 367 (2007)
93. W. Wegscheider, G. Schedelbeck, G. Abstreiter, M. Rother, M. Bichler, Phys. Rev. Lett. **79**, 1917 (1997)
94. G. Schedelbeck, W. Wegscheider, M. Bichler, G. Abstreiter, Science **278**, 1792 (1997)
95. J. Bauer, D. Schuh, E. Uccelli, R. Schulz, A. Kress, F. Hofbauer, J.J. Finley, G. Abstreiter, Appl. Phys. Lett. **85**, 4750 (2004)
96. E. Uccelli, M. Bichler, S. Nurnberger, G. Abstreiter, A. Fontcuberta i Morral, Nanotechnology **19**, 045303 (2008)
97. G.W. Bryant, G.S. Solomon (eds.), *Optics of Quantum Dots and Wires* (Artech House, London, 2005)
98. M. Bayer, M. Korkusinski, P. Hawrylak, T. Gutbrod, M. Michel, A. Forchel, Phys. Rev. Lett. **90**, 186801 (2003)
99. A. Hartmann, L. Loubies, F. Reinhardt, E. Kapon, Appl. Phys. Lett. **71**, 1314 (1997)
100. A. Hartmann, Y. Ducommun, L. Loubies, K. Leifer, E. Kapon, Appl. Phys. Lett. **73**, 2322 (1998)
101. A. Hartmann, Y. Ducommun, E. Kapon, U. Hohenester, E. Molinari, Phys. Rev. Lett. **84**, 5648 (2000)
102. E. Pelucchi, M. Baier, Y. Ducommun, S. Watanabe, E. Kapon, Phys. Stat. Sol. (b) **238**, 233 (2003)
103. E. Pelucchi, S. Watanabe, K. Leifer, B. Dwir, E. Kapon, Physica E **23**, 476 (2004)
104. Q. Zhu, K. Karlsson, E. Pelucchi, E. Kapon, Nano Lett **7**, 2227 (2007)
105. E. Pelucchi, S. Watanabe, K. Leifer, Q. Zhu, B. Dwir, P. DeLosRios, E. Kapon, Nano Lett. **7**, 1282 (2007)

106. S. Kiravittaya, B. Benyoucef, R. Zapf-Gottwick, A. Rastelli, O.G. Schmidt, Appl. Phys. Lett. **89**, 233102 (2006)
107. S. Kiravittaya, B. Benyoucef, R. Zapf-Gottwick, A. Rastelli, O.G. Schmidt, Physica E **40**, 1909 (2008)
108. O.G. Schmidt (ed.), *Lateral Alignment of Epitaxial Quantum Dots* (Springer, Berlin, 2007)
109. H. Heidemeyer, C. Müller, O.G. Schmidt, J. Cryst. Growth **261**, 444 (2004)
110. S. Kiravittaya, H. Heidemeyer, O.G. Schmidt, Physica E **23**, 253 (2004)
111. T. Takebe, M. Fujii, T. Yamamoto, K. Fujita, T. Watanabe, J. Vac. Sci. Technol. B **14**, 2731 (1996)
112. S. Kohmoto, H. Nakamura, T. Ishikawa, S. Nishikawa, T. Nishimura, K. Asakawa, Mat. Sci. Eng. B **88**, 292 (2002)
113. T. Takagahara, Surf. Sci. **267**, 310 (1992)
114. K. Shiraishi, H. Tamura, H. Takayanagi, Appl. Phys. Lett. **78**, 3702 (2001)
115. O.L. Lazarenkova, A.A. Balandin, Phys. Rev. B **66**, 245319 (2002)
116. E.M. Kessler, M. Grochol, C. Piermarocchi, Phys. Rev. B **77**, 085306 (2008)
117. S. Malik, C. Roberts, R. Murray, M. Pate, Appl. Phys. Lett. **71**, 1987 (1997)
118. S. Fafard, C.N. Allen, Appl. Phys. Lett. **75**, 2374 (1999)
119. H.S. Djie, B.S. Ooi, V. Aimez, Appl. Phys. Lett. **87**, 261102 (2005)
120. A. Rastelli, A. Ulhaq, S. Kiravittaya, L. Wang, A. Zrenner, O.G. Schmidt, Appl. Phys. Lett. **90**, 073120 (2007)

3

Optical Properties of Epitaxially Grown Wide Bandgap Single Quantum Dots

Gerd Bacher and Tilmar Kümmell

Summary. The strength of spatially resolved optical spectroscopy is demonstrated for revealing fascinating optical, electronic and magnetic properties of wide bandgap semiconductor single quantum dots based on II–VI and III-N compounds. Strong carrier confinement and high polarity result in large Coulomb and exchange energies and efficient photon and phonon coupling of excitons, charged excitons and biexcitons. Coherent state control as well as stimulated biexciton emission in individual quantum dots is achieved and superradiance in quantum dot ensembles is demonstrated. As major steps toward nonclassical light emitters, wide bandgap single quantum dots are shown to exhibit high temperature single photon emission as well as room temperature electroluminescence. In II–VI single quantum dots isoelectronically doped with magnetic ions, nanoscale magnetization can be optically probed and manipulated and in the limit of a single magnetic impurity embedded in a quantum dot, even the spin state of an individual magnetic atom in a solid state matrix can be addressed, which might be a promising step toward spin based quantum information processing.

3.1 Introduction

The self-organized epitaxial growth of quantum dots (QDs) based on wide bandgap II–VI and III-N semiconductor compounds allows the realization of a rich diversity of artificial materials with exciting optical, electronic and magnetic properties [1]. Due to the lattice mismatch between QD and barrier materials, nanoislands with typical lateral extensions in the 10 nm regime and heights of a only a few nm are formed by self-assembling, leading to a three-dimensional confinement of charge carriers. In contrast to the most popular III-As and III-P systems with their typical bandgaps in the red or near infrared regime, the material classes discussed here cover the visible or even the ultraviolet spectral range. Most important are self-organized CdTe [2–4] as well as CdSe [5–10] QDs, in part doped with magnetic ions [11–15], and embedded in a diversity of barrier materials [16–19], as well as (In,Ga)N QDs with (Ga,Al)N barriers [20–26].

Although the control of size, shape and density of the QDs is not yet as advanced as for InAs/GaAs or Ge/Si QDs, high quality QDs with excellent optical properties can be realized [1]. Unique properties like large Coulomb and exchange interactions [27, 28], strong phonon and photon coupling [18, 29–31], pronounced internal electric fields in wurtzite III-N QDs [32, 33] and the ability to dope especially the II–VI QDs isoelectronically with magnetic ions [11–15, 34] make these wide bandgap QDs quite interesting from both a basic physics and an application point of view. Concerning the latter one, ensembles of QDs have been successfully incorporated into electrically driven light emitters [35–39], and quite recently, the field of quantum information science has triggered a large interest in device concepts based on single quantum dots (SQDs). This includes single photon sources [40], where the short emission wavelength in the visible spectral range might be of interest for communication in free space or via plastic fibres [41], or even quantum computation schemes based on an early suggestion of Loss et al. [42], where the spin degree of freedom of individual particles (carriers, magnetic ions, etc.) in a SQD can be used to represent a Qubit. A thorough understanding of the optical, electronic and magnetic properties not only of QD ensembles, but in particular of individual QDs is therefore a key issue both for further pushing the development of high quality materials with well-designed properties and for stimulating new device concepts.

Optical spectroscopy has been shown to be a powerful tool for studying individual quantum dots. Due to its high sensitivity, photoluminescence (PL) spectroscopy is certainly one of the mostly used experimental techniques. The high spatial resolution required for addressing single quantum dots (SQDs) is usually obtained by a micro-PL setup in combination with lithographically defined mesas [7] or metal nanoapertures [11, 13, 32] for spatial selection and an energy dispersive element like a monochromator for spectral selection of an individual QD. Time-resolved linear [30, 43, 44] and nonlinear [45, 46] optics as well as the application of external magnetic [27, 28] or electric fields [47–50] are standard methods for investigating wide bandgap II–VI and III-N SQDs in the meanwhile. Moreover, it has been shown that such SQDs can be embedded into an electrically driven circuit, which, e.g., allows electroluminescence of an individual QD even up to room temperature [51] or the electrical control of nanomagnetism in a semiconductor [49].

Here, we review some recent results obtained from optical spectroscopy on epitaxially grown SQDs based on II–VI and III-N compounds. In an extension of our former review [1], we mainly focus here on some of the most outstanding properties of these materials: the strong carrier confinement and the ability to include magnetic ions into the matrix of the QD. As a consequence of the strong carrier confinement and the high polarity of the materials, large Coulomb and exchange energies as well as strong phonon and photon coupling are obtained. This issue is discussed in the first part of the article and results obtained for both, II–VI and III-N SQDs are presented. The subsequent section is devoted to the application potential of (In,Ga)N and CdSe SQDs for high temperature single photon sources. Optically triggered single photon emission up to $T = 200\,\mathrm{K}$ as well as electrically driven SQD electroluminescence at room temperature are shown. The last part of the article concentrates on magnetically doped QDs and thus on II–VI materials, where the exchange interaction between individual charge carriers and magnetic ions leads to SQDs with unique magnetic properties. We demonstrate optical monitoring and optical control of nanomagnetism in a semiconductor as well as the ability to optically probe the spin state of an individual magnetic ion in a solid state matrix.

3.2 Single Quantum Dots with Strong Carrier Confinement

Usually, the bandgap difference between QD and surrounding matrix are much larger in wide bandgap II–VI and III-N compound semiconductors as compared to narrow gap III-As or III-P materials. In addition, the QDs in the material systems considered here are quite small with diameters in the order of 10 nm and heights of a few nm. Both leads to a strong carrier confinement and by special growth procedures, like migration enhanced epitaxy [52] or atomic layer epitaxy [53], dot size and carrier confinement can be modified in a certain range. The broad palette of material combinations especially in II–VI materials paves a way to tailor the confinement energy further. For instance in the CdSe/ZnSe system, the barrier bandgap can be increased by an admixture of sulphur, leading to higher barriers especially for the holes. Mg admixture into the barrier material improves carrier confinement both in Te-based [54] and in Se-based [18, 19] systems. This indeed helps, e.g., to achieve high-temperature operation of SQD light emitters, as will be discussed later.

Due to the high polarity in wide-bandgap semiconductors, Coulomb and exchange energies are up to an order of magnitude higher than in typical III–V systems [27]. This enables on the one hand an easier experimental access to fundamental properties of semiconductor QDs, and on the other hand, excitonic effects show an enhanced robustness with respect to high temperatures. In this section we will first focus on some prominent experiments exploiting characteristic linear and nonlinear optical properties of wide-bandgap semiconductor SQDs, and then discuss some consequences for potential room temperature applications.

3.2.1 Exciton Complexes

Biexcitons

II–VI Materials

The biexciton (denoted in this article as XX or X_2) is a four-particle state. In its lowest energy state configuration, two electrons and two holes with antiparallel spins occupy the first quantized state of the conduction and the valence band in the QD, respectively. The biexciton state is therefore a singlet state with a total spin of $J = 0$. A direct optical transition to the QD ground state (i.e., no excess carriers, $J = 0$) is not possible due to the optical selection rules. Thus, the exciton state X represents the final state for the biexcitonic recombination. In II–VI semiconductors, as in III–V materials with a zincblende crystal lattice, Coulomb interaction leads to positive biexciton binding energies, i.e., the energetic distance between XX and X is smaller than the energy difference between the exciton state and the ground state. A typical optical fingerprint for the XX is therefore an additional PL line at the low energy side of the exciton emission X that exhibits a strong (quadratic) dependence on the excitation power.

This behavior is clearly visible in the left panel of Fig. 3.1. At low excitation density, the PL spectrum of CdSe/ZnSe SQDs consists of emission peaks stemming from exciton recombination of two individual QDs. With rising excitation density additional lines emerge, red shifted by about 24 meV with respect to the excitonic emission X and rapidly increasing in intensity, which can be attributed to biexciton emission X_2. The biexciton binding energy is obviously much larger than in III-As based QDs, where typical values of a few meV have been determined.

When having a closer look on the PL spectra presented in Fig. 3.1, some more information can be extracted. One should have in mind that in QDs, the light hole level is shifted to higher energies due to strain and confinement and thus, excitons are formed between electrons and heavy holes. The ground state of a heavy hole exciton in a SQD is a spin quadruplet, which can be

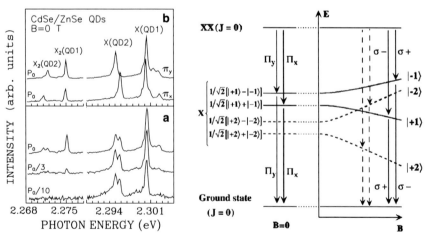

Fig. 3.1. *Left side:* Excitonic (X) and biexcitonic (X_2) emission from two individual CdSe/ZnSe SQDs for different excitation power P_0. The PL spectra shown in the *lower panel* are unpolarized, the data presented in the *upper panel* represent linearly polarized PL spectra (π_x and π_y, respectively) [27]. *Right side:* Energy level scheme for the biexciton–exciton cascade in a QD [55]

characterized by the z-component (=component in growth direction) of the total exciton spin, J_z. If the z-component of the electron spin, $s_z = \pm 1/2$, and the z-component of the total angular momentum (for simplicity also called "spin" in the following) of the heavy hole, $j_z = \pm 3/2$, are antiparallel, we get $J_z = s_z + j_z = \pm 1$ (the so-called "bright" exciton states), while for parallel spins of the particles $J_z = s_z + j_z = \pm 2$ (the "dark" exciton states).

In II–VI QDs the energy difference Δ_0 between bright and dark excitons that is given by the isotropic electron–hole interaction energy, amounts to about 1 meV and more [56,57] which is nearly an order of magnitude larger than in InAs/GaAs QDs. According to the selection rules, only the $|\pm 1\rangle$ states couple efficiently to the light field if no mixing between bright and dark states occurs. The anisotropic exchange interaction between electron and hole can lift the degeneracy of the spin doublets of the bright and the dark excitons, if the QD symmetry is lower than D_{2d}. In absence of a magnetic field this results in the following eigenstates of the excitons in a QD [27,55]

$$|X^{1,2}\rangle = (|+1\rangle \pm |-1\rangle)/\sqrt{2}, \qquad (3.1)$$

$$|X^{3,4}\rangle = (|+2\rangle \pm |-2\rangle)/\sqrt{2}. \qquad (3.2)$$

The resulting energy level scheme for the biexciton–exciton cascade is summarized in the right part of Fig. 3.1. Without magnetic fields, the optical transitions are expected to be linearly polarized because of the mixing of the spin states due to electron–hole exchange interaction. For high magnetic fields in Faraday geometry, the Zeeman energy dominates the exchange energy, finally leading to pure spin states as depicted in the right panel of Fig. 3.1, in agreement with experimental findings for both, CdSe/ZnSe and CdTe/ZnTe SQDs [27,28]. It should be noted that in magnetic fields also the dark states can contribute to the emission. This is generally valid for magnetic fields in Voigt geometry [58,59], but even in Faraday geometry dark states have been observed for strongly asymmetric InAs/GaAs [59] and CdTe/CdMgTe SQDs [55] due to a mixing with bright exciton states.

As can be seen in Fig. 3.1, the excitonic fine structure is reflected both in the exciton and in the biexciton recombination: SQD1 does not show a significant splitting of the exciton PL signal, while SQD2 exhibits a doublet with an energy separation of almost 1 meV indicating a reduced QD symmetry. Exactly the same behavior is observed in the corresponding biexciton lines. Moreover, the high energy component of the X emission in SQD2 (π_x-polarized) corresponds to the low energy component of the X_2 emission and vice versa, in agreement with the energy level scheme. All these effects are easily accessible in wide bandgap II–VI QDs because the characteristic energy splittings are significantly enhanced with respect to III-As semiconductor QDs. Recently, it was found that the exchange splitting can be manipulated by a post growth annealing step: In CdTe/ZnTe QDs the distribution of the splitting was reduced after exposing the QDs to temperatures of 420°C for

15 s [60]. In CdSe/ZnSe QDs, a slightly different experiment was performed: QDs embedded into small mesa structures were annealed for 30 s at temperatures between 100°C and 180°C. With this process, actually even the sign of the exchange splitting could be reversed [61].

Thanks to the large biexciton binding energy, II–VI QDs were the first, where the biexciton–exciton cascade could be traced directly in the time domain on a SQD level. Figure 3.2 depicts transient PL spectra (left) of both emission lines and the time-dependent intensity of the exciton and the biexciton signal (right) [30]. The biexciton emission shows a monoexponential decay with a time constant of 310 ps. The exciton reveals a more complex behavior: The onset of the exciton line is delayed, resulting in a "plateau-like" characteristics of the exciton decay curve. The excitation density in this experiment was set to a value where an average number of two electron hole pairs per excitation pulse in the SQD was generated. Model calculations taking into account the biexciton state, the bright and the dark exciton states and the "empty" QD (corresponding to a QD population with 2, 1, and 0 excitons, respectively), confirm that the exciton state is fed by the biexciton recombination causing the delayed onset and the "plateau-like" characteristics of the exciton emission dynamics.

This cascade-like process is of high importance when considering applications like single photon sources: Even if the quantum dot is "overpumped" by the excitation, the selection rules ensure a process where photons are emitted one-by-one and finally, each cascade ends with a single exciton recombination process at a well defined energy. While in InAs SQDs, contributions of the

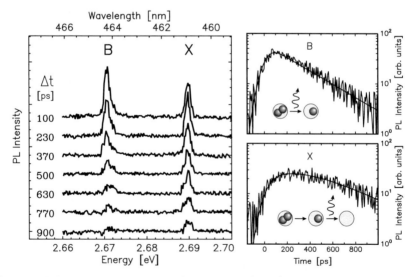

Fig. 3.2. *Left:* Transient PL spectra from a single CdSe/ZnSe QD showing the single exciton X and the biexciton transition (here denoted by B). *Right:* Decay curves for the exciton and the biexciton PL signal. After [30]

p-, d- or f-shell could be easily observed in PL, this seems to be suppressed in wide bandgap QDs of small size, probably due to Auger recombination. In photoluminescence excitation (PLE) experiments, however, higher QD states up to the d-shell could be verified in CdTe/ZnTe QDs [62], confirming the high optical quality and remarkable carrier confinement in these QDs.

III-Nitrides

During the last years, also SQDs based on III-nitrides have undergone detailed research. The situation is different here: (In,Ga)N/(Al,Ga)N quantum wells and quantum dots exhibit extraordinarily large internal electric fields in the MV/cm regime, at least in the wide-spread wurtzite phase. The internal fields have a considerable influence on both excitons and biexcitons, because they enforce a spatial separation between electrons and holes and therefore compete with the Coulomb attraction between the electrons and the holes in exciton complexes.

Considering the biexcitons, the internal fields can even result in a negative binding energy with values of $E_{XX} = -5\,\text{meV}$ for InGaN/GaN [32] and up to $E_{XX} = -30\,\text{meV}$ for GaN/AlGaN QDs [64]. Please note that we describe an antibinding state by a negative biexciton binding energy E_{XX}. Due to the strong three-dimensional carrier confinement in the QDs, these antibinding XX states are nevertheless observable. One indication of these states was given by Schömig et al. (see Fig. 3.3). With increasing excitation power, the emission from the QD ensemble shifts to higher energies, while

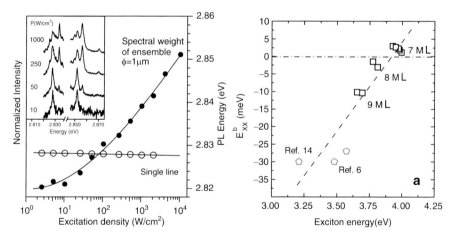

Fig. 3.3. *Left side:* Energy shift of a single InGaN/GaN QD emission line vs. excitation density compared to the energy shift of the spectral weight of a QD ensemble [32]. *Inset:* Power-dependent evolution of the PL spectrum stemming from two SQDs obtained with an aperture size of 250 nm at $T = 4\,\text{K}$. *Right side:* Dependence of E_{XX} on the exciton emission energy for GaN/AlN QDs of different vertical size [63]. The References 6 and 14 in the image correspond to [33] and [64], respectively

the PL energy of the exciton emission of an individual QD remains constant. Obviously, higher states and especially biexcitonic states with a negative binding energy, that exhibit a nonlinear power dependence, are populated, leading to the overall blue shift of the ensemble [32]. This is also clearly seen in the emission spectrum of a SQD (see inset of Fig. 3.3), where the emerging contribution of the biexciton becomes visible for increasing excitation power.

The actual biexciton binding energy depends strongly on the QD geometry, and this opens different ways to control it. For instance, it has been shown recently that E_{XX} in GaN/AlN QDs will change from a negative value to $E_{XX} > 0$, if the QDs become "small," i.e., the QD height falls below 7 ML [63]. Newest findings in very thin GaN QDs embedded into AlN nanowires that exhibit positive binding energies of up to 40 meV [65] support these results, however the exact reason for this extremely large positive biexciton binding energy is not yet clear up to now. A second approach for controlling the biexciton binding energy directly uses external electric fields to compensate for the internal fields. Jarjour et al. manipulated the electron–hole separation by contacting p-i-n structures with embedded InGaN QDs and applying voltages of up to 4 V [44]. This is sufficient for both, increasing the oscillator strength of the excitonic transition, which can be deduced from a reduced radiative lifetime, and changing the biexciton binding energy from $E_{XX} = -1.9$ meV in the unbiased case to $E_{XX} = +4.7$ meV, if a bias of 4 V is applied. In a similar way, the exciton binding energy can be manipulated, as can be seen from an atypical Stark effect: When applying a lateral electric field above $150\,\text{kV}\,\text{cm}^{-1}$, the expected redshift due to the band tilting is observed. However, for smaller fields, a distinct blueshift of the SQD signal is obtained, which is attributed to a reduction of the exciton binding energy by about 10 meV due to the electric field [50].

Charged Excitons

The great interest on charged states in SQDs results from the ability to control the emission from a single quantum object electrically [66]. For instance, charged quantum dots can store electrons with some spin information that can be read out independent of an initialization process [67].

Changing the charge state of a wide bandgap II–VI SQD has been first realized by contacting CdSe QD structures with a nanostructured metal mask that allows for the observation of the SQD emission through a nanoaperture with a diameter below 100 nm [48]. When applying a positive voltage to the metal top contact while the back contact applied to the n-doped substrate is grounded, the Fermi level is raised above the first quantized level in the conduction band of the QD and thus an electron will populate the QD. This can be seen very clearly in the left panel of Fig. 3.4. At large negative bias, the X emission from two individual QDs is visible, while in the case of positive bias, two additional lines emerge that can be attributed to a negatively charged exciton, i.e., a trion (X^-).

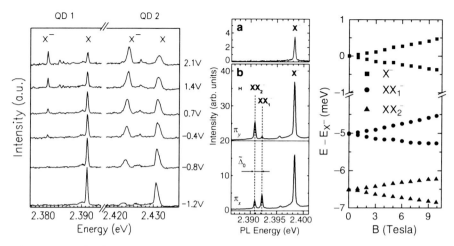

Fig. 3.4. *Left:* Characteristic PL spectra from two individual CdSe/ZnSe QDs for different voltages, showing the emission of neutral (X) and negatively charged excitons (X$^-$) [48]. *Center:* PL spectra from a negatively charged SQD, showing transitions corresponding to XX$^-$ and X$^-$ states [68]. *Right:* Zeeman splitting of the transitions shown in the center image [69]

For the trions one has to distinguish between the triplet trion (with parallel spins of the two carriers with the same charge) and the singlet trion where these spins are antiparallel. The final state for the recombination of the trion is a singly charged quantum dot. Due to the Coulomb interaction between the three charged particles, the recombination energy is shifted with respect to the neutral exciton line. For II–VI quantum dots, negatively [56, 70] as well as positively [28] charged trions have been observed, with binding energies spanning from 10 to 20 meV.

The triplet trion is an excited state: Due to Pauli blocking, the second electron (or hole) has to occupy the p-level. The spins can couple to a total spin of $J_z = \pm 1/2, \pm 3/2$, and $\pm 5/2$. It is difficult to observe an optical transition from this triplet trion to the ground state directly, because triplet trions usually are converted quite rapidly to the singlet state via spin flip. Akimov et al. found an interesting behavior of the PL spectra in weakly n-doped CdSe/ZnSe SQDs. These spectra consist of a single line, belonging to the negative trion X$^-$, and to *two* additional lines at the low energy side, emerging superlinearly with increasing excitation power in a biexciton-like manner (Fig. 3.4, center). These lines can be attributed to optical transitions from the charged biexciton to the trion [68,69]. The most probable transition is the recombination of an electron and a hole from the s-shells, which leaves the QD with another electron and a hole in the s-shell and an additional electron in the p-shell. This is the excited trion which is a triplet state in case of parallel electron spins. Because of the optical selection rules, only transitions to states with a total spin of $J_z = \pm 3/2$ and $J_z = \pm 1/2$ are allowed as the initial XX$^-$ state has a spin $J_z = \pm 1/2$. This results in two lines visible as XX$_1^-$ and XX$_2^-$ in the SQD PL spectrum. The separation between these lines is determined by the isotropic e–h-exchange interaction that – for neutral excitons – gives the splitting between dark and bright states. As both charged biexciton states described here are accessible in experiment, the splitting energy $\Delta_0 = 1.9$ meV can be extracted directly from the spectra.

It is interesting to note that the XX$^-$ transitions are linearly polarized, which again can be explained by the exchange interaction in asymmetric quantum dots. When applying a magnetic field in Faraday geometry, a Zeeman splitting is observed (see Fig. 3.4, right). From these experiments, the difference between the electron g-factors in the s-shell and in the p-shell as well as the ratio between the anisotropic and the isotropic part of the exchange interaction can be extracted [68].

The influence of magnetic fields on the different excitonic complexes is nicely summarized in Fig. 3.5 [28]. The left panel shows the PL energies in

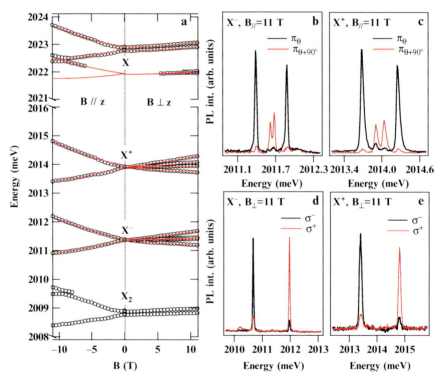

Fig. 3.5. Fine structure of exciton, biexciton, and positively and negatively charged exciton emission in a magnetic field for both Faraday and Voigt geometry [28]

dependence on the magnetic field for Faraday geometry (left) and Voigt geometry (right). The optical transitions from the neutral X and XX states show a distinct splitting at $B = 0\,\mathrm{T}$, which results from the electron hole exchange interaction, as discussed above. In case of the exciton, the splitting arises from the initial state of the transition, for the biexciton it comes from the final state of recombination. In contrast, the singly charged excitons reveal a different behavior: For X^-, the electron spins are antiparallel and will not interact with the hole spin; for X^+, the holes are paired in the same way and will not interact with the electron spin. For this reason, the charged states can clearly be identified by the missing zero-field splitting.

In Voigt geometry, the dark excitons can contribute to the PL signal, because the symmetry with respect to the quantum number J_z is broken [58, 59]. This can be seen immediately in case of the charged excitons, where the signal splits into four linearly polarized lines. For the neutral states, the dark states can be observed only at relatively high magnetic fields. This is in strong contrast to III–V quantum dots: Here, the dark excitons become visible at much lower magnetic fields, because the magnitude of the exchange energy is much smaller as compared to the Zeeman energy. From the energy splittings in the PL spectra as depicted in the right panel of Fig. 3.5, the g-factors parallel and perpendicular to the growth axis can be obtained. Leger et al. found that the hole g-factors for X^+ are enhanced with respect to neutral excitons or negative trions and attribute this to a weaker hole confinement in absence of a Coulomb attraction if no electrons are present in the QD [28].

3.2.2 Dynamics and Nonlinear Optics

Superradiance in Quantum Dots

Quantum dots are normally regarded as individual quantum objects that – in an ideal case – feel no mutual interaction; at least if they are separated significantly more than the exciton radius. This is normally the case for single layers of self organized QDs. However, it was found by Scheibner et al. [31] that the radiative exciton lifetime in a CdSe/ZnSe QD ensemble strongly depends on the excitation conditions. In the experiment the QDs were excited by linear polarized light either nonresonantly, i.e., above the barrier bandgap, or quasiresonantly, i.e., below the barrier bandgap.

Figure 3.6, left, shows the wavelength-dependent ratio between the radiative decay time under quasiresonant and under nonresonant excitation $\tau_{\mathrm{qr}}/\tau_{\mathrm{nr}}$. This ratio strongly depends on the energy and has a minimum at the maximum of the ensemble QD PL signal, i.e., at the spectral position, where the largest number of QDs are emitting. This phenomenon can be explained by the superradiance effect: The quasiresonantly excited QDs can be divided into subensembles that have the same ground state energy and spatial symmetry. These subensembles can couple radiatively, if they are excited quasiresonantly

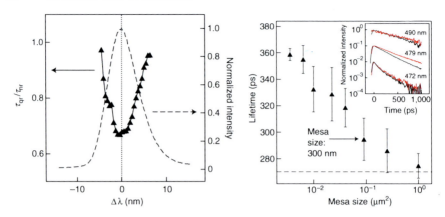

Fig. 3.6. *Left side:* Spectral dependence of the ratio τ_{qr}/τ_{nr} between quasiresonant and nonresonant radiative decay time. For comparison, the normalized PL spectrum is included as *dashed line*. *Right side:* Radiative lifetime at the maximum of the PL signal vs. mesa size. The *dashed line* indicates the lifetime for the unstructured part of the sample. *Inset:* Comparison of transient PL signals at different wavelengths from the unstructured part of the sample (*black*) and 60 nm mesas (*red*). From [31]

via a LO phonon replica and if they are spatially separated not more than approximately the effective wavelength in the material $\lambda_{\text{emission}}/n_{\text{ZnSe}} \approx 180$ nm. This electromagnetic coupling increases the radiation rate, thus leading to a shorter radiative recombination lifetime. In contrast, under nonresonant excitation, the initial polarization information is lost during relaxation into the QD ground state thus suppressing the superradiance effect.

In order to proof this mechanism, the number of radiatively coupled QDs was reduced by patterning the samples with mesa structures. The mesa edge length was varied between 1,000 and 60 nm. In Fig. 3.6, right (large panel), the lifetime is plotted in dependence on the mesa size. The most striking result is the clear reduction of the recombination rate corresponding to an increase of the lifetime for small mesas, i.e., small numbers of QDs coupled by the radiation field. It should be noted that artifacts like process induced damage or nonradiative recombination at the mesa surfaces would lead to a *reduction* of the lifetime with decreasing mesa size, in contrast the experimental data. Therefore, this finding confirms collective radiation effects between individual quantum dots over a comparatively long range of at least 150 nm.

Coherent Control and Stimulated Emission

An alternative access to biexciton formation is the excitation via a two photon process. In Fig. 3.7, left, the principle is shown: A picosecond laser pulse creates photons with exactly half the transition energy $|g\rangle \longrightarrow |b\rangle$. Absorption of two photons therefore populates the biexciton state. The time-integrated PL emission then shows all four transitions depicted in the scheme and the

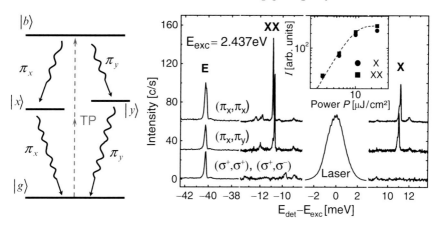

Fig. 3.7. *Left side:* Schematics of the exciton–biexciton system in a QD. For explanations, see text. From [71]. *Right side:* Laser and PL spectra from a SQD under pulsed excitation recorded in different polarization configurations. The excitation energy is 2.437 eV. *Inset:* Integral PL yield of exciton and biexciton recombination vs. excitation power [45]

exciting photon energy is exactly in the center between the X and the XX line (Fig. 3.7, right panel). The large separation between X and XX in CdSe/ZnSe SQDs makes this excitation by a ps pulse with a spectral width of about 2 meV possible. The components of the X and the XX transition are linearly polarized along π_x and π_y, respectively, and spectrally separated by about 0.3 meV. All four lines, however, are only visible under linearly polarized excitation, because in this case two photons with opposite spin can generate the biexciton with total spin $J = 0$. In contrast, under circular polarized excitation, no biexcitons are formed, because the sum of the angular momentum of the two photons does not allow an optical transition from the ground state with $J = 0$ to the biexciton $J = 0$ state.

This scheme enables two interesting experiments performed by Akimov, Henneberger and coworkers. First, the SQD is excited by a sequence of two ps laser pulses with a time delay that can be adjusted by a stabilized Michelson interferometer with a resolution of 0.03 fs [45]. This high temporal resolution allows for the two-photon coherent control of the biexciton state in a SQD. Figure 3.8, left part, shows the regime, where both laser pulses arrive nearly at the same time. Under these conditions, each pulse can contribute one photon to the two-photon excitation [72]. The optical interference between these two pulses is therefore reflected in the interferograms of both, excitons and biexcitons that exhibit oscillations with the laser frequency. In contrast, when the two pulses are clearly separated in time (see second and third column of the figure), the two-photon excitation occurs with the first *or* the second pulse. No optical interference is detected as there is no temporal overlap between the pulses. However, quantum

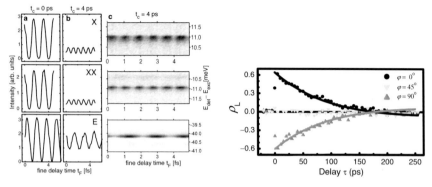

Fig. 3.8. *Left side:* Coherent control interferograms for exciton (X) and biexciton (XX) PL (*first and second row*) and for line E (*last row*), for pulses overlapping [*column* **a**] and separated in time [*column* **b**]. In *column* **c**, two-dimensional plots in the photon energy – delay time plane extracted from the same spectra as in *column* **b** are shown [45]. *Right side:* Evolution of the induced linear polarization degree ρ_L of the exciton emission as a function of the delay time between TP and ST pulse for different polarization configurations [71]

interference belonging to the transition from the ground state $|g\rangle$ to the biexcitonic state $|b\rangle$ can be seen. Because the energy difference between these two states is two times the laser energy, the interferograms show the *double* laser frequency. Due to decoherence mechanisms, the amplitude of the oscillation is strongly reduced with increasing time delay between both pulses. These findings very impressively show the two photon excitation process and define the time regime that enables a coherent control of the biexciton state. For comparison, the "E" line represents an excitonic transition that is excited via a single-photon process; in this case optical and quantum interference always have the same oscillation period.

In a second experiment, recombination via one of the two possible branches π_x and π_y is fostered by stimulated emission: The second picosecond laser pulse is now tuned to the energy of the transition between the $|b\rangle$ and the $|x\rangle$ and $|y\rangle$ state, and with a well-defined polarization, the transition between biexciton and exciton with exactly this polarization is stimulated [71]. The resulting population of the exciton sublevels is monitored via the linear polarization of the exciton recombination signal. The viewgraph in the right panel of Fig. 3.8 shows the induced linear polarization degree $\rho_L = (I_X - I_Y)/(I_X + I_Y)$ in dependence on the delay time τ between the two photon (TP) excitation and the stimulating (ST) pulse and for different polarization directions of the ST pulse. For $\varphi = 0$, i.e., the polarization of the ST excitation pulse is parallel to the X axis, one can see a significant polarization degree ρ_L at $\tau = 0$ that gradually decreases with τ, a clear hint for stimulated emission from the biexciton. The decrease of ρ_L with τ is related to the fact that for longer delay times spontaneous emission eventually dominates the stimulated emission. When $\varphi = \pi/2$, i.e., the ST pulse is polarized along the y axis, ρ_L has the opposite sign, as expected. For $\varphi = \pi/4$, no favorite recombination branch is selected, and therefore $\rho_L = 0$. Again, the large fine structure splitting of the exciton in wide bandgap II–VI materials allows one to observe stimulated emission of the biexciton in a SQD.

A different way to study coherent dynamics is the four wave mixing (FWM) technique, where the time dependence of coherent polarization is monitored by varying the time delay τ between two exciting pulses. In this way the phase coherence of excitons can be measured, providing information, e.g., on phonon scattering processes and their timescales. Patton et al. [46] succeeded in recording the FWM signal of one single CdTe quantum dot with a technique that provides spectrally resolved FWM signals with a femtosecond time resolution [73]. In Fig. 3.9, the signal from a SQD can be seen, consisting of a zero phonon line (ZPL) at 2.0083 eV that is superimposed on a broad acoustic phonon background. Due to the noise in the FWM signal, it is not possible to separate the phonon band belonging to the ZPL from the general background. However, the dephasing time T_2 of the ZPL line itself is accessible on the vertical axis (delay time τ): With increasing τ, the background is vanishing within a few picosecond and the ZPL signal quenches with a dephasing time of $T_2 = 13$ ps. This is significantly shorter than the radiative lifetime that amounts normally to a few 100 ps in CdTe QDs [74]. On the other hand the short dephasing time corresponds to a large homogeneous linewidth as estimated by FWHM $= 2\hbar/T_2$, which is comparable to the relatively large SQD linewidths of about 0.1 meV observed by PL spectroscopy in these structures. An interesting feature is in addition the beating behavior

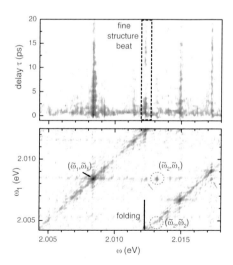

Fig. 3.9. *Top:* Spectrally resolved FWM intensity $I(\omega, \tau)$ vs. delay time τ. The logarithmic gray scale spans two orders of magnitude. *Bottom:* Two-dimensional spectrally resolved FWM intensity in logarithmic gray scale over three orders of magnitude. From [46]

visible for the line at 2.0123 eV. This beating with an oscillation period of $t_{\mathrm{osc}} = 14.5\,\mathrm{ps}$ indicates coherent coupling between energy states with a splitting of $h/t_{\mathrm{osc}} = 0.29\,\mathrm{meV}$ that might belong to the fine-structure split exciton states.

3.3 High Temperature Single Photon Sources

In the field of quantum information technology the availability of compact light sources emitting single photons on demand is a key issue. A semiconductor SQD is expected to exhibit superior efficiency and security for data transmission as compared to commercially available systems based on attenuated lasers, where efficiency is low and security against eavesdropping is principally limited due to a nonzero probability of multiphoton emission [40]. The challenge for a commercial use of SQDs is twofold: integration into an electrically driven circuit and operation under ambient conditions, i.e., at room temperature. Up to now, the InAs/GaAs material system [40, 75–77] represents the state of the art for semiconductor based SQD emitters, with the main drawback that their operation is limited to low temperatures, i.e., to laboratory conditions. In particular, electrically driven devices based on self-assembled SQDs have only been presented for operation temperatures below $T = 40\,\mathrm{K}$ [78–81]. Here, we will show that SQDs based on wide bandgap selenides and nitrides are able to overcome this low temperature restriction.

3.3.1 Toward High Temperature Emission in Single Quantum Dots

There are two important issues toward efficient high temperature single photon sources based on SQDs: First, the spectral line broadening due to phonon interaction should be comparable or even smaller than the biexciton binding energy as the biexciton–exciton cascade is widely regarded as a key element for reliable sources of single photons or entangled photons [82]. Second, nonradiative losses via defects within the QD or thermal carrier emission into the wetting layer or the barriers with subsequent nonradiative recombination therein have to be avoided.

In Fig. 3.10, temperature dependent SQD PL spectra are shown for different material systems. As expected, the temperature dependent bandgap shrinking results in a red shift of the SQD PL spectrum, while a distinct line broadening is obtained with increasing temperature. It is remarkable that in neither material system the PL lineshape can be described by a simple Lorentzian function, which would be expected for a pure homogeneous broadening due to a finite coherence time. Instead, pronounced wings at the low and the high energy tail of the spectra are observed. This was first observed and analyzed by Besombes et al. for CdTe/ZnTe SQD [29]. They demonstrated

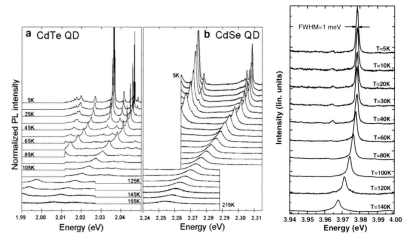

Fig. 3.10. *Left:* Evolution of the PL-spectrum from CdTe SQDs (**a**) and CdSe SQD (**b**), respectively, with temperature [83]. *Right:* Temperature dependent PL spectra of a cubic GaN/AlN SQD [84]

that the PL line consists of a zero-phonon line with Lorentzian lineshape in the center accompanied by acoustic phonon sidebands, i.e., the photon emission occurs together with the absorption or emission of acoustic phonons [83]. The phonon sidebands become more and more pronounced if the localization length decreases and thus they are particularly important for wide bandgap SQDs with their small size [54, 83–85].

It is remarkable that the optical emission of the SQDs seems to be limited to temperatures below about 150–200 K despite of the large band offset between the active QD layer and the barrier in these systems. Several efforts have been undertaken in order to reach efficient PL emission from wide bandgap QDs at room temperature. First, a high level of crystal quality for suppressing nonradiative losses via defects is required. Second, thermal emission of carriers out of the QD into the wetting layer or the barrier and subsequent nonradiative losses therein or at the sample surface and the substrate, respectively, has to be avoided. Progress has been achieved in the CdTe/ZnTe system by adding a certain amount of Mg into the barrier [16, 86] or by introducing a binary MgTe spacer layer between the QDs and the barrier [17]. In fact in these structures, a temperature independent recombination lifetime up to more than 200 K has been found, which indicates zero-dimensional carrier confinement over this whole temperature range [86].

A breakthrough for the selenide system was achieved quite recently. Embedding the active CdSe layer into a layer sequence of ZnSSe/MgS increases both the lateral and the vertical carrier confinement. This is found to result in a temperature independent recombination lifetime up to room temperature. In

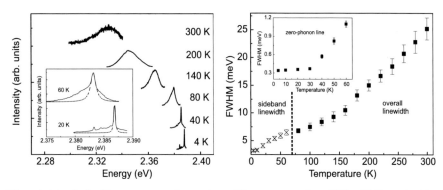

Fig. 3.11. *Left:* Temperature dependent PL spectra of a CdSe SQD embedded into ZnSSe/MgS barriers. In the *inset*, the low temperature lineshape consisting of the zero phonon line and the acoustic phonon sidebands is shown. *Right:* SQD PL linewidth FWHM vs. temperature. Data are taken from [18]

Fig. 3.11, left, the PL spectrum of a CdSe/ZnSSe/MgS SQD is depicted for different temperatures. A quite efficient SQD emission even at room temperature is obtained [18]. Due to the strong carrier confinement, pronounced acoustic phonon sidebands appear in the SQD PL spectra (see inset of Fig. 3.11). In the right part of Fig. 3.11, the SQD PL linewidth FWHM (full width half maximum) is plotted vs. temperature. In the low temperature regime, the zero phonon line as well as the acoustic phonon sidebands dominate the spectrum while with increasing temperature, exciton – LO phonon interaction becomes dominant. At room temperature, the SQD PL linewidth is about 26 meV.

A comparison between the SQD PL linewidth at elevated temperatures with the characteristic binding energies of biexcitons in the material system is quite instructive. In case of CdTe SQDs, typical biexciton binding energies in the order of 13 meV are found [56, 62] which is comparable to the expected SQD PL linewidth at about 200 K [83]. In hexagonal (In,Ga)N SQDs the biexciton binding energy is usually smaller or even negative [33, 63] due to the internal electric field, while the line broadening of a SQD due to phonon interaction is quite pronounced [32, 33] and thus the SQD linewidth usually exceeds the biexciton binding energy even at moderate temperatures. The situation is different for the CdSe/ZnSSe/MgS SQDs discussed here. The biexciton binding energy for Cd(S,Se) QDs is found to vary between 20 and almost 40 meV [27, 30, 87, 88] and thus may exceed the phonon induced linewidth broadening of about 26 meV even at room temperature. This makes these kind of SQDs especially attractive for applications as a solid state single photon source working at room temperature.

3.3.2 Single Photon Emitters Based on II–VI Compounds

A proof for the ability of wide bandgap SQDs to emit single photons can be obtained by photon-correlation measurements. In these kind of experiments, the SQD PL signal is sent to a Hanbury Brown and Twiss-type setup. The PL signal is divided into two channels and the probability that a photon hits the second detector at time $t+\tau$ if the first detector counted a photon at time t is given by the second order correlation function

$$g^2(\tau) = \frac{\langle I(t) \cdot I(t+\tau) \rangle}{\langle I(t)^2 \rangle}. \qquad (3.3)$$

An ideal single photon source is characterized by $g^2(\tau) = 0$, i.e., a single photon can enter either the first *or* the second channel. The result of these kind of measurements performed on a CdSe/ZnSSe SQD are depicted in Fig. 3.12, left. For these experiments performed by Sebald et al., a frequency doubled, mode-locked Ti–sapphire laser with a pulse width of 120 fs and a repetition rate of 82 MHz was used for excitation [85]. As can be seen in the figure, the number of correlations is strongly reduced at $\tau = 0$. This is a clear proof of single photon emission, while for a Poissonian light source the peak at $\tau = 0$ should be identical to the peaks at integer multiples of the laser repetition

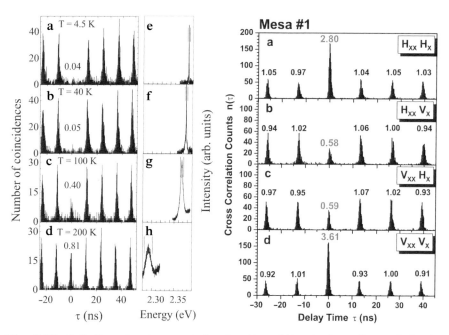

Fig. 3.12. *Left:* Second order correlation function of the single exciton emission in a CdSe/ZnSSe SQD under pulsed excitation at different temperatures. In the *right column*, the corresponding SQD PL spectra are depicted for each temperature [85]. *Right:* Polarization dependent cross-correlations between the exciton (X) and the biexciton (XX) emission of a CdSe SQD under nonresonant pulsed excitation. H/V indicate the applied orientation of polarization detection for XX ("start") and X ("stop"). From [90]

time. A measurable photon antibunching is even observed at $T = 200$ K and although $g^2(\tau) > 0.5$ due to some background, these data demonstrate the potential of this material system for high temperature single photon emitters. Similar results have been obtained in the meanwhile by other groups [89] and other II–VI materials like CdTe/ZnTe [74], but still restricted to lower operation temperatures.

Making use of the cascaded biexciton–exciton emission in SQDs, it is possible to generate polarization correlated, two-color photon pairs. In the right part of Fig. 3.12, the results of cross-correlation experiments on biexciton–exciton pairs are depicted [90]. The measurements were performed using linear polarizers in vertical (V) and horizontal (H) direction in order to account for the fine structure splitting of exciton states in CdSe SQDs due to electron–hole exchange interaction. In accordance to what is expected from the energy level scheme for the biexciton–exciton cascade (see Fig. 3.1), an antibunching is obtained for measurements with perpendicular polarization of the exciton and the biexciton emission whereas cross-correlation experiments performed with parallel polarization results in a distinct bunching behavior. This is a clear proof of a cascaded biexciton–exciton emission stemming from the same individual SQD. In order to generate entangled photon pairs the fine structure splitting due to the electron–hole exchange interaction has to be reduced below the limit of the homogeneous linewidth which might be possible, e.g., by a post growth annealing step [60, 61].

Besides the demonstration of single photon emission after optical excitation, the integration of wide bandgap SQDs into an electrically driven device is a key issue for applications. Although single CdSe nanocrystals have been successfully incorporated into an electrical circuit even showing room temperature operation [92], blinking and spectral diffusion as well as significant line broadening of the emission might limit their practical use quite seriously [93]. Recently, Arians et al. succeeded in embedding CdSe QDs into an electrically driven p-i-n diode and demonstrate electroluminescence from a SQD [94]. In order to avoid carrier losses due to thermal escape out of the SQD at elevated temperatures, the SQDs are embedded into ZnSSe/MgS barriers [18]. By using thin (1 nm) MgS barriers, an optimized trade-off between efficient carrier injection via tunneling and high quantum efficiencies at elevated temperatures is achieved.

Figure 3.13 shows electroluminescence (EL) spectra of a CdSe/ZnSSe/MgS SQD for various temperatures. It is quite remarkable that even at room temperature a pronounced EL signal stemming from one SQD is obtained. This has not been demonstrated for any other kind of self-assembled semiconductor QDs up to now and is a proof for the potential of II–VI SQDs for room temperature single photon emitters. The linewidth broadening due to phonon interaction is similar to what is observed for optically pumped SQD emission and thus even at room temperature comparable or even less than the biexciton binding energy. Looking at the threshold voltage U_t, defined as the voltage at the onset of the EL signal, a decrease with increasing temperature

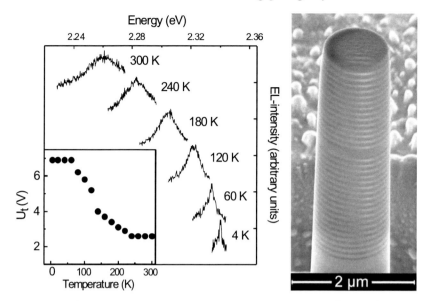

Fig. 3.13. *Left:* Temperature dependent EL spectra of a CdSe/ZnSSe/MgS SQD. In the *inset*, the threshold voltage U_t for the onset of the EL signal is plotted vs. temperature. The results are from [51]. *Right:* Scanning electron micrograph of a monolithic II–VI micropillar cavity [91] with a diameter of 1.2 µm

can be noticed. This is attributed to the thermal activation of the doping (in particular the p-doping) with increasing temperature. At room temperature, a surprisingly low value of $U_t = 2.6\,\mathrm{V}$ is obtained proving the quite efficient electrical carrier injection into the active SQD.

A further challenge for a practical single photon source operating at elevated temperatures is the light extraction efficiency of the device. Several attempts have been performed up to now to integrate self-assembled II–VI QDs into an optical cavity. This includes CdSe QDs embedded in ZnSe/ZnMgSSe microdiscs [95], in hybrid micropillars with SiO_2/TiO_2 distributed Bragg reflectors (DBRs) [96] and monolithic pillar microcavities using DBRs of MgS/ZnCdSe and focused ion beam etching (see Fig. 3.13, right) [91, 97]. While in particular in the latter approach, a tuning of the cavity modes as well as an enhanced spontaneous emission efficiency due to the Purcell effect have been demonstrated for QD ensembles, no results on self-assembled II–VI SQDs embedded into a microcavity have been published up to now. Moreover, the integration of a SQD coupled to a microcavity into an electrically driven device is still an unsolved issue.

3.3.3 Single Photon Emitters Based on Nitrides

An attractive alternative for high temperature single photon emitters working in the blue or even UV spectral range are III-N SQDs. This is in particular true as nitrides are nowadays well established in highly efficient light emitting or laser diodes, although QDs have not yet entered the market for this material system.

First measurements of the second-order coherence function on single hexagonal GaN/AlN QDs have been presented by Santori et al. [98]. As mentioned above, this material system exhibits negative biexciton binding energies due to the strong internal electric field. Figure 3.14 nicely shows the potential of III-N SQDs for high temperature single photon emission. In the left part of the figure, the PL spectrum of a SQD is depicted at $T = 200$ K. Low excitation power has been used in order to avoid any contribution of the biexciton emission. A nice antibunching at $\tau = 0$ is obtained even at $T = 200$ K, similar to what was observed before by Sebald et al. for CdSe/ZnSSe SQDs [85].

Although the development of commercially available light emitters based on nitrides is much more advanced as compared to II–VI materials, no results on electrically pumped GaN SQD emitters have been published up to now. In addition, only a few data on III-N microcavities containing QDs are available [99–101]. Jarjour et al. just recently reported on a cavity enhanced blue single photon emission from a InGaN/GaN SQD [101], where the QDs are embedded between GaN/AlN DBR layers at the bottom and SiO_x/TiO_x DBR mirrors on top of the structure. For practical applications at room temperature, however, a variety of challenges, including room temperature emission and integration into an electrically driven device, have to be addressed.

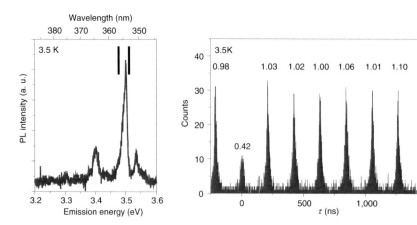

Fig. 3.14. PL spectrum (*left*) and correlation measurements (*right*) of a GaN/AlN SQD at $T = 200$ K. The spectral region between the *solid lines* in the PL spectrum indicates the portion of the spectrum that reaches the detectors for the correlation measurements after spectral filtering. From [33]

3.4 Spin Properties in II–VI Single Quantum Dots

Spintronics is an emerging field in physics with the intention of using the spin degree of freedom as an information carrier in semiconductor electronics [102, 103]. Since the spin of a particle is a quantum mechanical property, not only a discrete number of states, like the spin-up and the spin-down state of an electron, can be defined, but also a coherent superposition of them. This triggered already 10 years ago schemes for using the spin degree of freedom as a Qubit in quantum computation [42]. Here, SQDs are of strong interest because of the high stability of carrier spin states in such quasizero-dimensional quantum objects [67, 104].

Wide bandgap QDs based on II–VI compounds offer some exciting properties in this respect. First of all, magnetic ions like Mn^{2+} can be incorporated isoelectronically into the crystal matrix and the exchange interaction between charge carriers and the Mn^{2+} ions leads to a bulk of fascinating new aspects. Second, the small size of the QDs in these materials as compared to the "standard" InAs/GaAs system allows the study of the interaction between charge carriers and only a few hundreds of nuclear spins. Here, we will first present some selected spin properties in nonmagnetic II-VI SQDs with special emphasis on the preparation and the dynamics of spin states. Secondly, magnetically doped SQDs are discussed in the limit of both, ensembles of magnetic ions, where the SQD can be regarded as a semiconductor nanomagnet, and an individual Mn^{2+} ion embedded into a SQD.

3.4.1 Nonmagnetic Single Quantum Dots

Optical Preparation of Spin States

Time-resolved PL studies on QD ensembles already demonstrated long spin lifetimes in II–VI QDs [105] and in III-N QDs [106], in the latter case even up to room temperature. In these experiments, quasiresonant, polarized optical excitation generates electron–hole pairs with well defined spin orientation and the transient change of the PL polarization degree is indicative for the lifetime of the spin states. As shown experimentally, a pronounced polarization degree is only obtained, if linear polarized excitation is used, i.e., if the eigenstates of the bright excitons [see (3.1)] are excited.

As discussed before, this is due to the electron hole exchange interaction, which results in a coherent superposition of the bright spin-up exciton ($J_z = s_z + j_z = +1$ with $s_z = -1/2$ and $j_z = +3/2$) and the bright spin-down exciton ($J_z = -1$, consisting of $s_z = +1/2$ and $j_z = -3/2$) [27]. After an initial fast decrease of polarization during relaxation into the QD ground state, no further transient change of the polarization degree is found on the time scale of recombination. Thus, it is concluded that the coherent superposition of spin states in the QD ground state is not destroyed and thus neither the spin state

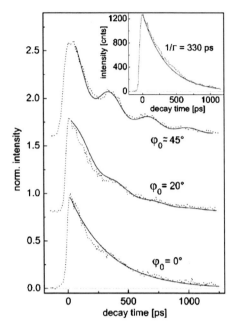

Fig. 3.15. PL transients of a single CdSe/ZnSe QD for quasiresonant excitation at the 1-LO phonon resonance. Three different polarization configurations are compared, where ϕ_0 represents the angle between the linear polarization of the laser beam and the [110] crystal orientation. In the *inset*, the PL transient after excitation in the wetting layer is shown for comparison. From [108]

of the electron nor the hole changes within the time window of the experiment, which is limited by the finite radiative recombination lifetime [105–107].

These findings can be used for optically preparing well-defined spin states on a SQD level. As shown in Fig. 3.15, photon beats are observed in time-resolved PL spectroscopy on a CdSe/ZnSe SQD, if one excites a superposition of the two spin-split eigenstates of the bright exciton. The beating period of 330 ps corresponds to a fine structure energy splitting of 13 µeV, not resolvable in the spectral domain [108].

The eigenstate symmetry can be changed by applying a sufficiently strong external magnetic field in Faraday geometry [27, 28]: If the Zeeman energy exceeds the electron hole exchange energy, pure spin-up and spin-down exciton eigenstates are obtained. Experiments performed by Mackowski et al. nicely demonstrate the impact of eigenstate symmetry on the optical preparation of spin states in a CdTe/ZnTe SQD [109] (see Fig. 3.16). While for zero external field, no polarized PL signal is found in case of circular polarized optical excitation, i.e., the spins randomize very rapidly, the situation changes if one performs the same experiment in an external magnetic field of $B_{\text{ext}} = 2.5\,\text{T}$. The distinct net circular polarization of the PL signal seen in the figure

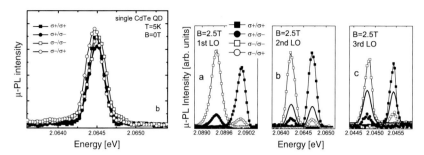

Fig. 3.16. *Left:* SQD PL-spectra obtained with an excess energy of 2 LO phonon energies at $B_{\text{ext}} = 0\,\text{T}$ for circular polarized excitation and detection. *Right:* PL-spectra at $B_{\text{ext}} = 2.5\,\text{T}$ of different CdTe/ZnTe SQDs with circular polarized excitation and detection. The excess energy was varied between 1 LO and 3 LO phonon energies [109]

indicates the ability of optically preparing pure spin-up and spin-down exciton states, depending on the polarization of the exciting laser, respectively, in an external magnetic field.

From these experiments it is obvious that in SQDs electron–hole pairs with a well-defined spin state can be prepared optically in the limit of quasiresonant excitation (e.g., via LO phonon cascades), if one considers the symmetry of the QD eigenstates. For practical applications in electrically driven devices, however, one may ask the question whether it is possible to inject spin polarized carriers into a SQD via a semiconductor heterointerface. Following the early concepts of spin injection into a semiconductor using a diluted magnetic semiconductor (DMS) as a spin aligner [110, 111], a ZnBeMnSe spin aligner was placed on top of a CdSe/ZnSe QD heterostructure. Via the quite efficient sp–d exchange interaction between charge carriers and Mn^{2+} ions, the carriers spins in the ZnBeMnSe spin aligner are completely polarized even in a weak or moderate external magnetic field.

A very simple experiment has been used in order to demonstrate spin injection into a SQD [112]: Unpolarized, quasiresonant excitation below the bandgap of the BeMnZnSe spin aligner and the ZnSe barrier generates unpolarized electron–hole pairs in the SQD. Due to a small, but finite relaxation from the upper to the lower Zeeman level, one obtains a small circular polarization degree $\rho_C = (I^{\sigma+} - I^{\sigma-})/(I^{\sigma+} + I^{\sigma-})$, which slightly increases with B_{ext} (see Fig. 3.17). If, however, the electron–hole pairs are generated in the ZnBeMnSe layer in case of UV excitation, the spins of the electrons and holes rapidly polarize in the DMS layer and occupy the lower Zeeman level in the spin aligner. The completely spin polarized electrons are transferred across the heterointerface into the SQD and in case of a successful spin injection, an increase of the SQD PL polarization degree as compared to below barrier excitation is expected.

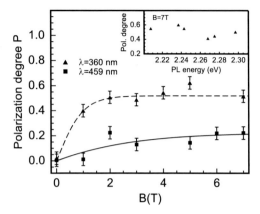

Fig. 3.17. Circular optical polarization degree ρ_C of the SQD emission vs. magnetic field for excitation above and below the DMS spin aligner. The *inset* shows the circular polarization degree of several other SQDs for above DMS excitation at $B_{\text{ext}} = 7\,\text{T}$. From [112]

The experimental data shown in Fig. 3.17 nicely confirm successful spin injection into a SQD. From a quantitative evaluation of the data, a spin injection efficiency of about 70% is obtained [112]. In the meanwhile, this concept of spin injection into a SQD across a heterointerface has been extended to hybrids of a DMS spin aligner and InAs SQDs, where, e.g., the role of band alignment and external voltage has been discussed [113] and even polarized electroluminescence has been achieved on a SQD level [114].

Spin Dynamics

In bulk semiconductors, spin–orbit interaction is known to severely limit the spin lifetimes of charge carriers. In particular the hole spin orientation is usually expected to randomize quite rapidly, e.g., due to valence band mixing. Localization on a nanometer scale in a QD suppresses spin–orbit interaction and indeed, a variety of available data demonstrate extremely long spin lifetimes of electrons in a QD at sufficiently strong magnetic fields [67, 104]. As the hyperfine interaction is expected to limit the zero-magnetic field relaxation time of carrier spins in a SQD [115–117], hole spins are currently in the focus of intensive research not only for the InAs/GaAs system [118, 119] but also for wide bandgap SQDs. This is because the role of hyperfine interaction with nuclear spins is expected to be reduced for holes as compared to electrons due to the p-character of the hole wavefunction [43].

In contrast to the experiments performed on neutral excitons discussed above, charged SQDs with resident electrons or holes provide an experimental tool for investigating the spin dynamics of electrons or holes separately. In a SQD charged with one excess electron, an optical excitation at low and

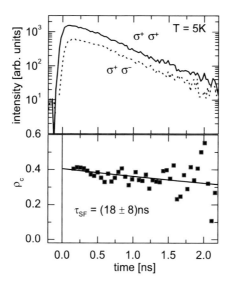

Fig. 3.18. *Upper part:* Decay characteristics of the trion PL signal after σ^+ excitation for both, σ^+ and σ^- detection measured at $T = 5$ K. *Lower panel:* Transient change of the circular polarization degree ρ_C. From [43]

moderate densities will result in the formation of a negative trion, where the spins of both electrons are aligned antiparallel. Therefore, there is no net electron–hole exchange interaction. Quasiresonant excitation below the energy of the trion triplet state with σ^+ polarized light will generate a heavy hole with $j_z = +3/2$. As can be seen in Fig. 3.18, a distinct circular polarization degree ρ_C of the trion PL signal in a SQD is found after picosecond, polarized optical excitation. A careful analysis of the data yields a spin lifetime of the heavy hole in the order of $\tau_s^h = 18$ ns in a CdSe/ZnSe SQD. With increasing temperature, τ_s^h decreases and at $T = 70$ K, a value of 0.8 ns is obtained.

In order to uncover the impact of hyperfine interaction on the carrier spin dynamics in wide bandgap SQD, the electron spin dynamics in CdSe/ZnSe SQDs have been studied by Akimov et al. again using the trion as an optical probe [120]. In this material system, the small hyperfine coupling constant and the small number of isotopes with finite nuclear spin make the Overhauser field between one and two orders of magnitude smaller than in the InAs/GaAs system. On the other hand, the small number of nuclei (a few 100 ... a few 1,000) leads to a Knight field, which is comparable to or even exceeds the one in InAs/GaAs QDs.

The idea of the experiment performed by Akimov et al. is shown in the left part of Fig. 3.19. Selective pumping a SQD by σ^+ polarized excitation generates trions with a spin-up hole ($j_z = +3/2$). The trion recombines by emitting again a σ^+ photon, or, after a hole spin flip, a σ^- photon. This pumping procedure generates a nonequilibrium electron spin population, provided

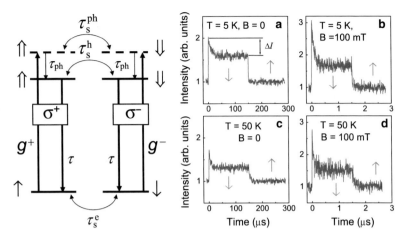

Fig. 3.19. *Left:* Optical transitions of a negatively charged QD with the characteristic time constants τ, τ_{ph} as the lifetime of the trion ground state and the LO-polaron state, respectively; τ_s^e, τ_s^h as the spin flip time of electron and heavy hole and g$^\pm$ as the optical generation rate for $\sigma^\pm$ excitation. *Right:* SQD electron spin transients under polarization switching excitation. The signal detection is in σ^+ polarization. Note the different x-axis for zero and finite magnetic field [120]

the electron spin flip time τ_s^e is much longer than the hole spin flip time τ_s^h. Switching to σ^- excitation, the opposite scenario starts which ends with a reversed spin occupation.

The experimental results are shown on the right part of Fig. 3.19. Data obtained at zero field and at a small external field of $B_{ext} = 0.1$ T are compared for two distinct temperatures. As can be seen, the transient response of the intensity, which reflects the time evolution of the electron spin, is roughly 2–3 orders of magnitude faster for $B_{ext} = 0$ as compared to $B_{ext} = 0.1$ T. From the data, it becomes obvious that at zero field hyperfine interaction with nuclear spins strongly modifies the spin dynamics of electrons and the strong Knight field results in a dynamical polarization of nuclear spins on a submillisecond time scale, i.e., much faster than for bulk semiconductors. This allows for a formation of a strong nonequilibrium dynamic nuclear polarization (DNP), as recently demonstrated from the same group [121], where signatures of DNP formation have been observed up to 100 K.

3.4.2 Magnetically Doped SQDs: Magnetism Meets Optics

In II–VI semiconductor QDs, magnetic ions like, e.g., Mn^{2+}, can be incorporated isoelectronically into the crystal matrix, either in the QD area or in the surrounding barrier. While first optical studies on a SQD level have been performed on "natural" CdMnTe/CdMgTe SQDs [122] or lithographically defined CdTe/CdMnTe SQDs [123], nowadays the research efforts mainly concentrate on self-assembled Cd(Mn)Se/Zn(Mn)Se [11, 12, 15, 34, 124, 125] or

Cd(Mn)Te/ZnTe [13, 14, 28, 49, 126] SQDs. As nonradiative losses via internal Mn^{2+} transitions are an important issue when trying to explore the properties of individual, magnetically doped QDs, great care have been taken to suppress this loss channel. This can be done either by limiting the concentration of magnetic ions or by preparing QDs with a bandgap below the energy required to promote a Mn^{2+} $3d$ electron to the first excited shell. Two limits are of particular interest: In case of a large amount of magnetic ions within the extension of the exciton wavefunction, the Mn^{2+} ion ensemble has to be treated like a paramagnetic system exposed to the carriers' exchange field, while in the limit of a single magnetic impurity in the QD, the discrete orientations of the manganese spin have to be considered.

Optically Probing Nanomagnetism

In Fig. 3.20, the PL spectra of individual CdMnTe/ZnTe (left) [14] and CdSe/ZnMnSe QDs (right) [12] are depicted. In the first example the magnetic ions are incorporated directly into the QD while in the second one they are nominally placed in the barrier. The extension of the exciton wavefunction into the barrier and, more important, the manganese segregation into the QD during growth induces a considerable overlap between the exciton wavefunction and the magnetic ions also in the latter case. It is obvious that with an decreasing concentration of magnetic ions, achieved in case of the CdMnTe QDs discussed here by a post growth annealing step, the PL linewidth of a SQD becomes significantly smaller. Moreover, by applying a magnetic field in Faraday geometry, one obtains a strong red shift and a pronounced linewidth narrowing with increasing B_{ext}.

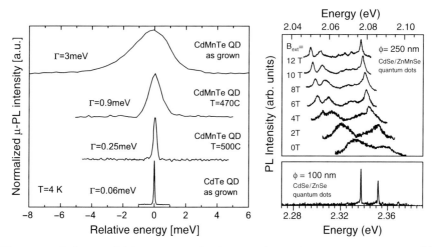

Fig. 3.20. *Left:* Zero field PL spectra for CdMnTe/ZnTe SQDs with different Mn^{2+} concentration [14]. *Right:* Magneto-PL spectra of individual CdSe/ZnMnSe QDs for different values of B_{ext}. For comparison, zero field data of CdSe/ZnSe reference SQDs are shown [12]

Magnetic Moment of a Single Quantum Dot

In order to understand these findings, one have to consider the magnetic ion system and its interaction with charge carriers in the QD. In DMS QDs containing lots of magnetic ions, the magnetization M of the paramagnetic Mn^{2+} system in a magnetic field B can be described by a modified Brillouin function

$$M(B,T) = x_{Mn} N_0 g_{Mn} \mu_B S_{eff}\ B_{5/2}\left(\frac{5\mu_B g_{Mn} B}{2 k_B T_{eff}}\right), \qquad (3.4)$$

where x_{Mn} is the manganese concentration, $g_{Mn} = 2$ is the g-factor of Mn^{2+} ions, and N_0 is the number of cations per unit volume. The effective spin $S_{eff} < 5/2$, and the effective temperature $T_{eff} = T + T_0$ take into account the antiferromagnetic interaction between neighboring Mn^{2+} spins.

At zero external field, optically generated electron–hole pairs are able to align the magnetic ion spins due to the *sp–d* exchange interaction, provided the formation time of such an "exciton magnetic polaron (EMP)" is shorter than the recombination lifetime [34]. The exchange energy is directly proportional to the magnetization M and can be written as

$$E_{exc}(B,T) = -\frac{\gamma(\alpha - \beta)}{2\mu_B g_{Mn}} M(B_{ext} + B_{exc}, T), \qquad (3.5)$$

where α and β are the exchange constants of the electrons and holes, respectively and γ (which is less than unity) takes into account the fact that only a part of the exciton wavefunction may actually overlap with the Mn^{2+} ion spins [122]. Here, the field $B = B_{ext} + B_{exc}$ includes both the external magnetic field B_{ext} and the exchange field B_{exc} describing the *sp–d* exchange interaction between the charge carriers and the spins of the magnetic ions. Note that due to the negligible in-plane g-factor of the heavy hole, both fields are parallel in Faraday geometry while this is not the case in Voigt geometry which complicates the situation [127].

In Fig. 3.21, the PL energy shift of a CdSe/ZnMnSe SQD is depicted vs. temperature and vs. external magnetic field applied in Faraday geometry. It is instructive to start the discussion in the limit of high temperatures and zero external magnetic field. Thermal disorder results in a completely randomized Mn^{2+} spin system. Reducing T, an EMP is formed due to the *sp–d* exchange interaction and at $T = 5$ K, we obtain a EMP binding energy of about 11 meV. If in addition an external magnetic field is applied, the Mn^{2+} spins become more and more aligned until saturation, which is evidenced by a further red shift of the PL energy, known as Giant Zeeman effect. The solid line is a fit according to (3.4) and (3.5). The perfect description of the experimental data supports that in case of equilibrium, i.e., a EMP formation time which is much shorter than the exciton recombination lifetime, the SQD PL energy variation directly reflects the exchange energy E_{exc} and thus the average magnetization

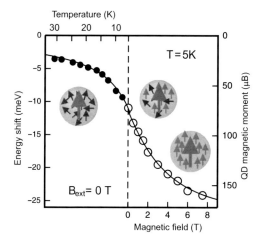

Fig. 3.21. PL energy shift ΔE (*left axis*) of a CdSe/ZnMnSe SQD vs. T and vs. B_{ext}. The *inset* schematically shows the orientation of the magnetic moments of the manganese ions (*small arrows*) relative to the exciton one (*large arrow*). The *right axis* gives the corresponding magnetic moment of the SQD under investigation. From [128]

of the SQD. Note that the Zeeman shift resulting from the band g-factor g_0 and the change of the bandgap with temperature are small and can therefore be neglected in this experiment.

The total magnetic moment of the SQD, μ_{QD}, which can be easily obtained by $\Delta E/B_{\text{exc}}$, is depicted on the right axis of Fig. 3.21. Here, the value of $B_{\text{exc}} = 2.6\,\text{T}$ extracted from the PL energy shift with T and B_{ext} is used. Starting from a magnetic moment of slightly above $70\,\mu_{\text{B}}$ at $T = 5\,\text{K}$ and $B_{\text{ext}} = 0$, the spin alignment occurring with increasing external magnetic field results in a total magnetic moment of the SQD of more than $150\,\mu_{\text{B}}$ at $B_{\text{ext}} = 8\,\text{T}$. Note that in case of a completely suppressed EMP formation, (3.5) is still valid for describing the Giant Zeeman shift of the SQD PL signal in an external field if $B_{\text{exc}} = 0$ is used. Quite recently, a magnetically doped SQD with a large positive g-factor and thus a Giant Zeeman shift has been combined with a nonmagnetic SQD exhibiting a small negative exciton g-factor. In such a SQD molecule, coherent tunnel coupling between two individual II–VI QDs could be achieved and the coupling is tunable by an external magnetic field [129].

Magnetic Fluctuations in a Single Quantum Dot

One of the most striking features in the emission spectra of magnetically doped SQDs is the large PL linewidth and its strong dependence on magnetic field (see Fig. 3.20). One should be aware that in this nanoscale magnetic system, the spin alignment due to sp–d exchange interaction competes with thermal disorder, which causes a temporally fluctuating magnetization. The exchange interaction between charge carriers and magnetic ions is mainly dominated by the p–d exchange between Mn^{2+} ions and heavy holes having a negligible in-plane g-factor. Thus, mainly longitudinal magnetic fluctuations are monitored in Faraday geometry, while longitudinal as well as transversal fluctuations contribute to the linewidth broadening of the SQD PL signal in Voigt geometry [127].

The statistical fluctuations of the magnetization M lead to a Gaussian shape of the SQD emission line with a full width at half maximum FWHM given by

$$\text{FWHM} = \sqrt{8\ln 2 \cdot \langle \delta(\mathbf{B}_{\text{exc}} \cdot \mathbf{M})^2 \rangle}, \tag{3.6}$$

where $\langle \delta(\mathbf{B}_{\text{exc}} \cdot \mathbf{M})^2 \rangle$ is the dispersion of $\mathbf{B}_{\text{exc}} \cdot \mathbf{M}$ and includes longitudinal and transversal fluctuations of the magnetization along and normal to $\mathbf{M}$, respectively. Taking into account the well-known dissipation fluctuation theorem $\langle \delta M^2 \rangle = (\mathrm{d}M/\mathrm{d}B)\, k_\mathrm{B} T / V_{\text{eff}}$, where $V_{\text{eff}} = \gamma V$ and V is the effective volume occupied by the exciton wavefunction, one can calculate the linewidth FWHM of the PL emission from a magnetically doped SQD [12, 127]. In the left part of Fig. 3.22 a comparison of experiment and theory for the FWHM measured on a CdSe/ZnMnSe SQD in both, Faraday and Voigt geometry is shown. A quite nice agreement is obtained.

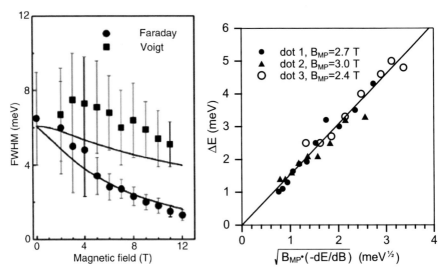

Fig. 3.22. *Left:* FWHM of a CdSe/ZnMnSe SQD vs. external magnetic field in Faraday and in Voigt geometry, respectively. Experimental data (*symbols*) and theory (*solid lines*) are compared. From [127]. *Right:* FWHM plotted for several CdSe/ZnMnSe SQDs vs. $(-B_{\text{exc}}\mathrm{d}E/\mathrm{d}B)^{1/2}$. Here, B_{exc} is denoted as B_{MP}. The *solid line* is a fit according to (3.7) [12]

In case of Faraday geometry, a quite simple relation between FWHM and the derivative of the Zeeman shift, dE/dB, can be extracted

$$\text{FWHM} = \sqrt{B_{\text{exc}} \cdot 8\ln 2 \cdot k_B T \left(-\left.\frac{dE}{dB}\right|_{B=B_{\text{ext}}+B_{\text{exc}}}\right)}. \quad (3.7)$$

As dE/dB can be easily determined from the energy shift of the SQD PL signal with magnetic field, (3.7) can be validated experimentally. This is done in Fig. 3.22, right, where experimental data from three different SQDs are compared to theory. The good agreement between experiment and theory again demonstrates the consistent picture developed above.

In analogy to the relation between the energy shift ΔE and the total magnetic moment μ_{QD} in the SQD, the linewidth FWHM can be related to a temporal variation of the magnetic moment, $\Delta \mu_{\text{QD}}$, by $\Delta \mu_{\text{QD}} = \text{FWHM}/2B_{\text{exc}}$. From the experimental data shown in Figs. 3.21 and 3.22, we extract a total magnetic moment of the SQD of $\mu_{\text{QD}} \pm \Delta \mu_{\text{QD}} = 70\,\mu_B \pm 17\,\mu_B$ at $B_{\text{ext}} = 0$ and $T = 5\,\text{K}$, which changes, e.g., to $\mu_{\text{QD}} \pm \Delta \mu_{\text{QD}} = 170\,\mu_B \pm 4\,\mu_B$ at $B_{\text{ext}} = 11\,\text{T}$. The optical spectroscopy on an individual magnetically doped semiconductor QD thus allows one to resolve magnetic moments of a few tens of Bohr magnetons and to trace fluctuations of only a few Bohr magnetons!

A more thorough theoretical analysis of the PL lineshape of magnetically doped SQDs has been performed by Wojnar et al. for CdMnTe/ZnTe QDs containing only a few tens of magnetic ions [126]. Using a muffin tin model for the QD shape and taking into account the exchange interaction between the exciton and the magnetic ions as well as the coupling to an external magnetic field, the PL intensity of the optical transition in a SQD is given by [126]

$$I(E) = \sum_{s_z, j_z} \exp(-E_I/k_B T) g(S) |\langle S, S_z, 3/2, j_z, 1/2, s_z | \sigma^{\pm} | S, S_z \rangle|^2 \quad (3.8)$$

with $S = 5/2$ and S_z the projection of the total Mn^{2+} magnetic moment on the direction of the magnetic field, $E = E_I - E_F$ with $E_I(E_F)$ the energy of the initial (final) state of the recombination process, and $g(S)$ the degeneracy of the quantum state of the N_{Mn} magnetic ions. The results of the calculations are presented together with corresponding experimental data in Fig. 3.23. As can be seen, both, the red shift as well as the linewidth narrowing with increasing external magnetic field can be reproduced. It is interesting to note that in smaller QDs containing less magnetic ions the magnetic field induced changes of the SQD PL signal are much less pronounced. This may be an indication of a stronger exchange interaction energy in small QDs as expected theoretically [130, 131] and thus a PL signal, which is less sensitive to an external field, or can simply be explained by the fact that a less amount of magnetic moments in the QD reduces its magnetization and thus the total exchange energy.

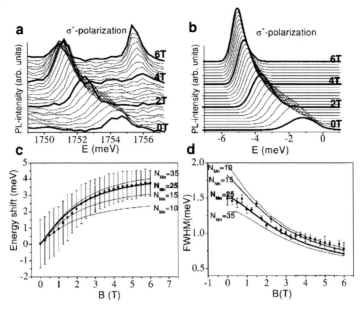

Fig. 3.23. Magneto-PL spectra of individual CdMnTe/ZnTe QDs in experiment (**a**) and theory (**b**). In the lower part of the figure, the energy shift ΔE (**c**) and the spectral width FWHM (**d**) of the SQD emission are depicted vs. magnetic field. Experiments (*symbols*) and theory (*lines*) are compared. For the calculations, a manganese concentration of $x_{Mn} = 0.0036$ and different values for the number of magnetic ions in the QD, N_{Mn}, have been used. The best fit is obtained for $N_{Mn} = 13$. From [126]

Even in the absence of EMP formation and without any external magnetic field, magnetic fluctuations of the manganese spin system result in a distinct SQD emission linewidth broadening. As each Mn^{2+} ion spin randomly fluctuates within its six possible spin projections $S_z = -5/2, \ldots, +5/2$, each photo-generated electron–hole pair experience a different spin configuration during its lifetime. Hundt et al. showed that in the absence of EMP formation the role of magnetic fluctuations increases with decreasing QD size and for a given dot size grows square-root-like with increasing number of Mn^{2+} ions inside the SQD [15]. In a recent work, a reduction of the impact of magnetic fluctuations on the SQD linewidth could be achieved by inserting a thin nonmagnetic spacer layer between the CdSe QDs and the ZnMnSe barrier [124, 125]. Interestingly, even for a quite small overlap γ between exciton wavefunction and magnetic ions in the barrier, a noticeable increase of the spin relaxation rate between bright spin-up and spin-down excitons is found. This is attributed to be most likely related to magnetic fluctuations, while apparently the spin transfer between bright and dark states, which would require a spin flip of either the electron or the hole, is still suppressed [125].

Optical Manipulation of Nanomagnetism

As shown above, rising the bath temperature results in a thermal demagnetization of a QD containing an EMP. Alternatively, the spin system can be heated via spin disordered hot carriers generated by optical excitation. This has been widely studied in DMS quantum wells [132–134] and even in QDs [15]. An enhanced temperature of the Mn^{2+} spin system due to laser excitation should be evidenced by both, a power dependent blue shift of the SQD PL signal and a broadening of the PL line provided an EMP is formed during the exciton lifetime.

In the absence of an external magnetic field one obtains the following relation for the temperature dependent PL linewidth FWHM of a SQD [128]

$$\text{FWHM} = \sqrt{8\ln 2 \cdot k_B T \cdot (T + T_0) \left.\frac{dE}{dT}\right|_{B_{\text{exc}}}}. \qquad (3.9)$$

In the upper part of Fig. 3.24, left, the PL energy shift of a CdSe/ZnMnSe SQD is plotted vs. excitation power and vs. bath temperature, respectively. Increasing the excitation power causes a pronounced blue shift of the PL signal, which amounts to almost 6 meV if one increases the excitation power from 0.1 to 10 mW. In the lower part of the figure, the variation of the FWHM with excitation power and temperature is shown. It becomes clear that the power-induced linewidth broadening is a consequence of the carrier-induced

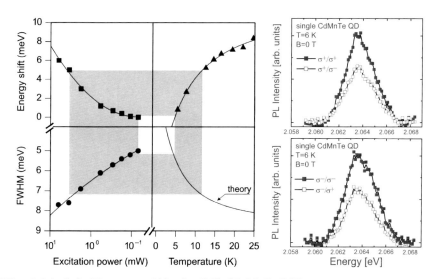

Fig. 3.24. *Left:* PL energy shift of a CdSe/ZnMnSe SQD vs. excitation power and temperature, respectively (*upper panel*). In the *lower panel*, FWHM is depicted vs. excitation power and vs. temperature, respectively [128]. *Right:* Polarized PL spectra of a CdMnTe/ZnTe SQD for σ^+ (*top*) and σ^- excitation (*bottom*), respectively. From [135]

spin heating of the manganese system. Increasing the excitation power from 0.1 mW to about 4 mW results in a PL energy shift of ≈5 meV and in an increase of the FWHM from 5 meV to more than 7 meV. This roughly corresponds to an increase of the Mn^{2+} spin temperature by about 7–8 K as can be seen by comparing the power dependent and the temperature dependent data. The data thus demonstrate the ability to reduce the magnetization of a magnetically doped SQD via an incoherent spin transfer from photoexcited hot carriers to localized magnetic moments.

Vice versa, in case of quasiresonant optical excitation with circularly polarized light, spin polarized electron–hole pairs can transfer their spin information directly to the magnetic system and thus increase the SQD magnetization. This shown in the right part of Fig. 3.24, where the PL spectra of a CdMnTe/ZnTe SQD are depicted for different excitation and detection schemes. The experiments have been performed in absence of an external magnetic field at $T = 6$ K. It is obvious that for cocircular polarized excitation and detection the SQD signal is significantly stronger than in case of cross-circular polarized excitation and detection. As schematically shown in the inset of the figure, this can be explained by assuming a transfer of spin information from the optically generated exciton to the magnetic ion system, i.e., an optical generation of SQD magnetization by laser excitation with suitable polarization [135].

3.4.3 Single Quantum Dots Doped with an Individual Mn^{2+} Ion

As discussed above, the Mn^{2+} spin ensemble in QDs doped with lots of magnetic ions can be described by the molecular field approximation and characterized by a magnetization. Nanooptics allows one to probe and to manipulate nanoscale magnetization in a semiconductor. In contrast, the situation completely changes if one considers a SQD doped with just one single magnetic impurity. This was first achieved by Besombes et al. in self-assembled CdTe/ZnTe/ZnMnTe QDs, where Mn intermixing during the growth allows for a sparse distribution of Mn^{2+} ions in the QD layer [13]. In the following, pioneering work on this issue has been performed by this group, which will be briefly discussed in the following.

Probing the Mn^{2+} Spin State by a Single Exciton

In Fig. 3.25, the PL spectra for an undoped CdTe SQD and a SQD doped with a single magnetic impurity are shown. For the undoped SQD, a spectrally narrow, single emission line is observed at $B_{ext} = 0$, which splits into the Zeeman doublet in an external magnetic field in Faraday geometry. Interestingly, six almost equidistant emission peaks are observed at zero field for the SQD doped with a single Mn^{2+} ion.

In order to explain these findings, one have to consider the spin states of the particles (electron, hole, Mn^{2+} ion) in the SQD. In a nonmagnetic QD, the

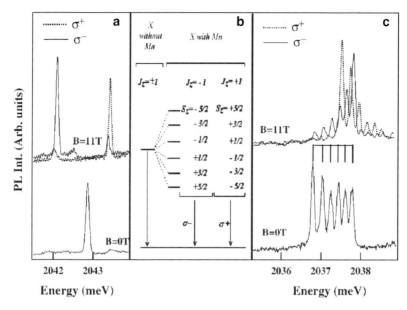

Fig. 3.25. Low temperature (5 K) PL spectra of a CdTe/ZnTe SQD (**a**) and a SQD doped with a single Mn^{2+} ion (**c**). Data obtained at $B_{\text{ext}} = 0$ and $B_{\text{ext}} = 11$ T are compared. In (**b**) the corresponding energy level scheme is illustrated. From [13]

bright heavy hole excitons contribute to the zero field PL spectrum and due to the anisotropic electron–hole exchange interaction, a linear polarized doublet is usually observed in PL. If the exchange splitting is small as compared to the experimental PL linewidth, a single emission peak is seen (see Fig. 3.25a). A Mn^{2+} ion with spin $S = 5/2$ and therefore six possible spin projections $S_z = \pm 5/2, \pm 3/2, \pm 1/2$ can interact with the carrier spins via the exchange interaction and the Hamiltonian of this system can be written as [13]

$$H_{\text{int}} = I_e \cdot \sigma \cdot S + I_h \cdot j \cdot S + I_{e-h} \cdot \sigma \cdot j, \tag{3.10}$$

where I_e (I_h) is the Mn-electron (hole) exchange integral, I_{e-h} the electron–hole exchange interaction and $\sigma(j)$ the magnetic moment of the electron (hole). Due to this exchange interaction the initial state of recombination (one electron, one hole, one Mn^{2+} ion) splits into six doubly degenerate states for the bright exciton ($s_z = 1/2$, $j_z = -3/2$, $S_z = \pm 5/2, \pm 3/2, \pm 1/2$ and $s_z = -1/2$, $j_z = +3/2$, $S_z = \pm 5/2, \pm 3/2, \pm 1/2$). As the final state of recombination (one Mn^{2+} ion) is a six times degenerate one, the SQD emission is expected to split into six distinct lines, in agreement with the experiment.

The energy splitting of the PL peaks is determined by the exchange integrals I_e and I_h and the electron–hole exchange interaction. It has been shown experimentally and theoretically that by reducing the overlap between the exciton wavefunction and the magnetic ion, the exchange splitting due to the

Mn-carrier interaction decreases [136] and the electron–hole exchange splitting becomes important. One obtains for the energy splitting of bright excitons for a given value of S_z [137]

$$\Delta E(S_z) = \sqrt{\delta_{e,h}^2 + (2|S_z|\delta_{Mn})^2} \qquad (3.11)$$

Here, $\delta_{e,h}$ is the anisotropic electron–hole exchange term and $2|S_z|\delta_{Mn}$ is the splitting induced by the Mn^{2+} only. Depending on the QD asymmetry and the overlap between the exciton wavefunction and the Mn^{2+} ion, the relative strength of both contributions in (3.11) varies.

In an external magnetic field the spin degeneracy in the initial and in the final state is further lifted. As can be seen in Fig. 3.25c, this results in 12 emission lines, six of them showing σ^+ and six of them σ^- polarization, if the magnetic field is applied in Faraday geometry. The behavior of the PL spectra of the magnetically doped SQD in a magnetic field in both, Faraday and Voigt geometry, is shown in more detail in Fig. 3.26.

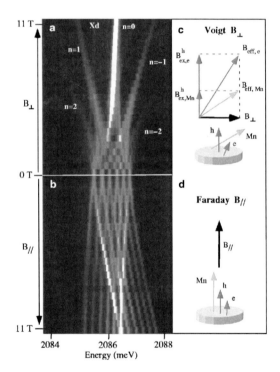

Fig. 3.26. Contour plots of the magneto PL signal of a CdTe SQD doped with one Mn^{2+} ion in Voigt (**a**) and Faraday (**b**) geometry. The *right part* of the figure schematically illustrates the spin orientation of the electron, the hole and the Mn^{2+} ion in a magnetic field. From [138]

In Faraday geometry, the quantization axis along the growth direction is conserved and the spin orientation of the electron, the hole and the Mn^{2+} ion is along the external field. In this configuration, the PL spectrum consists of 12 lines split by exchange and Zeeman interaction. In this geometry, the anisotropic part of the electron–hole exchange interaction in asymmetric QDs causes a mixing of bright exciton states with different Mn^{2+} spin states. Moreover, the electron–Mn^{2+} interaction couples bright and dark exciton states, which corresponds to a simultaneous spin flip of the electron and the Mn^{2+} ion. Both effects result in a distinct anticrossing behavior at elevated magnetic fields [13, 137].

In Voigt geometry, the situation is different. The Mn^{2+} spin is aligned along the magnetic field, i.e., perpendicular to the growth axis, while the spin orientation of the optically generated electron–hole pairs is along the growth axis. After optical excitation, the Mn^{2+} spin rotates away from its initial axis due to the hole exchange field and starts to precess around the effective field, which is given by the sum of the external field and the exchange field. In case of a longitudinal spin relaxation time T_1, which is shorter than the recombination lifetime, the exciton-Mn^{2+} spin complex reaches equilibrium and in this state, the projection of the Mn^{2+} spin on the external field B_{ext} differs from its original value. This requires an integer number n of spin flips, which are responsible for the replicas observed in the SQD PL spectra in Voigt geometry [138]. At high transverse magnetic fields the spin blockade is restored and only optical transitions in which the Mn^{2+} spin projection is conserved becomes visible. In addition, signatures of dark exciton states (in the figure labeled X_d) can be seen, similar to what is obtained in nonmagnetic SQDs (see Fig. 3.26).

Charged Excitons and Biexcitons Interacting with One Magnetic Ion

In analogy to what is known from nonmagnetic materials, SQDs can be populated by additional electrons and/or holes either by electrical injection or optical excitation [27, 48]. Embedding a SQD containing one magnetic ion in a Schottky diode structure, its charge state can be controlled by varying the bias voltage. As can be seen in Fig. 3.27, left, a rich diversity of individual PL peaks is obtained for neutral excitons, positively and negatively charged excitons and biexcitons, respectively. Apparently the exchange interaction between charge carriers and the magnetic ion dramatically modifies the characteristic emission spectrum of charged excitons and multiexcitons in a SQD doped with a single Mn^{2+} ion. While at negative and slightly positive voltages neutral and negatively charged exciton complexes dominate, a clear contribution of a positively charged exciton is found for a large positive bias.

Let us consider the case of a negatively charged exciton. The initial state of recombination consists of two electrons with antiparallel spin orientation and

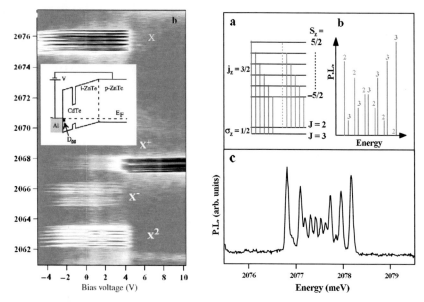

Fig. 3.27. *Left:* Contour plot of the PL intensity of a SQD doped with one magnetic ion embedded into a Schottky diode as a function of bias voltage. The recombination of the neutral exciton (X), the biexciton (X2) and positively (negatively) charged excitons X$^+$ (X$^-$) can be seen [49]. *Right:* Unpolarized emission spectrum of a negatively charged exciton X$^-$ coupled to a single Mn^{2+} ion (*lower panel*). In the *upper panel*, the energy level scheme and the corresponding optical transitions are shown schematically. From [139]

one hole exchange coupled to the Mn^{2+} ion, i.e., six doubly degenerate states for $j_z = 3/2$ and for $j_z = -3/2$, respectively. The final state of recombination consists of one electron coupled to the manganese ion. The twelve eigenstates of the electron-Mn^{2+} complex are split into a ground state septuplet (total spin $J = 3$) and a fivefold degenerate state with $J = 2$ (see Fig. 3.27, right). Taking into account the optical selection rules and the fact that the Mn spin is not affected by the optical transition, one obtains a multiplet of eleven lines, in quite nice agreement to what is found in experiment. Similar arguments can be used to explain the PL spectrum of a positively charged excitons containing a single magnetic ion. This ability of electrical control of the properties of a single magnetic ion in a solid state environment might even allow storing digital information on a single atom.

Simply by increasing the optical excitation power, the QD can be populated by two electrons and two holes in the *s*-shell. Due to the Pauli principle, the spin orientation of the two electrons and the two holes have to be antiparallel, i.e., the biexciton state is a spin singlet [27], which in first order is not expected to interact with the magnetic ions via exchange interaction. Due to the optical selection rules, a biexciton cannot annihilate directly into the

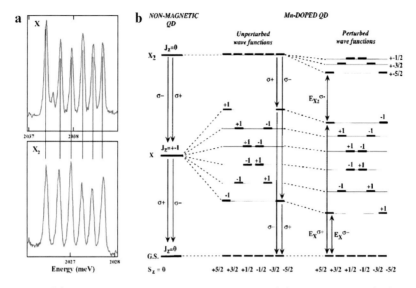

Fig. 3.28. (a) Fine structure splitting of the exciton (X) and biexciton (X_2) transition of a CdTe SQD doped with one Mn^{2+} at $B_{ext} = 0$. (b) Energy level scheme including the allowed optical transitions of the biexciton–exciton cascade in a nonmagnetic and a Mn^{2+} doped SQD. From [140]

ground state but has to recombine via the exciton state [30]. Thus, the final state of biexciton recombination is equivalent to the initial state of exciton recombination (see Fig. 3.2).

This is schematically shown in Fig. 3.28, where the energy level scheme of the biexciton–exciton cascade is illustrated for a nonmagnetic SQD and a SQD doped with one magnetic ion. In the ideal case of unperturbed carrier wavefunctions, the biexciton transition should exhibit the same fine structure splitting due to exchange interaction with the Mn^{2+} ions than the exciton transition. Indeed, as can be seen on the left part of Fig. 3.28, both, the exciton as well as the biexciton emission spectrum of a SQD doped with one magnetic ion exhibits six main emission peaks with identical energy splitting (the additional line observed in the PL spectrum of the exciton is related to the dark exciton transition). This interpretation is supported by magneto-PL studies: The mixing between bright and dark exciton states result in characteristic anticrossing features, which are observed for the high energy lines of the exciton transition in σ^- polarization and replicated for the low energy lines of the biexciton line in σ^+ polarization [141]. The slight deviation from an equidistant fine structure splitting is attributed to a lifting of degeneracy of the biexciton state due to a perturbation of the carriers' orbital wavefunction by the exchange coupling with the magnetic ion.

3.5 Conclusion

In summary, we reviewed some of the most fascinating properties of individual wide bandgap single quantum dots (SQDs) revealed by optical spectroscopy. The extraordinary large Coulomb and exchange energies result in a fine structure of the SQD emission spectra with large energy splittings. Linear as well as nonlinear optical properties like exciton superradiance or coherent control and stimulated emission of the biexciton in a single quantum dot could be shown. Well-defined spin states can be prepared in a SQD either by polarized optical excitation or by spin injection via a heterointerface and dynamical nuclear polarization on a submillisecond time scale is obtained. With respect to potential applications in (quantum) information science, two important aspects have been discussed. First, the strong inherent carrier confinement enables single photon emission at elevated temperatures and even room temperature electroluminescence of a SQD. Second, the incorporation of magnetic impurities into a II–VI SQD does allow one to optically probe and manipulate nanoscale magnetization in a semiconductor and even gives access to an optical and electrical control of the spin state of a single magnetic ion in a solid state environment.

Acknowledgments

The authors are indebted to R. Arians and S. Halm from Werkstoffe der Elektrotechnik, University Duisburg-Essen, H. Schömig, J. Seufert, R. Weigand, and A. Forchel from the Technische Physik at the University of Würzburg and to A.A. Maksimov and V.D. Kulakovskii from the Institute of Solid State Physics, Chernogolovka, for their collaboration. The research has been supported by the Deutsche Forschungsgemeinschaft.

References

1. G. Bacher, *Optical Spectroscopy on Epitaxially Grown II–VI Single Quantum Dots.* Topics in Applied Physics, vol 90 (Springer, Berlin, 2003), pp. 147–183
2. Y. Terai, S. Kuroda, K. Takita, T. Okuno, Y. Masumoto, Appl. Phys. Lett. **73**(25), 3757 (1998)
3. F. Tinjod, B. Gilles, S. Moehl, K. Kheng, H. Mariette, Appl. Phys. Lett. **82**(24), 4340 (2003)
4. S. Mackowski, L. Smith, H. Jackson, W. Heiss, J. Kossut, G. Karczewski, Appl. Phys. Lett. **83**(2), 254 (2003)
5. S. Xin, P. Wang, A. Yin, C. Kim, M. Dobrowolska, J. Merz, J. Furdyna, Appl. Phys. Lett. **69**(25), 3884 (1996)
6. F. Flack, N. Samarth, V. Nikitin, P. Crowell, J. Shi, J. Levy, D. Awschalom, Phys. Rev. B **54**(24), 17312 (1996)
7. T. Kümmell, R. Weigand, G. Bacher, A. Forchel, K. Leonardi, D. Hommel, H. Selke, Appl. Phys. Lett. **73**(21), 3105 (1998)

8. M. Rabe, M. Lowisch, F. Henneberger, J. Cryst. Growth **185**, 248 (1998)
9. D. Schikora, S. Schwedhelm, D. As, K. Lischka, D. Litvinov, A. Rosenauer, D. Gerthsen, M. Strassburg, A. Hoffmann, D. Bimberg, Appl. Phys. Lett. **76**(4), 418 (2000)
10. S. Ivanov, A. Toropov, S. Sorokin, T. Shubina, I. Sedova, A. Sitnikova, P. Kop'ev, Z. Alferov, H. Lugauer, G. Reuscher, M. Keim, F. Fischer, A. Waag, G. Landwehr, Appl. Phys. Lett. **74**(4), 498 (1999)
11. G. Bacher, H. Schömig, M. Welsch, S. Zaitsev, V. Kulakovskii, A. Forchel, S. Lee, M. Dobrowolska, J. Furdyna, B. Konig, W. Ossau, Appl. Phys. Lett. **79**(4), 524 (2001)
12. G. Bacher, A. Maksimov, H. Schömig, V. Kulakovskii, M. Welsch, A. Forchel, P. Dorozhkin, A. Chernenko, S. Lee, M. Dobrowolska, J. Furdyna, Phys. Rev. Lett. **89**(12), 127201 (2002). doi:10.1103/PhysRevLett.89.127201
13. L. Besombes, Y. Léger, L. Maingault, D. Ferrand, H. Mariette, J. Cibert, Phys. Rev. Lett. **93**(20), 207403 (2004). doi:10.1103/PhysRevLett.93.207403
14. S. Mackowski, H. Jackson, L. Smith, J. Kossut, G. Karczewski, W. Heiss, Appl. Phys. Lett. **83**(17), 3575 (2003). doi:10.1063/1.1622438
15. A. Hundt, J. Puls, F. Henneberger, Phys. Rev. B **69**(12), 121309 (2004). doi:10.1103/PhysRevB.69.121309
16. F. Tinjod, S. Moehl, K. Kheng, B. Gilles, H. Mariette, J. Appl. Phys. **95**(1), 102 (2004). doi:10.1063/1.1631755
17. S. Moehl, L. Maingault, K. Kheng, H. Mariette, Appl. Phys. Lett. **87**(3), 033111 (2005). doi:10.1063/1.2000335
18. R. Arians, T. Kümmell, G. Bacher, A. Gust, C. Kruse, D. Hommel, Appl. Phys. Lett. **90**(10), 101114 (2007). doi:10.1063/1.2710787
19. M. Funato, K. Omae, Y. Kawakami, S. Fujita, C. Bradford, A. Balocchi, K.A. Prior, B.C. Cavenett, Phys. Rev. B **73**(24), 245308 (2006). doi:10.1103/PhysRevB.73.245308
20. Y. Arakawa, T. Someya, K. Tachibana, Phys. Stat. Sol. (b) **224**(1), 1 (2001)
21. S. Tanaka, S. Iwai, Y. Aoyagi, Appl. Phys. Lett. **69**(26), 4096 (1996)
22. F. Widmann, B. Daudin, G. Feuillet, Y. Samson, J. Rouviere, N. Pelekanos, J. Appl. Phys. **83**(12), 7618 (1998)
23. B. Damilano, N. Grandjean, S. Dalmasso, J. Massies, Appl. Phys. Lett. **75**(24), 3751 (1999)
24. I. Krestnikov, N. Ledentsov, A. Hoffmann, D. Bimberg, A. Sakharov, W. Lundin, A. Tsatsul'nikov, A. Usikov, Z. Alferov, Y. Musikhin, D. Gerthsen, Phys. Rev. B **66**(15), 155310 (2002). doi:10.1103/PhysRevB.66.155310
25. O. Husberg, A. Khartchenko, D. As, H. Vogelsang, T. Frey, D. Schikora, K. Lischka, O. Noriega, A. Tabata, J. Leite, Appl. Phys. Lett. **79**(9), 1243 (2001)
26. F. Rol, B. Gayral, S. Founta, B. Daudin, J. Eymery, J. Gerard, H. Mariette, L. Dang, D. Peyrade, Phys. Stat. Sol. (b) **243**(7), 1652 (2006). doi:10.1002/pssb.200565406
27. V.D. Kulakovskii, G. Bacher, R. Weigand, T. Kümmell, A. Forchel, E. Borovitskaya, K. Leonardi, D. Hommel, Phys. Rev. Lett. **82**(8), 1780 (1999). doi:10.1103/PhysRevLett.82.1780
28. Y. Léger, L. Besombes, L. Maingault, H. Mariette, Phys. Rev. B **76**(4), 045331 (2007). doi:10.1103/PhysRevB.76.045331
29. L. Besombes, K. Kheng, L. Marsal, H. Mariette, Phys. Rev. B **63**(15), 155307 (2001)

30. G. Bacher, R. Weigand, J. Seufert, V. Kulakovskii, N. Gippius, A. Forchel, K. Leonardi, D. Hommel, Phys. Rev. Lett. **83**(21), 4417 (1999)
31. M. Scheibner, T. Schmidt, L. Worschech, A. Forchel, G. Bacher, T. Passow, D. Hommel, Nat. Phys. **3**(2), 106 (2007). doi:10.1038/nphys494
32. H. Schömig, S. Halm, A. Forchel, G. Bacher, J. Off, F. Scholz, Phys. Rev. Lett. **92**(10), 106802 (2004). doi:10.1103/PhysRevLett.92.106802
33. S. Kako, C. Santori, K. Hoshino, S. Goetzinger, Y. Yamamoto, Y. Arakawa, Nat. Mater. **5**(11), 887 (2006). doi:10.1038/nmat1763
34. J. Seufert, G. Bacher, M. Scheibner, A. Forchel, S. Lee, M. Dobrowolska, J. Furdyna, Phys. Rev. Lett. **88**(2), 027402 (2002). doi:10.1003/PhysRevLett.88.027402 10.1103/PhysRevLett.88.025501
35. M. Klude, T. Passow, R. Kroger, D. Hommel, Electron. Lett. **37**(18), 1119 (2001)
36. A. Gust, C. Kruse, E. Roventa, R. Kröger, K. Sebald, H. Lohmeyer, B. Brendemühl, J. Gutowski, D. Hommel, Phys. Stat. Sol. (c) **2**(3), 1098 (2005). doi:10.1002/pssb.2005646xx
37. T. Xu, A.Y. Nikiforov, R. France, C. Thomidis, A. Williams, T.D. Moustakas, Phys Stat. Sol. (a) **204**(6), 2098 (2007). doi:10.1002/pssa.200674834
38. Y. Su, S. Chang, L. Ji, C. Chang, L. Wu, W. Lai, T. Fang, K. Lam, Semicond. Sci. Technol. **19**(3), 389 (2004). doi:10.1088/0268-1242/19/3/016
39. S. Tanaka, J. Lee, P. Ramvall, H. Okagawa, Japn. J. Appl. Phys. **42**(8A), L885 (2003). doi:10.1143/JJAP.42.L885
40. A.J. Shields, Nat. Photonics **1**(4), 215 (2007). doi:10.1038/nphoton.2007.46
41. T. Matsuoka, T. Ito, T. Kaino, Electron. Lett. **36**(22), 1836 (2000)
42. D. Loss, D. DiVincenzo, Phys. Rev. A **57**(1), 120 (1998)
43. T. Flissikowski, I. Akimov, A. Hundt, F. Henneberger, Phys. Rev. B **68**(16), 161309 (2003). doi:10.1103/PhysRevB.68.161309
44. A.F. Jarjour, R.A. Oliver, A. Tahraoui, M.J. Kappers, C.J. Humphreys, R.A. Taylor, Phys. Rev. Lett. **99**(19), 197403 (2007). doi:10.1103/PhysRev Lett.99.197403. http://link.aps.org/abstract/ PRL/v99/e197403
45. T. Flissikowski, A. Betke, I. Akimov, F. Henneberger, Phys. Rev. Lett. **92**(22), 227401 (2004). doi:10.1103/PhysRevLett.92.227401
46. B. Patton, W. Langbein, U. Woggon, L. Maingault, H. Mariette, Phys. Rev. B **73**(23), 235354 (2006). doi:10.1103/PhysRevB.73.235354
47. J. Seufert, M. Obert, M. Scheibner, N. Gippius, G. Bacher, A. Forchel, T. Passow, K. Leonardi, D. Hommel, Appl. Phys. Lett. **79**(7), 1033 (2001)
48. J. Seufert, M. Rambach, G. Bacher, A. Forchel, T. Passow, D. Hommel, Appl. Phys. Lett. **82**(22), 3946 (2003). doi:10.1063/1.1580632
49. Y. Léger, L. Besombes, J. Fernandez-Rossier, L. Maingault, H. Mariette, Phys. Rev. Lett. **97**(10), 107401 (2006). doi:10.1103/PhysRevLett.97.107401
50. T. Nakaoka, S. Kako, Y. Arakawa, Phys. Rev. B **73**(12), 121305 (2006). doi:10.1103/PhysRevB.73.121305
51. R. Arians, A. Gust, T. Kümmell, C. Kruse, S. Zaitsev, G. Bacher, D. Hommel, Appl. Phys. Lett. **93**, 173506 (2008)
52. T. Passow, H. Heinke, T. Schmidt, J. Falta, A. Stockmann, H. Selke, P.L. Ryder, K. Leonardi, D. Hommel, Phys. Rev. B **64**(19), 193311 (2001). doi:10.1103/PhysRevB.64.193311
53. I.C. Robin, R. André, J.M. Gérard, Phys. Rev. B **74**(15), 155318 (2006). doi:10.1103/PhysRevB.74.155318. http://link.aps.org/abstract/PRB/v74/e155318

54. S. Moehl, F. Tinjod, K. Kheng, H. Mariette, Phys. Rev. B **69**(24), 245318 (2004). doi:10.1103/PhysRevB.69.245318
55. L. Besombes, K. Kheng, D. Martrou, Phys. Rev. Lett. **85**(2), 425 (2000). doi:10.1103/PhysRevLett.85.425
56. L. Besombes, K. Kheng, L. Marsal, H. Mariette, Phys. Rev. B **65**(12), 121314 (2002). doi:10.1103/PhysRevB.65.121314
57. J. Puls, M. Rabe, H. Wunsche, F. Henneberger, Phys. Rev. B **60**(24), R16303 (1999)
58. J. Puls, F. Henneberger, Phys. Stat. Sol. (a) **166**(3), 499 (1997). doi:yy.1002/pssa/200304005
59. M. Bayer, G. Ortner, O. Stern, A. Kuther, A.A. Gorbunov, A. Forchel, P. Hawrylak, S. Fafard, K. Hinzer, T.L. Reinecke, S.N. Walck, J.P. Reithmaier, F. Klopf, F. Schäfer, Phys. Rev. B **65**(19), 195315 (2002). doi:10.1103/PhysRevB.65.195315
60. K.P. Hewaparakrama, S. Mackowski, H.E. Jackson, L.M. Smith, W. Heiss, G. Karczewski, Nanotechnology **19**(12), 125706 (2008). doi:10.1088/0957-4484/19/12/125706
61. E. Margapoti, L. Worschech, A. Forchel, A. Tribu, T. Aichele, R. Andre, K. Kheng, Appl. Phys. Lett. **90**(18), 181927 (2007). doi:10.1063/1.2737131
62. L. Besombes, K. Kheng, L. Marsal, H. Mariette, Europhys. Lett. **65**(1), 144 (2004). doi:10.1209/epl/i2003-10055-9
63. D. Simeonov, A. Dussaigne, R. Butte, N. Grandjean, Phys. Rev. B **77**(7), 075306 (2008). doi:10.1103/PhysRevB.77.075306. http://link.aps.org/abstract/PRB/v77/e075306
64. S. Kako, K. Hoshino, S. Iwamoto, S. Ishida, Y. Arakawa, Appl. Phys. Lett. **85**(1), 64 (2004). doi:10.1063/1.1769586. http://link.aip.org/link/?APL/85/64/1
65. J. Renard, R. Songmuang, C. Bougerol, B. Daudin, B. Gayral, Nano Lett., **8**, 2092 (2008) http://pubs3.acs.org/acs/journals/doilookup?in_doi=10.1021/nl0800873
66. F. Findeis, M. Baier, A. Zrenner, M. Bichler, G. Abstreiter, U. Hohenester, E. Molinari, Phys. Rev. B **63**(12), 121309 (2001). doi:10.1103/PhysRevB.63.121309
67. M. Kroutvar, Y. Ducommun, D. Heiss, M. Bichler, D. Schuh, G. Abstreiter, J. Finley, Nature **432**(7013), 81 (2004). doi:10.1038/nature03008
68. I. Akimov, K. Kavokin, A. Hundt, F. Henneberger, Phys. Rev. B **71**(7), 075326 (2005). doi:10.1103/PhysRevB.71.075326
69. I.A. Akimov, A. Hundt, T. Flissikowski, F. Henneberger, Appl. Phys. Lett. **81**(25), 4730 (2002). doi:10.1063/1.1527694. http://link.aip.org/link/?APL/81/4730/1
70. T. Graham, A. Curran, X. Tang, J. Morrod, K. Prior, R. Warburton, Phys. Stat. Sol. (b) **243**(4), 782 (2006). doi:10.1002/pssb.200564654
71. I.A. Akimov, J.T. Andrews, F. Henneberger, Phys. Rev. Lett. **96**(6), 067401 (2006). doi:10.1103/PhysRevLett.96.067401. http://link.aps.org/abstract/PRL/v96/e067401
72. V. Blanchet, C. Nicole, M.A. Bouchene, B. Girard, Phys. Rev. Lett. **78**(14), 2716 (1997). doi:10.1103/PhysRevLett.78.2716
73. W. Langbein, B. Patton, Opt. Lett. **31**(7), 1151 (2006)

74. C. Couteau, S. Moehl, F. Tinjod, J. Gerard, K. Kheng, H. Mariette, J. Gaj, R. Romestain, J. Poizat, Appl. Phys. Lett. **85**(25), 6251 (2004). doi:10.1063/1.1842370
75. E. Waks, K. Inoue, C. Santori, D. Fattal, J. Vučković, G. Solomon, Y. Yamamoto, Nature **420**(6917), 762 (2002). doi:10.1038/420762a
76. C. Santori, D. Fattal, J. Vučković, G. Solomon, Y. Yamamoto, Nature **419**(6907), 594 (2002). doi:10.1038/nature01086
77. D. Press, S. Goetzinger, S. Reitzenstein, C. Hofmann, A. Loeffler, M. Kamp, A. Forchel, Y. Yamamoto, Phys. Rev. Lett. **98**(11), 117402 (2007). doi:10.1103/PhysRevLett.98.117402
78. Z. Yuan, B. Kardynal, R. Stevenson, A. Shields, C. Lobo, K. Cooper, N. Beattie, D. Ritchie, M. Pepper, Science **295**(5552), 102 (2002). doi:10.1126/science.1066790
79. M.B. Ward, T. Farrow, P. See, Z.L. Yuan, O.Z. Karimov, A.J. Bennett, A.J. Shields, P. Atkinson, K. Cooper, D.A. Ritchie, Appl. Phys. Lett. **90**(6), 063512 (2007). doi:10.1063/1.2472172
80. A. Lochmann, E. Stock, O. Schulz, R. Hopfer, D. Bimberg, V.A. Haisler, A.I. Toropov, A.K. Bakarov, A.K. Kalagin, Electron. Lett. **42**(13), 774 (2006). doi:10.1049/el:20061076
81. C. Monat, B. Alloing, C. Zinoni, L.H. Li, A. Fiore, Nano Lett. **6**(7), 1464 (2006). doi:10.1021/nl060800t
82. O. Benson, C. Santori, M. Pelton, Y. Yamamoto, Phys. Rev. Lett. **84**(11), 2513 (2000)
83. K. Kheng, S. Moehl, I.C. Robin, L. Maingault, R. Andre, H. Mariette, AIP Conf. Proc. **893**(1), 917 (2007). doi:10.1063/1.2730191. http://link.aip.org/link/?APC/893/917/1
84. F. Rol, S. Founta, H. Mariette, B. Daudin, L.S. Dang, J. Bleuse, D. Peyrade, J.M. Gerard, B. Gayral, Phys. Rev. B **75**(12), 125306 (2007). doi:10.1103/PhysRevB.75.125306
85. K. Sebald, P. Michler, T. Passow, D. Hommel, G. Bacher, A. Forchel, Appl. Phys. Lett. **81**(16), 2920 (2002). doi:10.1063/1.1515364
86. I. Robin, R. Andre, L. Dang, H. Mariette, S. Tatarenko, J. Gerard, K. Kheng, F. Tinjod, M. Bartels, K. Lischka, D. Schikora, Phys. Stat. Sol. (b) **241**(3), 542 (2004). doi:10.1002/pssb.200304268
87. T. Graham, B. Urbaszek, X. Tang, B. Bradford, C amd Cavenett, K. Prior, R. Warburton, Phys. Stat. Sol. (c) **1**(4), 755 (2004). doi:10.1002/pssb.200564654
88. U. Woggon, K. Hild, F. Gindele, W. Langbein, M. Hetterich, M. Grün, C. Klingshirn, Phys. Rev. B **61**(19), 12632 (2000). doi:10.1103/PhysRevB.61.12632
89. T. Aichele, V. Zwiller, O. Benson, I. Akimov, F. Henneberger, J. Opt. Soc. Am. **20**(10), 2189 (2003)
90. S. Ulrich, S. Strauf, P. Michler, G. Bacher, A. Forchel, Phys. Stat. Sol. (b) **238**(3), 607 (2003). doi:10.1002/pssb.200303185
91. H. Lohmeyer, J. Kalden, K. Sebald, C. Kruse, D. Hommel, J. Gutowski, Appl. Phys. Lett. **92**(1), 011116 (2008). doi:10.1063/1.2827574
92. H. Huang, A. Dorn, V. Bulovic, M.G. Bawendi, Appl. Phys. Lett. **90**(2), 023110 (2007). doi:10.1063/1.2425043
93. X. Brokmann, E. Giacobino, M. Dahan, J. Hermier, Appl. Phys. Lett. **85**(5), 712 (2004). doi:10.1063/1.1775280

94. R. Arians, T. Kümmell, G. Bacher, A. Gust, C. Kruse, D. Hommel, Physica E **40**(6), 1938 (2008). doi:10.1016/j.physe.2007.08.082
95. J. Renner, L. Worschech, A. Forchel, S. Mahapatra, K. Brunner, Appl. Phys. Lett. **89**(9), 091105 (2006). doi:10.1063/1.2345236
96. I. Robin, R. Andre, A. Balocchi, S. Carayon, S. Moehl, J. Gerard, L. Ferlazzo, Appl. Phys. Lett. **87**(23), 233114 (2005). doi:10.1063/1.2136433
97. H. Lohmeyer, C. Kruse, K. Sebald, J. Gutowski, D. Hommel, Appl. Phys. Lett. **89**(9), 091107 (2006). doi:10.1063/1.2338800
98. C. Santori, S. Gotzinger, Y. Yamamoto, S. Kako, K. Hoshino, Y. Arakawa, Appl. Phys. Lett. **87**(5), 051916 (2005). doi:10.1063/1.2006987
99. M. Arita, S. Ishida, S. Kako, S. Iwamoto, Y. Arakawa, Appl. Phys. Lett. **91**(5), 051106 (2007). doi:10.1063/1.2757596
100. K. Sebald, H. Lohmeyer, J. Gutowski, T. Yamaguchi, C. Kruse, D. Hommel, J. Wiersig, F. Jahnke, Phys. Stat. Sol. (b) **244**(6), 1806 (2007). doi:10.1002/pssb.200674827
101. A.F. Jarjour, R.A. Taylor, R.A. Oliver, M.J. Kappers, C.J. Humphreys, A. Tahraoui, Appl. Phys. Lett. **91**(5), 052101 (2007). doi:10.1063/1.2767217
102. S. Wolf, D. Awschalom, R. Buhrman, J. Daughton, S. von Molnar, M. Roukes, A. Chtchelkanova, D. Treger, Science **294**(5546), 1488 (2001)
103. D.D. Awschalom, M.E. Flatte, Nat. Phys. **3**(3), 153 (2007). doi:10.1038/nphys551
104. J. Elzerman, R. Hanson, L. van Beveren, B. Witkamp, L. Vandersypen, L. Kouwenhoven, Nature **430**(6998), 431 (2004). doi:10.1038/nature02693
105. M. Scheibner, G. Bacher, S. Weber, A. Forchel, T. Passow, D. Hommel, Phys. Rev. B **67**(15), 153302 (2003). doi:10.1103/PhysRevB.67.153302
106. D. Lagarde, A. Balocchi, H. Carrere, P. Renucci, T. Amand, X. Marie, S. Founta, H. Mariette, Phys. Rev. B **77**(4), 041304 (2008). doi:10.1103/PhysRevB.77.041304
107. M. Paillard, X. Marie, P. Renucci, T. Amand, A. Jbeli, J. Gerard, Phys. Rev. Lett. **86**(8), 1634 (2001)
108. T. Flissikowski, A. Hundt, M. Lowisch, M. Rabe, F. Henneberger, Phys. Rev. Lett. **86**(14), 3172 (2001)
109. S. Mackowski, T. Nguyen, T. Gurung, K. Hewaparakrama, H. Jackson, L. Smith, J. Wrobel, K. Fronc, J. Kossut, G. Karczewski, Phys. Rev. B **70**(24), 245312 (2004). doi:10.1103/PhysRevB.70.245312
110. R. Fiederling, M. Keim, G. Reuscher, W. Ossau, G. Schmidt, A. Waag, L. Molenkamp, Nature **402**(6763), 787 (1999)
111. Y. Ohno, D. Young, B. Beschoten, F. Matsukura, H. Ohno, D. Awschalom, Nature **402**(6763), 790 (1999)
112. J. Seufert, G. Bacher, H. Schömig, A. Forchel, L. Hansen, G. Schmidt, L. Molenkamp, Phys. Rev. B **69**(3), 035311 (2004). doi:10.1103/PhysRevB.69.035311
113. M. Ghali, R. Arians, T. Kümmell, G. Bacher, J. Wenisch, S. Mahapatra, K. Brunner, Appl. Phys. Lett. **90**(9), 093110 (2007). doi:10.1063/1.2710078
114. W. Loeffler, M. Hetterich, C. Mauser, S. Li, T. Passow, H. Kalt, Appl. Phys. Lett. **90**(23), 232105 (2007). doi:10.1063/1.2746405
115. I. Merkulov, A. Efros, M. Rosen, Phys. Rev. B **65**(20), 205309 (2002). doi:10.1103/PhysRevB.65.205309
116. A. Khaetskii, D. Loss, L. Glazman, Phys. Rev. Lett. **88**(18), 186802 (2002). doi:10.1103/PhysRevLett.88.186802

117. L.M. Woods, T.L. Reinecke, A.K. Rajagopal, Phys. Rev. B **77**(7), 073313 (2008). doi:10.1103/PhysRevB.77.073313
118. B.D. Gerardot, D. Brunner, P.A. Dalgarno, P. Ohberg, S. Seidl, M. Kroner, K. Karrai, N.G. Stoltz, P.M. Petroff, R.J. Warburton, Nature **451**(7177), 441 (2008). doi:10.1038/nature06472
119. A.J. Ramsay, S.J. Boyle, R.S. Kolodka, J.B.B. Oliveira, J. Skiba-Szymanska, H.Y. Liu, M. Hopkinson, A.M. Fox, M.S. Skolnick, Phys. Rev. Lett. **100**(19), 197401 (2008). doi:10.1103/PhysRevLett.100.197401
120. I.A. Akimov, D.H. Feng, F. Henneberger, Phys. Rev. Lett. **97**(5), 056602 (2006). doi:10.1103/PhysRevLett.97.056602
121. D.H. Feng, I.A. Akimov, F. Henneberger, Phys. Rev. Lett. **99**(3), 036604 (2007). doi:10.1103/PhyRevLett.99.036604
122. A. Maksimov, G. Bacher, A. McDonald, V. Kulakovskii, A. Forchel, C. Becker, G. Landwehr, L. Molenkamp, Phys. Rev. B **62**(12), R7767 (2000)
123. G. Bacher, T. Kümmell, D. Eisert, A. Forchel, B. Konig, W. Ossau, C. Becker, G. Landwehr, Appl. Phys. Lett. **75**(7), 956 (1999)
124. P. Dorozhkin, V. Kulakovskii, A. Chernenko, A. Brichkin, S. Ivanov, A. Toropov, Appl. Phys. Lett. **86**(6), 062507 (2005). doi:10.1063/1.1861954
125. E.A. Chekhovich, A.S. Brichkin, A.V. Chernenko, V.D. Kulakovskii, I.V. Sedova, S.V. Sorokin, S.V. Ivanov, Phys. Rev. B **76**(16), 165305 (2007). doi:10.1103/PhysRevB.76.165305
126. P. Wojnar, J. Suffczynski, K. Kowalik, A. Golnik, G. Karczewski, J. Kossut, Phys. Rev. B **75**(15), 155301 (2007). doi:10.1103/PhysRevB.75.155301
127. P. Dorozhkin, A. Chernenko, V. Kulakovskii, A. Brichkin, A. Maksimov, H. Schoemig, G. Bacher, A. Forchel, S. Lee, M. Dobrowolska, J. Furdyna, Phys. Rev. B **68**(19), 195313 (2003). doi:10.1103/PhysRevB.68.195313
128. G. Bacher, H. Schömig, M. Scheibner, A. Forchel, A. Maksimov, A. Chernenko, P. Dorozhkin, V. Kulakovskii, T. Kennedy, T. Reinecke, Physica E **26**(1-4), 37 (2005). doi:10.1016/j.physe.2004.08.019
129. G. Bacher, M.K. Welsch, A. Forchel, Y. Lyanda-Geller, T.L. Reinecke, C.R. Becker, L.W. Molenkamp, J. Appl. Phys. **103**(11), 113520 (2008). doi:10.1063/1.2937239. http://link.aip.org/link/?JAP/103/113520/1
130. T. Dietl, J. Spalek, Phys. Rev. B **28**(3), 1548 (1983)
131. A. Golnik, J. Ginter, J. Gaj, J. Phys. C **16**(31), 6073 (1983)
132. B. König, I. Merkulov, D. Yakovlev, W. Ossau, S. Ryabchenko, M. Kutrowski, T. Wojtowicz, G. Karczewski, J. Kossut, Phys. Rev. B **61**(24), 16870 (2000)
133. D. Keller, D. Yakovlev, B. Konig, W. Ossau, T. Gruber, A. Waag, L. Molenkamp, A. Scherbakov, Phys. Rev. B **65**(3), 035313 (2002). doi:10.1103/PhysRevB.65.035313
134. M. Kneip, D. Yakovlev, M. Bayer, A. Maksimov, I. Tartakovskii, D. Keller, W. Ossau, L. Molenkamp, A. Waag, Phys. Rev. B **73**(3), 035306 (2006). doi:10.1103/PhysRevB.73.035306
135. S. Mackowski, T. Gurung, H. Jackson, L. Smith, G. Karczewski, J. Kossut, Appl. Phys. Lett. **87**(7), 072502 (2005). doi:10.1063/1.2011785
136. L. Maingault, L. Besombes, Y. Léger, C. Bougerol, H. Mariette, Appl. Phys. Lett. **89**(19), 193109 (2006). doi:10.1063/1.2387116
137. Y. Léger, L. Besombes, L. Maingault, D. Ferrand, H. Mariette, Phys. Rev. Lett. **95**(4), 047403 (2005). doi:10.1103/PhysRevLett.95.047403

138. Y. Léger, L. Besombes, L. Maingault, D. Ferrand, H. Mariette, Phys. Rev. B **72**(24), 241309 (2005). doi:10.1103/PhysRevB.72.241309
139. Y. Léger, L. Besombes, L. Maingault, J. Fernandez-Rossier, D. Ferrand, H. Mariette, Phys. Stat. Sol. (b) **243**(15), 3912 (2006). doi:10.1002/pssb.200671506
140. H. Mariette, L. Besombes, C. Bougerol, D. Ferrand, Y. Léger, L. Maingault, J. Cibert, Phys. Stat. Sol. (b) **243**(14), 3709 (2006). doi:10.1002/pssb.200642256
141. L. Besombes, Y. Léger, L. Maingault, D. Ferrand, H. Mariette, J. Cibert, Phys. Rev. B **71**(16), 161307 (2005). doi:10.1103/PhysRevB.71.161307

4

Coherent Electron Spin Dynamics in Quantum Dots

Manfred Bayer, Alex Greilich, and Dmitri R. Yakovlev

Summary. The coherent spin dynamics of electrons confined in quantum dots is discussed. A new measurement technique, mode-locking of electron spin precession by and with a pulsed excitation laser is used to address the coherence, which otherwise would be masked in ensemble studies by dephasing. The background of nuclei leads to a refocusing such that all optically excited electron spins become synchronized with the laser. With this tool spin coherence times in the microseconds range are demonstrated at cryogenic temperatures. The mode-locking can be tailored by the laser excitation protocol such that strong signals at arbitrary times can be generated in Faraday rotation experiments.

4.1 Introduction

Recently the coherent dynamics of elementary excitations in semiconductor heterostructures has attracted considerable interest for applications in quantum information processing, even though it is challenging to identify two-level systems, which are well isolated from the environment so that their coherence is retained for long enough times. This demand has directed interest toward semiconductor quantum dots because of their discrete energy level structure, due to which they bear some resemblance to atoms found in nature. In particular interest has moved toward spin excitations in semiconductors [1–3], motivated in particular by the observation of very long electron spin coherence times T_2 already for bulk semiconductors [4]. Further, it has been shown that the spin relaxation mechanisms which are effective in higher-dimensional systems are strongly suppressed in quantum dots. For electrons, for example, only the spin–orbit coupling and the interaction with the nuclear background is effective, while for holes also the interaction with the nuclei is suppressed. The interest in quantum dot spins was enhanced further by the demonstration of very long electron spin relaxation lifetimes, T_1, in the milliseconds-range [5]. This has raised hopes that T_2, which may theoretically last as long as $2T_1$ [6], could be similarly long.

In this chapter we discuss the coherent spin dynamics of electrons confined in semiconductor quantum dots. To monitor this dynamics, the spins are oriented optically by circularly polarized pulsed laser excitation and their subsequent coherent precession about an external magnetic field is detected. Destruction of spin coherence by scattering leads to a change of the phase of the precession, so that through a phase measurement detailed insight in the spin dynamics can be taken. In that way an ensemble of carrier spins is addressed, which complicates the task due to variations of the precession frequencies, which are particularly prominent for the quantum dot case. However, a remarkable tool to overcome this limitation has been discovered in our experiments. The spin ensemble can be synchronized by and with a periodic train of laser pulses, thus providing a locking of several electron spin precession modes. On a time scale of a microsecond the excitation protocol selects only a fraction of dots, namely those which satisfy the phase synchronization conditions (PSC). However, on a longer time scale of seconds up to minutes the nuclear fields in the quantum dots are rearranged such that all dots contribute to the coherent signal.

Spin Coherence and Spin Dephasing Times

Application of an external magnetic field **B** generally leads to a splitting of the spin states of an electron given by $g\mu_B B$, where g is the g-factor of the electron. In the following we use the heterostructure growth direction z as spin quantization axis, even if a magnetic field is applied in a direction different from that. The z-axis coincides with the optical axis in our spectroscopic studies (see below). Circularly polarized light then injects electrons in one of the spin basis states. If the magnetic field is applied along z (Faraday configuration), the electron spin orientation is therefore stable until relaxation occurs. If the field is applied normal to the optical axis (Voigt configuration), the spins are created in a superposition of the spin split eigenstates and as a consequence the spins undergo a precession with frequency $\Omega = g\mu_B B/\hbar$.

The dynamics of the spins after optical orientation can be described by two phenomenological times constants, the relaxation times T_1 and T_2 for the spin components longitudinal and transverse to B, respectively [8]. The Faraday configuration allows one to study the longitudinal relaxation T_1, which describes relaxation from the upper to the lower spin split state, for example. This inelastic process requires energy exchange with a surrounding bath, from which energy can be absorbed or into which energy can be emitted. The acoustic phonons represent, for example, such a bath as do also free carriers. Coupling to this latter bath is, however, strongly suppressed due to the quantization of the kinetic energy in quantum dots.

On the other hand, the Voigt configuration allows one to study the transverse relaxation time T_2 which is the time a spin is precessing about the magnetic field until a phase changing scattering event occurs. We will refer to this time as *spin coherence time*. While inelastic scattering will destroy the phase coherence anyway, also elastic scattering through interaction with a bath may contribute to destruction of the phase coherence. Therefore T_2 is typically much shorter than the longitudinal relaxation time T_1, and the relation between them is established through:

$$\frac{1}{T_2} = \frac{2}{T_1} + \frac{1}{T_2^\star}, \qquad (4.1)$$

where the second term on the right-hand side comprises the elastic scattering channels.

Measurements of the transverse relaxation time can be rather easily achieved if performed on a single spin or a homogeneous ensemble of spins over times during which the experimental conditions do not change. The relaxation times are masked when they are performed on an inhomogeneous ensemble of electron spins. The origin of the inhomogeneity may be, for example, a dispersion of precession frequencies at fixed magnetic field due to g-factor variations. Then the spins in the ensemble "run quickly out of phase" after initial spin orientation. Depending on the dispersion, the macroscopic ensemble response looses its coherence then considerably faster than the individual spin coherence time T_2, while the coherence on the individual spin level is still retained. The time during which the ensemble coherence is lost is called the *dephasing time* T_2^*. By introducing the inhomogeneity related relaxation time T_2^{inh} one can establish a connection between the dephasing and the relaxation time:

$$\frac{1}{T_2^*} = \frac{1}{T_2} + \frac{1}{T_2^{\text{inh}}}. \qquad (4.2)$$

For completeness we note for our case of single charged quantum dots, that two ensemble factors contribute to the dephasing. The g-factor dispersion of the electrons in the ensemble dominate T_2^{inh} in strong magnetic fields. In zero field and in low fields T_2^{inh} is limited by variations of the fluctuation fields of the nuclei, with which the electron interacts through the hyperfine interaction.

The fluctuations in an ensemble measurement often allow only measurement of T_2^* unless special tools such as spin-echo techniques [9] are applied. This is also true for measurements on single quantum dots, for which the low spectroscopic signal requires multiple repetitions of the experiments during which the nuclear background varies. Here we report on a novel technique which allows one to extract T_2 from an ensemble measurement. This technique is based on mode-locking of electron spin coherence, as presented in Sect. 4.2.1.

The spin coherence of electronic states has been studied in semiconductor structures of higher dimensionality such as bulk-like thin films and quantum wells [10, 11]. The spin dephasing time of an electron has been found to vary over a wide range. A T_2^* time of 300 ns has been measured in bulk GaAs at liquid helium temperature [12]. For quantum wells the longest spin dephasing times reported so far are 10 ns for GaAs/(Al,Ga)As [13] and 30 ns for CdTe/(Cd,Mg)Te [14, 15] in structures with a very diluted two-dimensional electron gas.

The experimental data have been obtained by time-resolved pump–probe Faraday rotation with picosecond-resolution [17–19]. A scheme of the Faraday rotation experiment is given in Fig. 4.1a. To have spectral resolution in the millielectronvolt-range the experiments have been performed with laser pulses with a duration of about 1.5 ps, corresponding to about 1.5 meV spectral width. The light pulses were emitted from a mode-locked Ti:Sapphire oscillator generating pulses at a repetition rate of 75.6 MHz, i.e., with a period of 13.2 ns between the pulses. This repetition rate could be reduced by a pulse picker. The laser intensity can be characterized by the dimensionless laser pulse area Θ, defined by $\Theta = \int \mathbf{d} \cdot \mathbf{E}(t) dt / \hbar$, where $\mathbf{d}$ is the matrix element for the optically excited dipole transition and $\mathbf{E}$ is the electric field of the laser pulse.

The basic features of the experiment are the following: the quantum dot heterostructure is excited along its growth axis z by a train of circularly polarized pump pulses, which are energetically resonant with the ground state

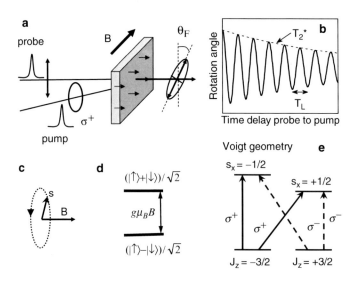

Fig. 4.1. (a) Experimental geometry for pump–probe Faraday rotation. The pump is circular polarized and the probe is linearly polarized. (b) Scheme of Faraday rotation signal due to electron spin precession about a transverse magnetic field. (c) Classical description of electron spin precession. (d) Electron spin states split by a magnetic field. The quantum mechanical description of spin precession is based on excitation of a coherent superposition of the two states. (e) Energy level diagram for heavy-hole exciton transitions in quantum dots in a magnetic field along the x-axis (Voigt geometry)

optical transition. The pulses inject spin oriented electrons, if allowed by Pauli principle due to the resident electron spin. The electron spin ensemble builds up a macroscopic magnetization, which is tested by a much weaker, linearly polarized probe pulse with frequency identical to the pump frequency hits. The rotation of the polarization plane of the transmitted probe pulse is analyzed as function of the delay between the pump and probe pulses. The probe beam itself does not affect the magnetization and was carefully checked through the linearity of the detected probe signal on the probe intensity.

An external magnetic field **B** is applied normal to the structure growth direction, along the x-axis. This field orientation leads to a precession of electron spins optically oriented along the z-axis about the field with a frequency given by $\Omega \equiv \Omega_x = g_x \mu_B B/\hbar$, where g_x is the electron g-factor along the field. Using our notations, the plane of the oscillatory precessional motion lies in the z–y plane. The rotation angle of the probe polarization is proportional to the magnetization along the optical axis z, therefore maps out the oscillations in the z–y plane and is sensitive to scattering events which lead to destruction of the magnetization.

A typical Faraday rotation trace recorded on a reference quantum well is shown in Fig. 4.1b. The signal contains information about the (static) spin splitting and the (dynamic) spin coherence, and can be analyzed by an exponentially damped harmonic function $\exp(-t/T_2^\star) \cdot \cos(\Omega t)$. In general, experimentally observed signals may look more complicated as for example two types of carriers or localized and free carriers, which differ in g-factor and dephasing time, may contribute.

To make the selection rules for the optically excited transitions more transparent, the scheme in Fig. 4.1e sketches how the coherent superposition of two electron spin basis states is excited in a quantum dot subject to an external magnetic field. The optical transitions involve almost pure heavy-holes (even though the hole g-factor in the dot plane is nonzero) at the top of the valence band and electrons at the bottom of the conduction band. Therefore the hole g-factor is neglected for simplicity and the hole spin projection on the field vanishes. The spin projection of electron and hole onto the optical axis are given by $S_z = \pm 1/2$ and $J_z = \pm 3/2$, respectively. Therefore σ^+ polarized light, for example, can inject a hole with $J_z = -3/2$ and an electron with $+1/2$. σ^- polarized light involves the electron and hole with opposite spin orientations.

4.2 Spin Coherence in Singly Charged Quantum Dots

For an electron confined in a quantum dot, most of the spin relaxation mechanisms efficient in higher dimensional systems are suppressed. This holds for all mechanisms involving the spin–orbit interaction and also for carrier–carrier scattering for a singly charged dot. However, the hyperfine interaction of the electron with the dot nuclei is enhanced because of the carrier localization.

To suppress the unavoidable inhomogeneity of a quantum dot ensemble spectroscopic techniques have been improved to an extent that single dot studies become possible. Still the spectroscopic response of a single quantum dot is weak, partly also because of the lack of sensitive detectors, for example. Therefore such experiments often require minutes long measurement times. Here we report on experiments of ensembles of singly charged quantum dots, so that the signal is strong but one cannot avoid the inhomogeneities.

Samples The (In,Ga)As/GaAs quantum dot sample contained 20 layers of dots separated from each other by 60 nm GaAs barriers [16]. The density of dots in each layer was about 10^{10} cm^{-2}. 20 nm below each dot layer an n-doping δ-sheet with a Si-dopant density roughly equal to the dot density was placed. The As-grown InAs/GaAs sample shows ground state luminescence at a wavelength around 1.2 μm at cryogenic temperatures. After thermal annealing for 30 s at 960°C, which causes intermixing of the In and Ga atoms, the ground state emission is shifted to 0.89 μm and therefore is in the sensitivity range of a Si-detector. Transmission electron microscopy studies have shown that after overgrowth the As-grown dots are still about dome-shaped and are rather large with a diameter of about 25 nm and a height of about 5 nm. Thermal annealing increases these parameters. From Faraday rotation studies [16] we estimate that about 75% of the dots are occupied by a single electron, while 25% contain no residual charge.

A photoluminescence spectrum of this sample measured under excitation into the wetting layer at 1.46 eV is shown in Fig. 4.2a. The emission band has

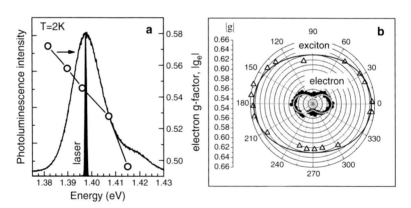

Fig. 4.2. (a) Photoluminescence spectrum of the studied (In,Ga)As/GaAs quantum dot sample. The *filled trace* gives the spectrum of the excitation laser used in the Faraday rotation experiments, which could be tuned across the inhomogeneously broadened emission band. The *symbols* give the electron in-plane g-factor along the $[1\bar{1}0]$ direction across this band. (b) In-plane angular dependence of the electron (*circles*) and exciton (*triangles*) g-factors obtained from circular dichroism experiments. *Lines* are fits to the data [20]. $B = 5$ T. Angle zero corresponds to a field orientation along the x-direction which is defined by the [110] crystal axis [20]

a full width at a half maximum of 15 meV. Faraday rotation signals recorded on this sample are shown in Fig. 4.3a. Quantum beats with at least two different frequencies are clearly observed, resulting in a modulation of the signal at short delay times. After 300 ps, which is close to the exciton lifetime, the modulation vanishes, and a monotonic decay of the beats amplitude is seen. The decay time due to dephasing strongly depends on magnetic field, decreasing from $T_2^* = 3$ ns at 1 T to 0.5 ns at 6 T.

The long-lived component of the Faraday rotation signal is caused by spin precession of the resident electrons in the singly charged quantum dots [16]. From the magnetic field dependence of the precession frequency we extract an electron g-factor $|g| = 0.54$ [20]. The modulation of the signal at early times, which is caused by interference of the resident electron signal with signal from excitons in charge neutral quantum dots, allows us to derive also the hole g-factor $|g_{h\perp}| = 0.15$ along the magnetic field.

Spectral Dependence of the Electron g-Factor

The energy dispersion of the electron g-factor within the dot ensemble has been measured by varying the excitation energy across the emission band for fixed field orientation. Figure 4.2a shows that the g-factor decreases monotonously from 0.57 on the low energy flank to 0.49 at the high energy flank. This variation can be understood if one makes the assumption that the main effect of the confinement is an increase of the bandgap E_g. The deviation of g_e from the free electron g-factor $g_0 = 2$, determined from $\mathbf{k} \cdot \mathbf{p}$ calculations, is given by [21, 22]:

$$g_e = g_0 - \frac{4m_0 P^2}{3\hbar^2} \frac{\Delta}{E_g (E_g + \Delta)}. \tag{4.3}$$

Here m_0 is the free electron mass, P is the matrix element describing the coupling between valence and conduction band, and Δ is the spin–orbit splitting of the valence band. The decrease of the g-factor modulus with increasing emission energy could be then only explained if the g-factor has a negative sign. This argument is supported by measurements of the dynamic nuclear polarization [20], similar to those described in [23]. This allows us to determine also signs for the exciton and hole g-factors.

4.2.1 Mode Locking of Spin Coherence in an Ensemble of Quantum Dots

Ensemble dephasing does not lead to destruction of the coherence on a single spin level, but masks it due to the rapid accumulation of phase differences among the different spins with increasing delay. The T_2 time may be obtained by elaborated spin-echo techniques [9]. A less complicated and robust

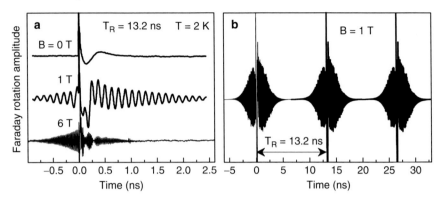

Fig. 4.3. (a) Pump–probe Faraday rotation signals at different magnetic fields in singly charged (In,Ga)As/GaAs quantum dots. The pump power density is 60 W cm^{-2}, the probe density is 20 W cm^{-2} [24]. (b) Faraday rotation signal recorded for a longer delay range in which three pump pulses were located

measurement scheme would be therefore highly desirable. Such a scheme may be also useful for the processing of quantum information, including initialization, manipulation, and read-out of a coherent spin state.

For that purpose we take a close-up look at the Faraday rotation traces at negative delays, as shown in Fig. 4.3a. Long-lived electron spin coherence is seen not only at positive delays, but surprisingly strong coherent signal is seen also for negative delays which must also arise from electrons due to the same oscillation period as for positive delays. The amplitude of the spin beats increases when approaching zero delay $t = 0$.

Figure 4.3b shows the signal when scanning the delay over a larger time interval, in which three pump pulses, separated by 13.2 ns from each other, are located. At each pump pulse arrival electron spin coherence is created, which is completely dephased after a few nanoseconds. Before the next pump arrival coherent signal from the electrons reappears. This negative delay signal can occur only because the coherence of the electron spin in each individual quantum dot prevails for much longer times than T_R, i.e., if $T_2 \gg T_R$.

Spin Coherence Time of the Electron Spin

Independent of the origin of the coherent signal at negative delays, its observation opens a way towards measuring the single spin coherence time T_2 in an ensemble: This can be achieved by increasing the separation T_R between the pump pulse into a range comparable with T_2, so that spin phase scattering events occur resulting in a decrease of the Faraday rotation signal. The corresponding data are shown in Fig. 4.4, where the Faraday rotation amplitude detected at a fixed negative delay shortly before the next pump pulse arrival is shown as function of T_R. The result of model calculations shown by the line

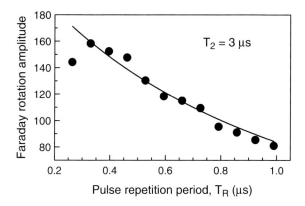

Fig. 4.4. Faraday rotation amplitude at a negative delay of -80 ps as function of the separation between subsequent pump pulses measured at $B = 6$ T for $T = 6$ K. The line shows calculations with the spin coherence time $T_2 = 3\,\mu$s as fit parameter [24]

allow us to extract the coherence time of a single dot, $T_2 = 3.0 \pm 0.3\,\mu$s at $B = 6$ T, which is four orders of magnitude longer than the ensemble dephasing time $T_2^* = 0.4$ ns.

T_2 as function of temperature is plotted in Fig. 4.5 by the squares [35]. For the applied experimental conditions ($B = 2$T) the measured spin coherence time T_2 is about $0.6\,\mu$s at cryogenic temperatures. T_2 remains constant with temperature increment up to 15 K. However, we find a surprisingly sharp drop of T_2 down to $0.25\,\mu$s at 20 K. At 30 K the coherence time is estimated to be 30 ± 10 ns, and drops further into the 2 ns range at $T = 50$ K. Under these conditions it has become comparable to T_2^*, as the corresponding data in Fig. 4.5 shows.

Mechanism of Spin Synchronization

To understand the reappearance of the spin coherence signal at negative delays, we consider excitation of a single quantum dot by a periodic train of circularly polarized laser pulses with intensity π. The first effect of this excitation is a synchronization of the electron spin precession. The degree of spin synchronization is defined by $P(\omega_e) = 2|S_z(\omega_e)|$, where the z-component of the electron spin vector, $S_z(\omega_e)$, is taken at the moment of pulse arrival. If the pulse period, T_R, is equal to an integer number N times the electron spin precession period, $2\pi/\omega_e$, the effect of the π-pulse is an almost complete electron spin alignment along the light propagation direction z [24] at the moments of pulse arrival. The synchronization degree for π-pulse application is given by $P_\pi = \exp(-T_R/T_2)/[2 - \exp(-T_R/T_2)]$. In our case P_π is close to unity, corresponding to 100% synchronization, because $T_R \ll T_2$. This synchronization occurs for every quantum dot in the ensemble.

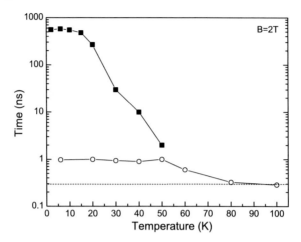

Fig. 4.5. Decoherence time T_2 (*squares*) and dephasing time T_2^* (*full circles*) vs. temperature at $B = 2\,\text{T}$. *Dotted line* marks the exciton/trion lifetime

The optically excited ensemble with a broad distribution of g-factors and therefore precession frequencies contains subsets of quantum dots whose precession frequencies fulfill a PSC such that the precession frequency is a multiple of the laser repetition rate $\omega_R = 2\pi/T_R$:

$$\omega_e = 2\pi N/T_R \equiv N\omega_R. \tag{4.4}$$

The equidistantly spaced PSC modes selected from the distribution of electron precession frequencies are shown by the dashed lines in Fig. 4.6. For not too small magnetic fields the electron spin precession frequency is much higher than the laser repetition rate of 75.6 MHz. Therefore the synchronized spins undergo a large number of full 2π rotations ($N \gg 1$) between two pump pulses. In addition the width of optically excited frequencies is much larger than the laser repetition rate, so that a multiplet of subsets with different N become synchronized. This is illustrated by the scheme in Fig. 4.6. The left handed side shows three precession modes satisfying the PSC (4.4) with $N = 4, 6, 8$, while the case of a spin precessing with a frequency different from the PSC is shown on the right side. The number of PSC modes M can be estimated by dividing the number of excited modes $\Delta\omega \propto \mu_B \Delta g B/\hbar$ by the distance between adjacent synchronized modes $2\pi/T_R$: $M = \mu_B \Delta g B T_R/\hbar$.

The electron spins in the PSC dots (left panel) all run into phase again when the next pump pulse is approached, resulting in a macroscopically measurable Faraday magnetization, despite of their very different phases between the pulses. Quantitatively, their contribution to the spin polarization of the ensemble at a time t after orientation by a pump pulse is given by $-0.5\cos(N\omega_R t)$. Summing over all oscillating terms of the synchronized spin subsets leads to constructive interference of their contributions to the Faraday

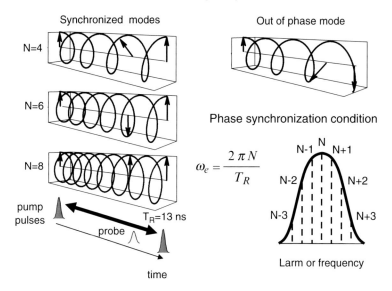

Fig. 4.6. Scheme for the phase synchronization of electron spin coherence with a periodic train of laser pulses. On the *left* modes satisfying the PSC (4.4) are shown. On the *right* a non-PSC dot is shown. The *arrows* indicate the orientation of the electron spins in a quantum dot. The *thick solid line in the bottom right panel* shows the distribution of electron precession frequencies optically excited in the dot ensemble. The PSC modes selected from this distribution are shown by the *dashed lines*

rotation signal around the pump pulse arrival times. The rest of quantum dots represents a dephased background with zero average magnetization and therefore does not contribute to signal at times $t \gg T_2^*$, even though also the spins in these dots still precess coherently.

π-pulse excitation is not critical for the electron spin synchronization by a resonant periodic laser pulse train, but synchronization can be achieved for any intensity. The pulses create a coherent superposition of a trion state and an electron state in a quantum dot, leading to long-lived coherence of the resident electron spins, because the coherence is not affected by the radiative decay of the trion component. Each pulse of, for example, σ^+ polarized light changes the electron spin projection along the light propagation direction by $\Delta S_z = -(1 - 2|S_z(t \to t_n)|)W/2$, where $t_n = nT_R$ is the time of n-th pulse arrival, and $W = \sin^2(\Theta/2)$ [16, 24]. Consequently, the pulse train orients the electron spin opposite to the light propagation direction, and it also increases the degree of electron spin synchronization P. Application of $\Theta = \pi$-pulses (corresponding to $W = 1$) leads to a more than 99% degree of electron spin synchronization already after a dozen of pulses. However, if the electron spin coherence time is long enough ($T_2 \gg T_R$), an extended train of pulses leads also to a high degree of spin synchronization in case of $\Theta \ll 1$ ($W \approx \Theta^2/4$).

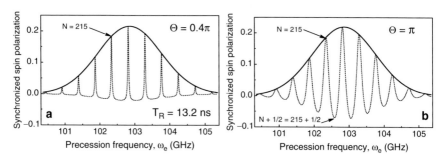

Fig. 4.7. Calculated spectra of phase synchronized electron spin precession modes created by a train of circularly polarized pulses calculated for pulse area $\Theta = 0.4\pi$ and π at the moments of pulse arrival. $T_R = 13.2$ ns. (a) At low pumping intensity the pulse train synchronizes the electron spin precession in a very narrow frequency range around the frequencies fulfilling the PSC $\omega_e = 2\pi N/T_R$. (b) π-pulses broaden the phase synchronized modes. In addition, electron spins with polarization opposite to the synchronized modes contribute significantly at frequencies between the PSC modes. Parameters of calculations: $B = 2$ T, $|g_e| = 0.57$, $\Delta g_e = 0.005$ and $T_2 = 3\,\mu$s [24]

The effect of the pump intensity on the synchronized spin polarization distribution is shown in Fig. 4.7 for $\Theta = 0.4\pi$ and π. The total density of optically excited electron spin precession modes is shown by the solid line, representing the envelop for the synchronized spin polarization. The quasidiscrete structure of the distribution created by the pulse train (the dashed lines) is the most important feature, which allows us to study the spin coherence of a single quantum dot in an ensemble: The continuous density of spin precession modes causes fast dephasing of the FR signal after spin orientation by the pump pulse. However, the gaps in the density of precession modes due to spin synchronization with the laser facilitate the observed constructive interference at negative delays. These gaps are created by mode locking of the electron spin precession with the periodic laser pulse sequence.

Requirements for Quantum Dot Ensemble

The presented data immediately provoke the question what properties an inhomogeneous quantum dot ensemble should have for demonstrating mode-locking. Naturally homogeneous quantum dot ensembles would be optimal for quantum information processing. However, fabrication of such ensembles cannot be foreseen on the basis of technology, which always gives a sizeable inhomogeneity. Accepting the resulting unavoidable distribution of electron g-factors mode-locking might be the tool to exploit spin coherence during the single spin decoherence time. Simultaneously the mode-locking technique has the advantage of a strong spectroscopic response, as it involved many quantum dots. Further, it gives flexibility with respect to changing the experimental conditions such as the laser excitation protocol (e.g., wavelength, pulse duration and repetition rate) which changes the PSC. In response to such a change, the ensemble involves other quantum dot subsets in the spin synchronization. However, a very broad distribution of electron g-factors would lead to a very fast dephasing in the ensemble, making it difficult to exploit the spin synchronization as it persists only over very short time ranges around the pump pulses.

Control of the Ensemble Spin Synchronization

A high degree of control over the synchronized spin coherence in an ensemble of singly charged dots can be obtained through tailoring the periodic excitation protocol. For that purpose, no longer a train of single pulses but a train of pump pulse doublets with fixed separation between the two pulses in the doublet is used. Free adjustable parameters in the laser protocol are the separations between the pulses as well as their polarization.

Two Pump Pulse Excitation Protocol

Each pulse emitted by the laser is now split into two pulses with a fixed delay $T_\mathrm{D} < T_\mathrm{R}$ between them. Both pumps are circular copolarized and have the same intensities. The resulting Faraday rotation traces for $T_\mathrm{D} = 1.84\,\mathrm{ns}$ are shown in Fig. 4.8a. When the quantum dots are exposed to either the first (top trace) or the second (mid trace) of the two pump pulses, the Faraday rotation signals are identical except for the shift by T_D. The signal changes drastically under excitation by the train of pulse doublets (bottom trace): Around the arrival of pump 1 the same Faraday rotation response is observed as before in the one-pump experiment. Around pump 2 also qualitatively the same signal is observed but with a considerably enhanced amplitude.

Even more remarkable are the echo-like features showing up before the first and after the second pump pulse. These bursts are symmetrically shaped with the same decay and rise times T_2^*. The temporal separation between them is a multiple of T_D. Note also that the bursts show no additional modulation as seen at positive delays when a pump is applied. This modulation lasts only over times identical to the exciton lifetime and arises from interference of signal from excitons in charge neutral dots with the electron signal in the charged dots. At the times of the bursts the excitons have recombined so that only resident electron signal is detected.

The electron spins in the quantum dot ensemble have been clocked by introducing a second frequency, which is determined by the laser pulse separation T_D. More specifically, the electron spin distribution has been reshaped by the pulse doublet excitation which has picked subensembles which are locked on two frequencies such that the electron precession frequency are multiples of the two rates $2\pi/T_\mathrm{D}$ and $2\pi/(T_\mathrm{R} - T_\mathrm{D})$. This clocking results in multiple bursts in the Faraday rotation response. The change of the precession mode

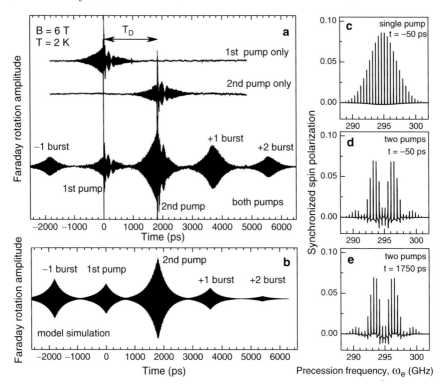

Fig. 4.8. Control of the electron spin synchronization in (In,Ga)As/GaAs quantum dots by a trains of pump pulse doublets with a repetition time $T_R = 13.2$ ns between the doublets. The shift in time between the pulses in the doublet is $T_D = 1.84$ ns. (**a**) Faraday rotation signal measured for separate action of either the first or the second pump pulse (the two *upper curves*) as well as for joint action of both pumps (the *bottom curve*). The pumps were σ^+ copolarized. (**b**) Model calculation of the Faraday rotation signal in the two pump pulse experiment with parameters $\Theta = \pi$ and $\gamma = 3.2$ GHz. Panels (**c–e**) illustrate the modification of the PSC mode spectrum when going from a single pump pulse to a two pump pulse excitation protocol. Parameters in the calculation: $\Theta = 0.4\pi$, $\gamma = 3.2$ GHz, $|g_e| = 0.57$ and $\Delta g_e = 0.004$. Panel (**c**) is for the single pump protocol, measured right before the pump pulse. Panels (**d**) and (**e**) are for two pump pulses measured before pump 1 and pump 2, respectively [24]

spectrum when going from pump singlet to doublet excitation is shown in Fig. 4.8c–e. The model calculations of the Faraday rotation signal in panel (b) are in agreement with the experimental signals.

Burst Shaping by Changing the Delay Between Pump Pulses

Figure 4.9 shows Faraday rotation single excited by a pump–pulse doublet train. The separation between the doublets is $T_R = 13.2$ ns, and the two pulses in the doublet have the same intensity and polarization. The separation T_D

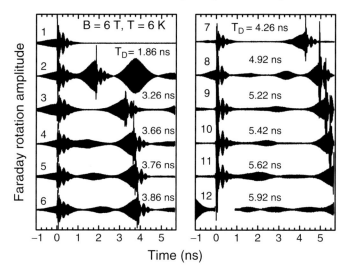

Fig. 4.9. Faraday rotation traces measured as function of delay between probe and first pump pulse at time zero. A second pump pulse was applied with a delay T_D relative to the first one. The T_D is indicated at each trace. The *top left trace* is a reference without second pump [25]

between the pulses in each doublet was varied between $\sim T_R/7$ and $\sim T_R/2$. With these variations the signal changes strongly, depending on whether the delay time T_D is commensurate with the repetition period T_R, $T_D = T_R/i$ with $i = 2, 3, 4, \ldots$, or incommensurate, $T_D \neq T_R/i$. For commensurability the signal shows strong periodic bursts in the Faraday rotation signal with a period equal to T_D, as seen for $T_D = 1.86\,\text{ns} \approx T_R/7$. Commensurability is also achieved for $T_D = T_R/4 \approx 3.26\,\text{ns}$ and $T_D = T_R/3 \approx 4.26\,\text{ns}$.

When T_D and T_R are incommensurate, the Faraday rotation signal shows bursts of spin quantum beats between the two pulses of the pump doublet, in addition to bursts outside of the doublet. One can see a single burst midway between the pumps for $T_D = 3.76$ and $5.22\,\text{ns}$. Two bursts, each equidistant from the closest pump and also equidistant from one another, appear at $T_D = 4.92$ and $5.62\,\text{ns}$. Three equidistant bursts occur at $T_D = 5.92\,\text{ns}$.

Although the time dependencies of the Faraday rotation signals look very different for commensurate and incommensurate T_D and T_R, in both cases they result from constructive interference of synchronized spin precession modes [25]. Quantifying the phase synchronization for the electron spin under pulse doublet excitation leads to the two conditions involving the time intervals T_D and $T_R - T_D$:

$$\omega_e = 2\pi NK/T_D,$$
$$\omega_e = 2\pi NL/(T_R - T_D), \qquad (4.5)$$

where a common integer N has been extracted in the multiplicity of the electron spin precessions as compared to the two effective repetition rates resulting from the two characteristic time separations. K and L are integers. Combinations of the two equations imposes limitations on the T_D values:

$$T_D = [K/(K+L)]T_R, \qquad (4.6)$$

which for $T_D < T_R/2$ (a constraint without limitation of generality) leads to $K < L$. By this PSC the positions of all bursts in the signals in Fig. 4.9 can be explained. For commensurability, one has $K=1$ so that $T_D = T_R/(1+L)$. In this case constructive interferences should occur with a period T_D as seen for $T_D = 1.86$ ns ($L = 6$).

For incommensurability the number of bursts between the pulses and the delays at which they appear can be tailored. There should be just one burst, when $K=2$, because then the constructive interference has period $T_D/2$. In experiment, a single burst is indeed seen for $T_D = 3.76$ ns ($L=5$) and 5.22 ns ($L=3$), see Fig. 4.9. Two bursts are seen for $T_D = 4.92$ and 5.62 ns, corresponding to $K=3$ and $L=5$ and 4, respectively. Finally, the Faraday rotation signal with $T_D = 5.92$ ns shows three bursts between the pumps, which is described by $K=4$ and $L=5$.

Stability of Mode-Locking

The Faraday rotation bursts due to spin mode-locking show a remarkable stability against variations of the magnetic field from 1 to 10 T. The appearance of bursts, however, changes with field strength. The bursts are squeezed due to a decrease of T_2^* with increasing field, but the delays at which the bursts appear remain unchanged. Also the burst amplitude does not vary strongly in this field range. The stability with respect to variations of the magnetic field is a consequence of the mode-locking mechanism, which is not connected to specific properties of the quantum dots, e.g., their electron spin precession frequency. The periodic pump pulse protocol always can select proper subsets of quantum dots which satisfy the PSC, even for strongly varied experimental conditions.

4.2.2 Nuclei Induced Frequency Focusing of Spin Coherence

In this section we describe an effect which arises from the hyperfine interaction of the electron spin confined in a quantum dot with the nuclear spins. An in-depth consideration of hyperfine interaction effects can be found in the next chapter. The spatial carrier confinement enhances the hyperfine interaction with the lattice nuclei, leading to spin decoherence and dephasing [26, 36]. This problem may be overcome by polarizing the nuclear spins [27], but the high degree of polarization required, close to 100%, to suppress decoherence has not been achieved so far.

However, as we will show here, the hyperfine interaction, rather than being detrimental, can be also utilized. Namely it can remove also the background of dephased electrons which do not fulfill the phase synchronization from the very beginning so that the initially continuous mode spectrum of electron spin precession is fully transformed into a comb-like spectrum consisting of a few discrete modes only. The information on this mode distribution can be stored in the nuclear system for tens of minutes because of the long nuclear memory times [10, 28–30].

The Faraday rotation technique involving pump pulse singlet or doublet excitation was applied. Corresponding signals measured are shown Fig. 4.10a. Surprisingly, the signal pattern created by the two pulse protocol is memorized over several minutes, as the lowest trace shows. One would expect that blocking of the second pulse in a pump doublet would destroy the periodic burst pattern on a microsecond time scale according to the electron spin coherence time, T_2, in these dots [24]. Only the signal around the first pump should remain over the scanned range of pump–probe delays. The middle trace was recorded after the sample was illuminated for ~ 20 min by the pump-doublet train. Immediately after this measurement, the second pump was blocked and a measurement using only the single pump train was started (bottom trace). Contrary to the expectations, the signal shows qualitatively all characteristic for a pump doublet protocol. A strong signal ("burst 0") appears around the delays where the second pump was originally located. Further signals, denoted by "burst 1" and "burst 2", are also observed. The system, therefore, remembers for minutes its previous exposure to a pump doublet protocol.

Additional Faraday rotation traces were recorded in a short delay range around "burst 0" for different times after closing the second pump (Fig. 4.10b). The decay kinetics was measured at a fixed delay of 1.857 ns (corresponding to maximum Faraday rotation signal) vs. the time after switching off pump 2 (curve in Fig. 4.10c). A strong signal is seen even after 40 min. The observed dynamics is well described by a bi-exponential dependence on elapsed time t, $a_1 \exp(-t/\tau_1) + a_2 \exp(-t/\tau_2)$ with a memory time τ_1 of a minute and $\tau_2 = 10.4$ min. The decay, however, critically depends on the light illumination conditions. When the system is held in complete darkness (both pumps and probe blocked), no relaxation occurs at all on an hour time scale. This is shown by the circles in Fig. 4.10c, which give the Faraday rotation amplitude when switching on pump 1 as well as probe after a dark period t.

The observed long memory of the excitation protocol must be imprinted in the dot nuclei, for which long spin relaxation times up to hours or even days have been reported in high magnetic fields [28, 29]. The nuclei in a particular dot must have been aligned along the magnetic field through the hyperfine interaction with the electron during exposure to the pump train. This alignment, in turn, changes the electron spin precession frequency, $\omega_e = \omega_e + \Omega_{N,x}$, where the nuclear contribution, $\Omega_{N,x}$, is proportional to the nuclear polarization. The slow decay dynamics of the Faraday rotation signal indicates that

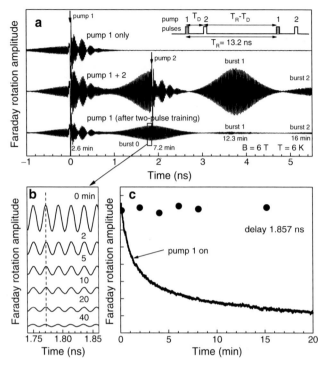

Fig. 4.10. (a) Faraday rotation traces measured on an ensemble of singly-charged (In,Ga)As/GaAs quantum dots. Details of the optical excitation protocol are given in the sketch. The *top trace* was measured using a train of pump pulse singlets. The *middle trace* was excited by a pump pulse doublet protocol with the second pump delayed by $T_D = 1.86$ ns relative to the first one. A measurement over the whole delay time range took about 20 min. The *lowest trace* was taken for a single pump pulse excitation protocol with pump 2 closed. Recording started right after measurement of the *middle trace*. Some times at which the different bursts were measured are indicated. The pump and probe power densities were 50 and 10 W cm^{-2}. (b) Faraday rotation signals measured over a small delay range at the maximum of "burst 0" for different times after closing the second pump, while pump 1 and the probe were always on. (c) Relaxation kinetics of the Faraday rotation amplitude at a delay of 1.857 ns after switching off pump 2. Before this measurement, the system was treated for 20 min by pump doublet excitation. The curve was measured with pump 2 blocked at $t = 0$. The *circles* show the signal for different times in complete darkness (both pumps and probe were blocked). $B = 6$ T and $T = 6$ K [31]

the optical excitation stimulates a nuclear adjustment such that the number of dots, for which the electron spin precession frequencies satisfy the PSC is increased.

The nuclear polarization is changed by electron–nuclei spin flip-flop processes due to the hyperfine interaction [32]. Such processes, however, are suppressed in a strong magnetic field due to the energy mismatch between the electron and nuclear Zeeman splittings by about three orders of magnitude. Flip-flop transitions, which are assisted by phonons compensating this mismatch, have a low probability due to the phonon-bottleneck in quantum dots [33, 34]. This explains the robustness of the nuclear spin polarization in darkness (Fig. 4.10c).

Consequently, resonant optical excitation of the singlet trion state in the quantum dots becomes the most efficient mechanism in the nuclear spin polarization dynamics. The excitation process rapidly turns "off" the hyperfine field of a resident electron acting on the nuclei, and the field is subsequently turned "on" again by the trion radiative decay. Thereby it allows a flip-flop process during switching without energy conservation.

The nuclear spin-flip rate resulting from this mechanism is proportional to the rate of optical excitation of the electron, $\Gamma_l(\omega_e)$. According to the selection rules, the probability of exciting the electron to a trion by σ^+ polarized light is proportional to $1/2 + S_z(\omega_e)$, where $S_z(\omega_e)$ is the component of the electron spin polarization along the light propagation direction taken at the moment of pump pulse arrival. Therefore, the excitation rate is given by: $\Gamma_l(\omega_e) \sim [1/2 + S_z(\omega_e)]/T_R$. For electrons satisfying the PSC, $S_z(\omega_e) \approx -1/2$, so that the excitation probability is very low due to Pauli blocking [24]. Due to the long spin coherence time, T_2, the excitation rate is reduced by two orders of magnitude to $1/T_2$ for these electrons as compared to $1/T_R + 1/T_2$ for the rest of electrons [31].

Due to $\Gamma_l(\omega_e)$, the nuclear relaxation rate shows a strong periodic dependence on ω_e, with the period determined by the PSC of the particular applied excitation protocol: $2\pi/T_R$ for the single pulse train and $2\pi/T_D$ for the double pulse train protocol (Fig. 4.11a). The huge difference in nuclear spin flip rates explains tends $\Omega_{N,x}$ in each dot tends to reach the mode-locking condition by undergoing a random walk process. In dots for which the PSC is not fulfilled, the nuclear contribution to ω_e changes randomly due to the light stimulated flip-flop processes. The typical change $\Delta\Omega_{N,x}$ of this contribution to ω_e is limited by statistical fluctuations of the nuclear spin polarization. For the studied (In,Ga)As dots, $\Delta\Omega_{N,x}$ lies on a gigahertz scale [31] which is comparable with the separation between the phase synchronized modes $2\pi/T_R \sim 0.48\,\text{GHz}$, making this readjustment of the precession frequency rather efficient. As a result, the nuclear contribution occasionally drives an electron into a PSC mode, where its precession frequency is virtually frozen on a minutes time scale. This leads to the frequency focusing in each dot and to accumulation of dots, for which the confined electron spins match the PSC.

The frequency focusing modifies the spin precession mode density of the dot ensemble (Fig. 4.11b and close-up in panel Fig. 4.11c). Without focusing, the density of electron spin precession modes is Gaussian with a width: $\Delta\omega_e = [(\Delta\Omega_{N,x})^2 + (\mu_B \Delta g_e B/\hbar)^2]^{1/2}$, where Δg_e is the g-factor dispersion. Frequency focusing modifies the original continuum density to a comb-like distribution. Eventually the whole ensemble participates in a coherent precession locked on

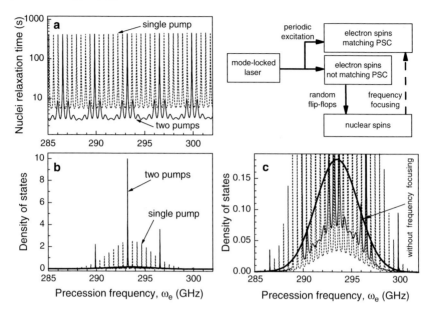

Fig. 4.11. Scheme of the nuclear induced frequency focusing of electron spin precession modes. Periodic resonant excitation by a mode-locked circularly polarized laser synchronizes the precessions of electron spins whose frequency satisfy the PSC. The optical excitation also leads to fast nuclear relaxation time in quantum dots, which do not satisfy the PSC, through optically stimulated flip-flop processes among the electron and a nuclear spin. The resulting change of the effective nuclear magnetic field modifies the electron spin precession frequency until it becomes frozen upon reaching the mode-locking condition. (**a**) Average spin relaxation time of the As nuclei vs. the electron spin precession frequency calculated for singlet pump (*dashed*) and doublet pump (*solid*) excitation. (**b**) Density of electron spin precession modes in an ensemble of singly charged dots after nuclear focussing, calculated for singlet pump (*dashed*) and doublet pump (*solid*) excitation. The *thick black line* shows the initial optically excited density of electron spin precession modes, whose width is determined by the electron g-factor dispersion and the nuclear polarization fluctuations. Panel (**c**) is a closeup of panel (**b**) for better visibility of the details. Calculations were done for $B = 6\,\text{T}$, $|g_e| = 0.555$, $\Delta g_e = 0.0037$, $\gamma = 1\,\text{GHz}$, $T_R = 13.2\,\text{ns}$, $T_D = T_R/7\,\text{ns}$, and $T_2 = 3\,\mu\text{s}$ [31]

a few frequencies only. This suggests that a laser protocol (defined by pulse sequence, width, and rate) can be designed such that it focuses the electron-spin precession frequencies in the dot ensemble to a single mode. It this case the spin coherence of the ensemble would dephase with the single electron coherence time, i.e., $T_2^\star = T_2$.

The focusing of electrons onto PSC modes is directly manifested by the Faraday rotation signals in Fig. 4.10a, as it causes comparable amplitudes before and after the pump pulses. Calculations show that, without frequency

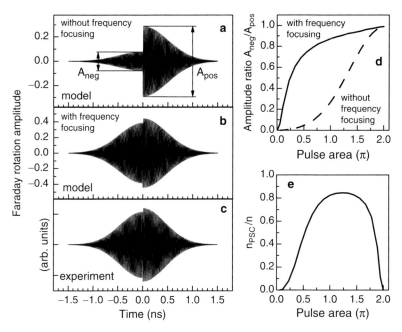

Fig. 4.12. Panels (**a**), (**b**) and (**c**) show Faraday rotation traces of an ensemble of singly charged QDs subject to a single pump pulse excitation protocol with pulse area $\Theta = \pi$ at a magnetic field $B = 6\,\mathrm{T}$. (**a**) FR traces calculated with the density of electron spin precession modes unchanged by the nuclei. (**b**) FR traces calculated with the density of electron spin precession modes modified by the nuclei. (**c**) Experimentally measured Faraday rotation signal after extracting the contribution of excitons in charge neutral quantum dots. (**d**) Calculated ratio of the signal amplitudes, $A_{\mathrm{neg}}/A_{\mathrm{pos}}$, with (*solid*) and without (*dashed*) including the nuclear rearrangement, as function of pump pulse area. (**e**) Dependence of the relative number of electrons, n_{psc}/n, in a QD ensemble involved in the mode-locked precession as function of pump pulse area. For the parameters in the calculation see Fig. 4.11 [31]

focusing, the amplitude at negative delays, A_{neg}, would not exceed 30% of the positive delay signal amplitude, A_{pos} (Fig. 4.12a). The strong optical pump pulses with pulse area Θ used in the present experiment involve all quantum dots in the ensemble, and their total contribution should make the Faraday rotation signal much stronger after the pulse than before, when only mode-locked electrons are relevant. However, the nuclear adjustment increases the negative delay signal to more than 90% (Fig. 4.12b) of the positive delay signal. The large experimental value of $A_{\mathrm{neg}}/A_{\mathrm{pos}}$ (panel (c)) confirms that in our experiment almost all electrons within the optically excited dot ensemble become involved in the coherent spin precession. Calculations of the pump intensity dependence of the ratio $A_{\mathrm{neg}}/A_{\mathrm{pos}}$ show that the nuclear focusing increases the ratio of electrons involved in the coherent spin precession relative to their total number, n_{psc}/n, almost to unity, even at rather low excitation densities (Fig. 4.12e,d).

As almost all the electrons in the ensemble are driven into a coherent spin precession, the exciting laser acts as a metronome which establishes robust macroscopic quantum bits, each characterized by a fixed precession frequency. These macroscopic quantum bits exist in dephasing free subspaces, which may open new promising perspectives for exploiting the coherent properties of ensembles of charged QDs during the single electron spin coherence time T_2.

4.2.3 Conclusion

The results presented in this section show that the shortcomings which are typically attributed to quantum dot spin ensembles may be overcome by elaborated laser excitation protocols. The related advantages arise from the robustness of the phase synchronization in a quantum dot ensemble: (1) a strong detection signal with relatively small noise, and (2) an insensitivity with respect to changes of external parameters like optical excitation protocol and magnetic field strength.

Acknowledgments

This chapter is a result of our collaboration with E.A. Zhukov, I.A. Yugova, F. Hernandez, R. Oulton, Al.L. Efros, A. Shabaev, I.A. Merkulov, D. Reuter, and A.D. Wieck.

References

1. D. Loss, D.P. DiVincenzo, Phys. Rev. A **57**, 120 (1998)
2. A. Imamoglu, D.D. Awschalom, G. Burkard, D.P. DiVincenzo, D. Loss, M. Sherwin, A. Small, Phys. Rev. Lett. **83**, 4204 (1999)
3. D.D. Awschalom, D. Loss, N. Samarth (eds.), *Semiconductor and Quantum Computation* (Springer, Heidelberg, 2002)
4. J.M. Kikkawa, D.D. Awschalom, Science **287**, 473 (2000)
5. J.M. Elzerman, R. Hanson, L.H. Willems Van Beveren, B. Witkamp, L.M.K. Vandersypen, L.P. Kouwenhoven, Nature **430**, 431 (2004); M. Kroutvar, Y. Ducommun, D. Heiss, M. Bichler, D. Schuh, G. Abstreiter, J.J. Finley, Nature **432**, 81 (2004)
6. W.A. Coish, D. Loss, Phys. Rev. B **70**, 195340 (2004)
7. J.R. Petta, A.C. Johnson, J.M. Taylor, E.A. Laird, A. Yacoby, M.D. Lukin, C.M. Marcus, M.P. Hanson, A.C. Gossard, Science **309**, 2180 (2005)
8. A. Abragam, *The Principles of Nuclear Magnetism* (Oxford Science, Oxford, 1961)
9. C.P. Slichter, *Principles of Magnetic Resonance* (Springer, Berlin, 1996)
10. D.D. Awschalom, N. Samarth, in *Semiconductor Spintronics and Quantum Computation*, ed. by D.D. Awschalom, D. Loss, N. Samarth (Springer, Berlin, 2002), pp. 147–193

11. I. Zutic, J. Fabian, S. Das Sarma, Rev. Mod. Phys. **78**, 323 (2004)
12. R.I. Dzhioev, B.P. Zakharchenya, V.L. Korenev, D. Gammon, D.S. Katzer, JETP Lett. **74**, 182 (2001)
13. R.I. Dzhioev, V.L. Korenev, B.P. Zakharchenya, D. Gammon, A.S. Bracker, J.G. Tischler, D.S. Katzer, Phys. Rev. B **66**, 153409 (2002)
14. E.A. Zhukov, D.R. Yakovlev, M. Bayer, G. Karczewski, T. Wojtowicz, J. Kossut, Phys. Stat. Sol. (b) **243**, 878 (2006)
15. H. Hoffmann, G. V. Astakhov, T. Kiessling, W. Ossau, G. Karczewski, T. Wojtowicz, J. Kossut, L. W. Molenkamp, Phys. Rev. B **74**, 073407 (2006)
16. A. Greilich, R. Oulton, E.A. Zhukov, I.A. Yugova, D.R. Yakovlev, M. Bayer, A. Shabaev, Al.L. Efros, I.A. Merkulov, V. Stavarache, D. Reuter, A. Wieck, Phys. Rev. Lett. **96**, 227401 (2006)
17. J.J. Baumberg, D.D. Awschalom, N. Samarth, H. Luo, J.K. Furdyna, Phys. Rev. Lett. **72**, 717 (1994)
18. N.I. Zheludev, M.A. Brummell, A. Malinowski, S.V. Popov, R.T. Harley, D.E. Ashenford, B. Lunn, Solid State Commun. **89**, 823 (1994)
19. J.M. Kikkawa, D.D. Awschalom, Phys. Rev. Lett. **80**, 4313 (1998)
20. I.A. Yugova, A. Greilich, E.A. Zhukov, D.R. Yakovlev, M. Bayer, D. Reuter, A.D. Wieck, Phys. Rev. B **75**, 195325 (2007)
21. P.Y. Yu, M. Cardona, *Fundamentals of Semiconductors* (Springer, Berlin, 1996)
22. I.A. Yugova, A. Greilich, D.R. Yakovlev, A.A. Kiselev, M. Bayer, V.V. Petrov, Yu.K. Dolgikh, D. Reuter, A. D. Wieck, Phys. Rev. B **75**, 245302 (2007)
23. D. Paget, G. Lampel, B. Sapoval, V.I. Safarov, Phys. Rev. B **15**, 5780 (1977)
24. A. Greilich, D.R. Yakovlev, A. Shabaev, Al.L. Efros, I. A. Yugova, R. Oulton, V. Stavarache, D. Reuter, A. Wieck, M. Bayer, Science **313**, 341 (2006)
25. A. Greilich, M. Wiemann, F.G.G. Hernandez, D.R. Yakovlev, I.A. Yugova, M. Bayer, A. Shabaev, Al.L. Efros, D. Reuter, A.D. Wieck, Phys. Rev. B **75**, 233301 (2007)
26. A.V. Khaetskii, D. Loss, L. Glazman, Phys. Rev. Lett. **88**, 186802 (2002)
27. W.A. Coish, D. Loss, Phys. Rev. B **70**, 195340 (2004)
28. V.L. Berkovits, A.I. Ekimov, V.I. Safarov, Sov. Phys. – JETP **38**, 169 (1974)
29. D. Paget, Phys. Rev. B **25**, 4444 (1982)
30. R. Oulton, A. Greilich, S. Yu. Verbin, R.V. Cherbunin, T. Auer, D.R. Yakovlev, M. Bayer, V. Stavarache, D. Reuter, A. Wieck, Phys. Rev. Lett. **98**, 107401 (2007)
31. A. Greilich, A. Shabaev, D.R. Yakovlev, Al.L. Efros, I.A. Yugova, D. Reuter, A.D. Wieck, M. Bayer, Science **317**, 1896 (2007)
32. M.I. Dyakonov, V.I. Perel, in *Optical Orientation*, ed. by F. Meier, B.P. Zakharchenja (North-Holland, Amsterdam, 1984)
33. A. Khaetskii, Yu.V. Nazarov, Phys. Rev. B **61**, 12639 (2000)
34. M. Kroutvar, Y. Ducommun, D. Heiss, M. Bichler, D. Schuh, G. Abstreiter, J.J. Finley, Nature **432**, 81 (2004)
35. F.G.G. Hernandez, A. Greilich, F. Brito, M. Wiemann, D.R. Yakovlev, D. Reuter, A.D. Wieck, and M. Bayer, Phys. Rev. B **78**, 041303(R) (2008)
36. I.A. Merkulov, Al.L. Efros, and M. Rosen, Phys. Rev. B **65**, 205309 (2002)

5

Quantum Dot Nuclear Spin Polarization

Patrick Maletinsky and Atac Imamoglu

Summary. We present an all-optical study of the coupled electron–nuclear spin system in an individual, self-assembled quantum dot (QD). Nuclear spins are polarized by optical orientation of a quantum dot electron and subsequent transfer of angular momentum to the nuclear spins. We find the coupling between the electron and the nuclear spins to be highly nonlinear, leading to hysteresis of the nuclear spin polarization in external magnetic fields. Furthermore, we present a time-resolved study of the nuclear spin polarization where we show that the nuclear spin lifetime depends drastically on the charging state of the QD. If a single electron occupies the QD, nuclear spin polarization decays within a few milliseconds. This decay time is extended to a few minutes in the absence of the QD electron, indicating that the QD nuclei constitute a mesoscopic nuclear spin system which is very well isolated from its environment.

5.1 Introduction

The QD nuclear spin ensemble investigated in this work is addressed in experiments which consist of two basic steps. First, optical excitation of a QD with circularly polarized light is used to dynamically polarize the nuclear spins in the QD. Second, the nuclear spin polarization is detected by measuring the energy-shift of a QD emission line due to the effective magnetic field of the spin polarized nuclei.

When an electron spin system is driven out of thermal equilibrium through an external agent, electron spin relaxation drives the electrons back to a thermal state. Since this relaxation is partly happening via the nuclear spin reservoir, angular momentum is transferred to the nuclei and a net nuclear polarization can be established [1]. An efficient way of achieving this situation is optical excitation of spin polarized electrons in bulk semiconductors. Optically induced dynamical nuclear spin polarization (DNSP) was first demonstrated in silicon [2] and was later on studied extensively for nuclei close to paramagnetic impurities in GaAs [3].

Individual, optically active, self assembled QDs present an excellent system for studying optically induced DNSP in more depth, thereby revealing subtleties of the electron–nuclear spin system that were experimentally not accessible before. Several aspects distinguish the QD system from its bulk counterpart mentioned above: The narrow QD emission lines enable a direct measurement of electron nuclear interaction energy. This is not possible in bulk systems where typical widths of emission lines are an order of magnitude larger than the electronic energy shifts induced by polarized nuclei. In addition, the possibility of addressing a single QD has the advantage of removing effects of sample inhomogeneities and crosstalk between individual islands of spin polarized nuclei. Due to the different atomic composition and strain distribution of the QD as compared to its surrounding host material, the ensemble of $\sim 10^4$–10^5 QD nuclear spins can be considered as truly isolated from the environment. Therefore, the coupled electron–nuclear spin system of a QD is an implementation of a well isolated system of a single electron spin, coupled to a slowly varying, small nuclear spin reservoir, i.e., the central spin problem.

More recent interest in the dynamics of QD nuclear spins has arisen from the experimental and theoretical findings that the slow fluctuations of the nuclear spins constitute the dominant source of decoherence of QD electron spins. Since these fluctuations have long correlation times, the resulting electron-spin decoherence is non-Markovian and hence the evolution of the electron spin is highly complex [4–6]. Controlling the nuclear spin fluctuations, for instance by producing a substantial nuclear spin polarization or by performing a series of projective measurements on the nuclear spins [7], could potentially suppress this decoherence mechanism [8].

Optical orientation of QD nuclear spins has been demonstrated experimentally by a few groups [9–16]. However, the degree of DNSP achieved in these experiments has been limited to $\sim$10–20% in low external magnetic fields and did not exceed 60% in high fields. In order to reach even higher degrees of DNSP, a detailed analysis of both the formation dynamics and the limiting factors of DNSP is required. Since DNSP is a balance between nuclear spin polarization by the QD electron and nuclear spin depolarization, a measurement of the corresponding timescales as well as an identification of the factors that influence the nuclear spin dynamics is of great interest and relevance. Inherent properties of the QD nuclear spin system, like the respective role of nuclear spin diffusion, quadrupolar interactions and trapped excess QD charges can be investigated using time-resolved measurements of the nuclear spins. Furthermore, experimental determination of the nuclear spin decay time directly yields the correlation time of the fluctuations of the Overhauser field along the axis in which the nuclei are polarized – a crucial quantity for understanding the limits of electron spin coherence in QDs [17].

In this work, we give an overview of our experimental assessment of the above mentioned points. After describing the sample investigated in this work and the experimental setup of our spectroscopy system in some detail in

Sect. 5.2, we give a short review of the theory essential for the interpretation of our experimental findings in Sect. 5.3. We then turn to our experimental results, starting with the low field behavior of DNSP in Sect. 5.4, where we describe the first measurement of the Knight field of a single electron confined to a QD. The nonlinear nature of the coupled electron–nuclear spin system is most pronounced in strong external magnetic fields, where DNSP may have a hysteretic response to the applied field, as we will show in Sect. 5.5. Finally, in Sect. 5.6 we focus on time-resolved measurements of DNSP, both in low and in high external magnetic fields. These measurements allowed us to identify the dominant nuclear spin relaxation mechanisms in self-assembled QDs which we discuss in detail at the end of this section.

5.2 Experimental Methods

The experimental results presented in this work have been obtained on individual, self assembled InGaAs QDs. These QDs have been embedded in gated structures that allow for deterministic QD charging with individual excess charges (electrons or holes) [18]. The sample was grown by molecular beam epitaxy on a (100) semiinsulating GaAs substrate. The QDs are spaced by 25 nm of GaAs from a 40 nm doped n^{++}-GaAs layer, followed by 30 nm GaAs and 29 periods of AlAs/GaAs (2/2 nm) superlattice barrier layer, and capped by 4-nm GaAs. A bias voltage is applied between the top Schottky and back ohmic contacts to control the charging state of the QDs. The low density of QDs ($<0.1\,\mu\text{m}^{-2}$) allows us to address a single QD using a microphotoluminescence (μ-PL) setup.

Our standard μ-PL setup is based on the combination of a ZrO$_2$ solid immersion lens in Weierstrass configuration, directly fixed onto the sample, and a PL collection lens of numerical aperture ∼0.1 mounted outside the cryostat. Depending on the discussed experiment, the sample was placed either in a helium-bath cryostat equipped with a superconducting magnet, reaching a maximum magnetic field strength of 10 T, or in a helium flow cryostat. There, variable low magnetic fields of $B_{\text{ext}} < 20\,\text{mT}$ or, alternatively, fixed magnetic fields of $B_{\text{ext}} \approx 200\,\text{mT}$ were applied using external Helmholtz coils or a permanent magnet, respectively. In the case of the flow cryostat, the collection lens was replaced by a microscope objective with a numerical aperture of 0.26, resulting in an enhanced photon collection efficiency.

The spectroscopy system consists of a spectrometer of 0.75 m focal length and a liquid-nitrogen cooled charge coupled device (CCD) camera. The spectral resolution of this system is limited to ∼30 μeV by the finite CCD pixel separation. However, the precision to which the emission energy of a given spectral line can be determined, can be increased to ∼2 μeV, by calculating a weighted average of the emission line over the relevant CCD pixels [19]. Alternatively, by using a scanning Fabry–Perot interferometer of 62 μeV free spectral range and a finesse ≥70 as a narrow-band frequency filter in front of the spectrometer, a spectral resolution <1 μeV can be achieved [14].

The PL polarization and spin splitting are studied by resonantly exciting a single QD in one of its excited ("p-shell") states at $T = 5$ K. The PL spectral lines associated with different charging states of a single QD [18] can be identified from the PL intensity contour plot as a function of the bias voltage and emission energy (Fig. 5.1a). The emission line coming from neutral exciton (X^0) recombination exhibits a fine-structure splitting of ~20 μeV due to the anisotropic electron–hole exchange interaction (AEI) [20]. The emission from the negatively (positively) charged trion X^{-1} (X^{+1}) arises from optical excitation of QD charged with a single electron (hole). The photons emitted upon X^{-1} (X^{+1}) recombination are red (blue) shifted by ~4.6 meV (~1.9 meV) with respect to the X^0 line.

Unless stated otherwise, the discussed experiments have been performed on the negatively charged exciton X^{-1} at the center of its PL stability region with respect to gate voltage. In this regime, electron cotunneling to the nearby reservoir was shown to be minimized [21] and the QD is occupied by a single electron in its ground state. Optical excitation is performed in a resonant way into the "p-shell," which lies approximately one LO phonon energy above the emission energy of X^{-1} ($E_0 = 1.3155$ eV).[1] The excitation power is fixed close to saturation of the observed emission line. We found that these conditions lead to a maximal preservation of PL light polarization ($|\rho_c^\pm| \approx 75\%$ at $B = 0$ T) after excitation with circularly polarized light.

The polarization of the excitation laser and of the PL are denoted as ($\sigma^\alpha, \sigma^\beta$), where σ^α and σ^β are the excitation and detection polarization, respectively. The index α or β takes one of four values: linear polarization along the crystal axes ($x : [1\bar{1}0], y : [110]$) or circular polarization ($\pm$: light helicity ± 1). The degree of circular polarization is defined as $\rho_c^\pm \equiv (I^\pm - I^\mp)/(I^+ + I^-)$, where I^β denote the intensity of PL in the ($\sigma^\pm, \sigma^\beta$) configuration. The polarization characteristics of the system is calibrated using the strongly polarized emission of the LO phonon of the GaAs substrate (Fig. 5.1a) [23]. The combined fidelity of polarization preparation and detection was found to be better than 98%.

Circularly polarized, resonant p-shell pumping of a single electron (hole) charged QD generates optically oriented trions with hole (electron) spin $J_z = +3/2$ ($S_z = -1/2$) or $J_z = -3/2$ ($S_z = +1/2$), under σ^+ and σ^-

---

[1] We exclude resonant ground state excitation by simultaneous LO phonon emission through various experimental observations: We observe several distinct QD excitation resonances at different energies. The GaAs LO phonon line for these resonances can thereby lie above or below the corresponding excitonic emission energy. The observed resonances have widths on the order of 100 μeV, in accordance with [22]. Only a few of these resonances lead to the high degree of PL circular polarization and to the nuclear spin polarization effects that will be discussed later. These facts corroborate or model of resonant excitation into the "p-shell."

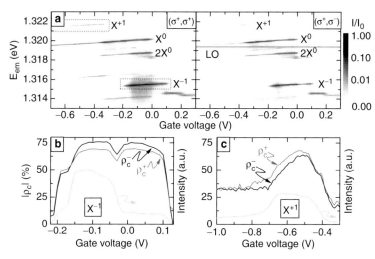

Fig. 5.1. Photoluminescence from a single charge-tunable QD. (**a**) Contour plot of the PL intensity as a function of applied bias voltage under circularly polarized excitation and co- (cross-) circularly polarized detection (($\sigma^+,\sigma^{+(-)}$), respectively. The intensity is normalized with respect to the X^{-1} peak-intensity of $I_0 = 3 \times 10^4$ counts s^{-1}. The excitation laser energy is tuned to a p-shell resonance of the singly charged QD at 1.356 eV, ∼36.6 meV above the bulk GaAs LO phonon line indicated in (**a**). Panels (**b**) and (**c**) show the integrated intensities of X^{-1} and X^{+1} emission in the (σ^+,σ^+) configuration (*light gray*), as well as the degree of circular PL polarization $\rho_c^\pm$ under $\sigma^\pm$ excitation. For (**c**), the laser is tuned into a p-shell resonance for X^{+1}

pumping, respectively [12, 22]. The intradot excitation ensures maximal carrier spin preservation during relaxation, which is confirmed by the high degree of circular polarization of the PL emission lines (Fig. 5.1b,c). The initial state of X^{+1} is composed of two holes in a singlet-state and one electron. Therefore, σ^+ (σ^-) polarized PL from X^{+1} indicates that the optically created electron was in the $S_z = -1/2$ ($S_z = 1/2$) state. Analogously, for X^{-1}, circular polarization of the emitted light reflects both the spin of the hole in the QD before photon emission and the initial spin of the residual QD electron after photon emission. A high degree of circular polarization of X^{-1}-emission thus indicates a highly spin polarized residual electron in the QD.

5.3 Electron and Nuclear Spin Systems in a Single Quantum Dot and Implications for Electron Spin Coherence

Circularly polarized excitation and the analysis of PL light polarization are powerful experimental tools for manipulating and measuring the spin polarization of the electron spin system in a single QD. The high degree of QD

PL polarization described in Sect. 5.2 indicates that electron spin relaxation mechanisms in the investigated QDs are relatively weak. Since the electron spin system is in thermal contact with the nuclear spin system through the hyperfine interaction, spin polarization is transferred from one to the other, thereby cooling the nuclear spin system. At the same time, the coupling of the nuclear spins to their environment will heat up the nuclear spins, leading to a finite nuclear spin temperature in a dynamical equilibrium. In this section, we give an overview of the basic spin interactions and relaxation mechanisms of the QD electron and nuclear spins and discuss the coupling mechanism between the two spin systems. Finally, we discuss how this coupling affects the coherence properties of the electron spin.

The total Hamiltonian $\hat{H}$ for the electron–nuclear spin system can be written as[2]

$$\hat{H} = \hat{H}_Z^{el} + \hat{H}_Z^{nuc} + \hat{H}_{dip} + \hat{H}_{HF}. \tag{5.1}$$

$\hat{H}_Z^{el}$ and $\hat{H}_Z^{nuc}$ denote the electron and nuclear Zeeman Hamiltonian, respectively, while $\hat{H}_{dip}$ denotes the nuclear dipole–dipole interactions and $\hat{H}_{HF}$ the hyperfine interaction, which couples the electron and nuclear spins. We will discuss the individual terms of this Hamiltonian in more detail in the following paragraphs.

5.3.1 Electron Spin System

The electron spin system considered in this work consists of a residual QD electron in the case of X^{-1} and of the (metastable) photo-generated QD electron in the case of X^{+1}. Information about the electron spin is obtained by measuring the circular polarization of PL light.

When placed in a magnetic field, the spin of a QD electron experiences the Zeeman interaction, described by the Hamiltonian

$$\hat{H}_Z^{el} = g_{el}^* \mu_B \hat{\mathbf{S}}_{el} \cdot \mathbf{B}, \tag{5.2}$$

where μ_B is the Bohr magneton, g_{el}^* the effective electron g-factor and $\hat{\mathbf{S}}_{el}$ the electron spin operator ($\hat{\mathbf{S}}_{el} \equiv \frac{1}{2}\hat{\boldsymbol{\sigma}}$, with the Pauli-matrices $\hat{\boldsymbol{\sigma}}$).

For a consistent notation with respect to the sign of the g-factors, we also note the Zeeman Hamiltonian for holes, which we write as $\hat{H}_Z^h = -g_h^* \mu_B \hat{\mathbf{S}}_h \cdot \mathbf{B}$ [24], with the effective hole g-factor g_h^* and the hole spin operator $\hat{\mathbf{S}}_h$. In this representation, the heavy-hole wave functions $|\pm \frac{3}{2}\rangle$ convert to pseudospins $|\pm \frac{1}{2}\rangle$. The experimentally found g-factors g_{el}^* and g_h^* are both negative in our system.

Due to the tight confinement of electrons in self assembled QDs, the electron spin is well protected from spin-relaxation by electron–phonon coupling

---

[2] In this work, we only consider singly charged excitons. Exchange interactions therefore play no role for the energies of the recombining excitons [20].

via spin–orbit interaction, which is otherwise very effective in a solid state environment. This typically extends electron T_1 times in QD systems up to 1 s in moderate magnetic fields [25]. However, the coupling of the QD electron to the nearby electron reservoir introduces an additional – albeit controllable – decay channel for the electron spin. The QD electron can make a virtual transition to the Fermi sea of the reservoir and be replaced by a spin-flipped electron from the same reservoir. The rate of these cotunneling events depends critically on the tunnel barrier between the QD and the electron reservoir and therefore in the details of the QD heterostructure. For the structures used in this work, the resulting T_1 time of the electron has an upper bound of ~3 ns at gate voltages in the center of the 1e$^-$ stability region and decreases rapidly toward the edges of this region [26]. This tunability of the electron spin lifetime will be an essential feature in the measurements of the nuclear spin lifetime, as will be shown later.

5.3.2 Nuclear Spin System

The QDs considered in this work consist of InAs with an admixture of Ga due to material diffusion during the growth process. The nuclear spins present in our QDs therefore consist of three nuclear species with their naturally occurring isotopes: ^{115}In (95.3%), ^{113}In (4.7%), ^{75}As, ^{69}Ga (60.1%), ^{71}Ga (39.9%) – their natural abundances are given in brackets. Indium has a total spin of 9/2 while all other nuclei have spin 3/2. In the following, we will only consider the most abundant nuclear species for simplicity and neglect the at least ~10% admixture of Ga in our dots.

Each nucleus i is characterized by its spin I^i and the gyromagnetic ratio γ_i, describing the response to a static external magnetic field $\mathbf{B}$:

$$\hat{H}_Z^{\text{nuc}} = -\gamma_i \hbar \hat{\mathbf{I}}^i \cdot \mathbf{B}, \qquad (5.3)$$

where $\hat{\mathbf{I}}^i$ is the (dimensionless) spin-operator of the ith nucleus. The gyromagnetic ratios for In and As are $\gamma_{^{115}\text{In}} = 9.365\,\text{MHz}\,\text{T}^{-1}$ and $\gamma_{^{75}\text{As}} = 7.315\,\text{MHz}\,\text{T}^{-1}$.

Interactions of nuclear spins dominate the nuclear energy spectrum at low magnetic fields and lead to spin-transport in high fields. The simplest, and in semiconductors usually dominant nuclear spin–spin interaction is the dipolar coupling between two nuclear spins i and j. It can be written as [27]:

$$\hat{H}_{\text{dip}}^{i,j} = \frac{\mu_0 \hbar^2 \gamma_i \gamma_j}{4\pi r_{ij}^3} \left(\hat{\mathbf{I}}^i \cdot \hat{\mathbf{I}}^j - 3 \frac{\left(\hat{\mathbf{I}}^i \cdot \mathbf{r}_{ij}\right)\left(\hat{\mathbf{I}}^j \cdot \mathbf{r}_{ij}\right)}{r_{ij}^2} \right), \qquad (5.4)$$

where $\mathbf{r}_{ij}$ is the vector of length r_{ij} joining the two nuclei, and μ_0 is the permeability of free space. The dipolar Hamiltonian of the total nuclear spin system is then a sum over all nuclear spin pairs: $\hat{H}_{\text{dip}} = \sum_{i<j} \hat{H}_{\text{dip}}^{i,j}$.

A common decomposition of this Hamiltonian divides $\hat{H}_{\text{dip}}^{i,j}$ into spin-conserving ("secular") and nonconserving ("nonsecular") parts [27]. While the former are responsible for nuclear spin diffusion within the lattice, the latter can lead to depolarization of nuclear spins in low magnetic fields. The strength of the interaction $\hat{H}_{\text{dip}}$ is usually characterized by a "local field" B_{loc}, which is the effective magnetic field generated on the site of a nucleus by its neighboring nuclear spins. For bulk GaAs, B_{loc} is on the order of 0.1 mT [3]. It can be shown that the nonsecular terms of $\hat{H}_{\text{dip}}$ only contribute to the evolution of the nuclear spin systems for magnetic fields $B_{\text{ext}} \leq B_{\text{loc}}$. Below these fields, the nonsecular nuclear dipole–dipole interactions depolarize the nuclear spins very effectively on a timescale $T_2 \approx 10$–$100\,\mu\text{s}$.[3]

Nuclear spin relaxation in III–V compounds has been experimentally investigated in detail using standard NMR techniques [29]. The resulting T_1 times in InAs and GaAs were on the order of 1,000 s for Ga and As and roughly 200 s for In at a temperature of 4 K. These values however were shown to be limited by nuclear spin relaxation by paramagnetic impurities, a mechanism absent in individual QDs. The remaining relevant nuclear spin relaxation mechanism is quadrupolar relaxation. This process results from nuclear transitions induced by the coupling of the nuclear quadrupole moment to phonon-generated electric field gradients at the nuclear site. The corresponding phonon-induced relaxation rate scales with temperature as T^2 and dominates over the temperature-independent relaxation by paramagnetic impurities for $T > 20$ K. We therefore estimate the nuclear T_1-time in our QDs to be further reduced by two orders of magnitude compared to the reported values [29]. Even though the nuclear T_1-time could be reduced again by the strong lattice deformations in self-assembled QDs, we assume nuclear spin–lattice relaxation to be completely negligible for our experiments. The only remaining mechanism leading to a decay of nuclear spin polarization in a QD could be spin-diffusion out of the QD into the surrounding bulk material. The spin-diffusion constant in GaAs has been measured experimentally to be $D = 10^{-13}\,\text{cm}^2\,\text{s}^{-1}$ [30]. The typical timescale for diffusion out of a QD with a diameter $d \approx 20$ nm is therefore $d^2/D \approx 1$ min. However, the different nuclear species and local lattice structure within the QD compared to its surrounding bulk material should further reduce the diffusion constant and increase the diffusion time. Experimental work on nuclear spin diffusion between quantum wells has shown that this reduction amounts at least to a factor of 10 [31].

---

[3] We note that this is not a decoherence process in the sense of loss of information from the nuclear spins to a reservoir, resulting in a decay of the off diagonal elements of the nuclear spin density matrix. Rather, due to nuclear spin–spin interactions, these elements all evolve at different frequencies spread over an energy interval $\Delta E \approx \gamma B_{\text{loc}}$. Then, as far as the expectation values of observables are concerned, the off-diagonal elements of the nuclear density matrix can be taken equal to zero after a time $T_2 \simeq \hbar/\Delta E$ [28]. After this time, the nuclear spin state is fully described by its mean energy or equivalently, its nuclear spin temperature. The relaxation time of this nuclear spin temperature is denoted as T_1.

5.3.3 Hyperfine Interaction

The dominant contribution to the coupling between the electron- and the nuclear-spin systems originates from the Fermi contact hyperfine interaction. For an electron in a QD and in first order perturbation theory, this interaction can be written as [3, 4, 32]:

$$\hat{H}_{\mathrm{hf}} = \frac{\nu_0}{8} \sum_i A_i |\psi(\mathbf{R}_i)|^2 \hat{\mathbf{S}}_{\mathrm{el}} \cdot \hat{\mathbf{I}}^i, \tag{5.5}$$

where ν_0 is the volume of the InAs-crystal unit cell containing eight nuclei, $\psi(\mathbf{r})$ is the electron envelope wave function,[4] and $\mathbf{R}_i$ is the location of the ith nucleus. $A_i = \frac{2}{3}\mu_0 g_0 \mu_B \hbar \gamma_i |u(\mathbf{R}_i)|^2$ is the hyperfine coupling constant and g_0 the free electron g-factor. A_i depends on the value of the electron Bloch function $u(\mathbf{R}_i)$ at the nuclear site. For all the nuclei in our system it is positive and on the order of 50 μeV (i.e., $A_{\mathrm{In}} = 56\,\mu\mathrm{eV}$ and $A_{\mathrm{As}} = 46\,\mu\mathrm{eV}$ [33]). We note that only electrons in the conduction band couple to the nuclear spins through (5.5). For carriers in the valence band of III–V semiconductors, this interaction vanishes due to the p-type symmetry of $u(\mathbf{R}_i)$ [34].

With the identity $\hat{\mathbf{S}}_{\mathrm{el}} \cdot \hat{\mathbf{I}}^i = I_z^i S_z + \frac{1}{2}(I_+^i S_- + I_-^i S_+)$, where $S_\pm$ and $I_\pm^i$ are the electron and nuclear spin raising and lowering operators, respectively, (5.5) can be decomposed into two parts [35]: A dynamical part ($\propto I_+^i S_- + I_-^i S_+$), allowing for the transfer of angular momentum between the two spin systems, and a static part ($\propto I_z^i S_z$), affecting the energies of the electron and the nuclear spins.

The dynamical contribution leads to a thermal equilibration of the electron and the nuclear spin systems. Neglecting other spin relaxation mechanisms and polarization due to thermalization in the external magnetic field, the mean nuclear spin polarization $\langle I_z^i \rangle$ along the quantization axis z is linked to the electron spin polarization S_z through the Curie-law like relation [36]:

$$\langle I_z^i \rangle = I^i B_{I^i}(x), \quad \mathrm{with} \quad x = I^i \ln\left(\frac{1+2S_z}{1-2S_z}\right). \tag{5.6}$$

B_{I^i} is the Brillouin function of order I^i. For small electron spin polarizations $S_z \ll 1/2$, (5.6) can be expanded to $\langle I_z^i \rangle = 4/3 I^i (I^i + 1)\langle S_z \rangle$, while for $S_z \approx 1/2$, $\langle I_z^i \rangle = I^i$.

The static part of the hyperfine interaction leads to the notion of the "effective magnetic fields," either seen by the electron due to spin polarized nuclei (Overhauser field $\mathbf{B}_{\mathrm{nuc}}$), or by the nuclei due to a spin polarized electron (Knight field $\mathbf{B}_{\mathrm{el}}$). Here, we only consider their projection along the z-axis, which we denote as B_{nuc} and B_{el}, respectively. The corresponding Knight field operator $\hat{B}_{\mathrm{el}}^i$ is

$$\hat{B}_{\mathrm{el}}^i = -\frac{1}{\hbar \gamma_i} \frac{\nu_0}{8} A_i |\psi(\mathbf{R}_i)|^2 \hat{S}_z. \tag{5.7}$$

---

[4] In our convention, $|\psi(\mathbf{r})|^2$ is normalized to $\frac{8}{\nu_0}$.

Its expectation value $B_{el}^i = \langle S|\hat{B}_{el}^i|S\rangle$ depends on the electron spin state $|S\rangle$ and on the exact location of the nucleus i. For a fully polarized electron bound to a shallow donor[5] in GaAs, the maximal value of B_{el}^i has been estimated to be 13 mT for Ga and 22 mT As. These values, however, are further reduced to $f_{el}B_{el}^i$ if the QD is occupied by a single electron only in a finite fraction f_{el} of the total measurement time [3].

Analogously, the Overhauser field operator can be written as

$$\hat{B}_{\text{nuc}} = \frac{1}{g_{el}^*\mu_B}\frac{\nu_0}{8}\sum_i A_i\,|\psi(\mathbf{R}_i)|^2\,\hat{I}_z^i, \tag{5.8}$$

with the expectation value $B_{\text{nuc}} = \langle\mu|\hat{B}_{\text{nuc}}|\mu\rangle$ for a given nuclear spin state $|\mu\rangle$. This effective field leads to a total electron Zeeman splitting in the presence of both nuclear and external magnetic fields:

$$\Delta E_{el}^Z = g_{el}^*\mu_B(B_{\text{ext}} + B_{\text{nuc}}). \tag{5.9}$$

The electronic energy shift due to spin polarized nuclei (ΔE_{OS}) is referred to as the Overhauser shift (OS). For a fully polarized nuclear spin system in bulk InAs, the OS due to the In and As nuclei amounts to $\Delta E_{\text{OS}}^{\text{In}} = \frac{1}{2}\frac{9}{2}56\,\mu\text{eV} = 126\,\mu\text{eV}$ and $\Delta E_{\text{OS}}^{\text{As}} = \frac{1}{2}\frac{3}{2}46\,\mu\text{eV} = 35\,\mu\text{eV}$, respectively.

Singly charged excitons are ideal candidates for a spectroscopic study of DNSP in QDs. In these charge complexes, exchange interactions between the carriers play no role [20] and the magnetic field dispersion of the spin splittings of excitonic recombination lines is solely due to Zeeman interaction of the spins with (effective) magnetic fields. For X^{-1}, the total Zeeman splitting of the PL recombination line thus amounts to

$$\Delta E_{X^{-1}}^Z = -g_h^*\mu_B B_{\text{ext}} - g_{el}^*\mu_B(B_{\text{ext}} + B_{\text{nuc}}), \tag{5.10}$$

where g_{el}^* and g_h^* are the electron- and hole g-factors, respectively.

Exciting the QD with linearly polarized light creates residual electrons in a superposition of spin up and down, resulting in no nuclear polarization and therefore in $B_{\text{nuc}} = 0$. Thus, comparing the Zeeman splittings of X^{-1} under linearly- and circularly polarized excitation ($\Delta E_{X^{-1}}^{Z,\text{lin.}}$ and $\Delta E_{X^{-1}}^{Z,\sigma^\pm}$, respectively) gives a direct measure of ΔE_{OS} and B_{nuc}:

$$\Delta E_{\text{OS}} = \Delta E_{X^{-1}}^{Z,\sigma^\pm} - \Delta E_{X^{-1}}^{Z,\text{lin.}} = -g_{el}^*\mu_B B_{\text{nuc}}. \tag{5.11}$$

The analysis of the spin splittings for X^{+1} is analogous and will not be given here.

5.3.4 Hyperfine-Induced Electron Spin Decoherence

As we argued in Sect. 5.3.1, QD electron spins are well protected from relaxation and their evolution is governed by the Zeeman interaction of the electron

---
[5] With a Bohr radius of 10 nm – comparable to our QD confinement length.

with (effective) magnetic fields. Besides the externally applied magnetic field, the electron experiences the strong nuclear magnetic field B_{nuc} (cf. (5.8)). This field can attain values of several teslas for fully polarized nuclei and can therefore significantly contribute to the QD electron spin dynamics. If the nuclear spins are not explicitly prepared in a specific spin state, each nuclear spin points in a random direction and the mean nuclear spin polarization of the ensemble of nuclear spins is a random variable with a Gaussian distribution, having a width $\propto \sqrt{N}$. The fluctuating nuclear field can therefore be estimated to be $\Delta B_{\text{nuc}} \simeq A/\sqrt{N}g_{\text{el}}^*\mu_B$, which is on the order of $10\,\text{mT}$ for typical QD sizes with $N \approx 10^5$.

In low external magnetic fields, $B_{\text{ext}} \approx \Delta B_{\text{nuc}}$, the evolution of the direction of the electron spin is therefore mostly determined by its interaction with the nuclear spins. The transverse components of B_{nuc} with respect to the electron spin direction lead to coherent Larmor precession. Equivalently, the flip-flop terms in (5.5) drive transitions between the electron spin up and down states, thereby coupling these two states and defining a new quantization direction for the electron spins.

In a magnetic field $B_{\text{ext}} \gg \Delta B_{\text{nuc}}$, where the electron Zeeman splitting largely exceeds the hyperfine coupling strength with the random nuclear field, electron–nuclear flip-flop events become energetically forbidden and the components of the nuclear field transverse to B_{ext} can be neglected. However, the longitudinal components of the nuclear field fluctuations lead to a fluctuation of the energies of the electron spin states. This fluctuation in turn leads to a decoherence of the electron spin with a corresponding T_2^* time [32]:

$$T_2^* = \frac{\hbar}{g_{\text{el}}^*\mu_B\sqrt{2\Delta B_{\text{nuc}}^2/3}}. \tag{5.12}$$

With $\Delta B_{\text{nuc}} \approx 10\,\text{mT}$, T_2^* is of the order of $3\,\text{ns}$, in accordance with recent transport measurements in a QD system [37]. We note that this apparent decoherence on a timescale T_2^* only comes from the fact that real experiments constitute an average of many measurements, each of which is realized under another nuclear spin configuration. In each run, the electron therefore has another Larmor precession frequency, which combined with the experimental averaging leads to an apparent damping of the electron spin precession. For a given nuclear spin state, however, the evolution of the electron–nuclear spin system is perfectly coherent as long as the direction and magnitude of the nuclear field are not altered. This averaging effect can be eliminated by using spin-echo techniques which unwind the effect of a static nuclear field on the electron spins. Under such conditions, the electron spin dephasing time in electrostatically defined QDs was found to be $T_2 \approx 1\,\mu\text{s}$ at $B_{\text{ext}} = 100\,\text{mT}$ – two orders of magnitude longer than T_2^* [37].

The remaining electron spin decoherence is a consequence of the evolution of the nuclear spin system and in particular of the z-component B_{nuc}^z of the nuclear field. This evolution was neglected in deriving (5.12) and the arguments leading to the finite T_2^* time were given based on the picture of the "frozen fluctuations" of the nuclear field. In order to understand the pure dephasing processes for a QD electron (characterized by a time T_2), it is therefore necessary to understand the mechanisms which cause the evolution of ΔB_{nuc} and the timescale at which this evolution happens.

In Sect. 5.3.2 it was shown that nuclear spin lattice relaxation is negligible in the temperature range relevant to the experiments discussed here. The only interactions relevant for the evolution of the nuclear field are therefore the nuclear dipole–dipole interactions (5.4) and the hyperfine interaction with conduction band electrons (5.5). Their main effect in changing B_{nuc}^z are flip-flop processes between different nuclei (i and j), which change the nuclear field because the two coupled nuclei might have different interaction strengths with the electron spin (i.e., $|\psi(\mathbf{R}_i)|^2 \neq |\psi(\mathbf{R}_j)|^2$).

Nuclear dipolar interaction strengths are characterized by the local dipolar field B_{loc} (cf. Sect. 5.3.2). The evolution of a nuclear spin due to interactions with its neighbors therefore happens at a rate corresponding to the nuclear Larmor precession rate in the field B_{loc}, i.e., at ~ 1–10 kHz. For external magnetic fields exceeding B_{loc}, the only relevant contribution for the evolution of B_{nuc}^z are the secular terms of the dipole–dipole interaction which lead to diffusion and redistribution of the nuclear spins. The corresponding rate of change of B_{nuc}^z due to dipolar interactions is therefore further reduced. The time required to change a given nuclear field by a magnitude ΔB_{nuc} has been estimated to be ~ 0.01–10 s [38].

The role of the hyperfine interaction for the evolution of the nuclear spin system is twofold. First, the Knight field B_{el} of a QD electron leads to precession of the nuclear spins around the electron spin at a rate given by the nuclear Larmor frequency in a field B_{el}. Being on the order of 1 mT [14], the Knight field leads to nuclear spin evolution on timescales comparable to the corresponding estimates for the dipolar interactions. This evolution, however gets suppressed in external magnetic fields exceeding B_{el}, where electron–nuclear flip-flop processes vanish in first order. The second effect of the hyperfine interaction on the nuclear spins are electron-mediated, long ranged nuclear spin–spin interaction. This second order process termed "indirect interaction" consists of a virtual electron–nuclear spin-flip followed by a spin-"flop" of the electron with another nuclear spin. The resulting decorrelation time of the nuclear magnetic field has been shown to scale as $N^{3/2} \Delta E_{\text{el}}^Z / A^2$ for $A/\Delta E_{\text{el}}^Z \ll 1$ [39], with a rough estimate of 1 ms for $N = 10^5$, $A = 50\,\mu\text{eV}$ and $\Delta E_{\text{el}}^Z = 200\,\mu\text{eV}$.

Calculating the electron spin dephasing rate caused by the slow but random nuclear field fluctuations occurring on a timescale of $\sim 100\,\mu\text{s}$ turns out to be a difficult task [40]. This rate not only depends on the timescale of nuclear field fluctuations, but also on the correlations of these fluctuations as well as on the width of the initial nuclear field distribution. Still, attempts have been made to calculate this quantity including the interaction mechanisms discussed above [4, 41]. The results of these calculations lead $T_2 \approx 1$–$100\,\mu\text{s}$, in rough agreement with experimental results [37].

Since most solid state systems contain nuclei with nonzero spin, the only way to further increase the electron spin coherence time is to manipulate the nuclear spins. The goal of such a manipulation is then to suppress fluctuations of the mean nuclear spin, thereby leaving the electron to interact coherently with a static nuclear field. Two methods for suppressing nuclear spin fluctuations have been proposed up to now. They involve creating a very high degree of nuclear spin polarization [4] or, alternatively, repeated projective measurements of the nuclear field [7], causing a "quantum Zeno effect" on the nuclear spins.

5.4 Optical Pumping of Nuclear Spins in Low Fields: The Role of the Knight Field

In low external magnetic fields, the evolution of the nuclear spin system is governed by the nuclear dipole–dipole interactions, characterized in strength by a local magnetic field B_{loc} (cf. Sect. 5.3.2). If the total magnetic field seen by the nuclei is smaller than B_{loc}, nuclear angular momentum is not a conserved quantity. Instead, the nuclear spins evolve into a highly entangled many body state due to their dipolar coupling. In order to observe any nuclear spin polarization, it is therefore necessary to apply a magnetic field exceeding B_{loc}. This fact has been observed experimentally in different QD systems [13, 42] as well as in various bulk NMR experiments [28]. Remarkably, in a situation of tight electron confinement, the Knight field B_{el} can attain values exceeding the local field B_{loc}, thereby "stabilizing" the nuclear spin polarization and allowing for the observation of nuclear spin polarization even in the absence of an externally applied magnetic field.

5.4.1 Nuclear Spin Cooling in the Knight Field of the QD Electron

Creating a Knight field strong enough to suppress the nonsecular terms of the dipole–dipole interaction requires a sizable QD electron spin polarization. We realize this situation by exciting our QDs resonantly in one of their excited ("p-shell") states as described in Sect. 5.2.

Figure 5.2a shows the emission spectrum of X^{-1} at zero external magnetic field obtained by using a scanning Fabry–Perot interferometer under linearly and circularly polarized excitation. Under σ^x-polarized laser excitation, no fine structure splitting is observed, confirming the absence of nuclear spin polarization. Exciting the QD with $\sigma^{\pm}$-polarized light, spin doublets with a $\sim 13\mu\mathrm{eV}$ splitting appear. For X^{-1}, PL peaks that are cocircular with the excitation laser have lowest energy for both σ^+ and σ^- excitation (Fig. 5.2a), indicating that the direction of the effective magnetic field causing the observed splitting is determined by the direction of the QD electron spin. We therefore attribute the observed splitting to DNSP and further test this hypothesis by performing an analogous measurement using X^{+1} trion excitation

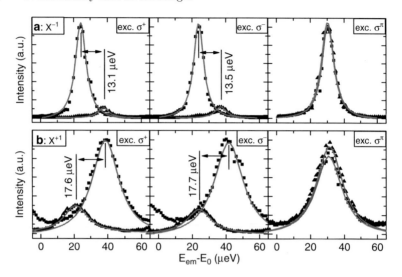

Fig. 5.2. Spin splitting induced by the Overhauser field. High-spectral-resolution PL spectra measured with a Fabry–Perot scanning interferometer (spectral resolution ~1 µeV) under $B_{\text{ext}} = 0$. The polarizations of the excitation laser are denoted in the figure. PL is detected co- and cross-polarized polarization with respect to the excitation (*squares* and *triangles*, respectively). Under circularly polarized excitation of X^- (a) as well as of X^{+1} (b), a significant nuclear spin polarization develops. In contrast, linearly polarized excitation does not lead to nuclear spin polarization. An energy offset of $E_0 \approx 1.3155\,\text{eV}$ (1.3215 eV) is subtracted from the X^- (X^+) data

(Fig. 5.2b). For X^{+1}, the observed energy sequence in PL emission is reversed, indicating that for a given excitation polarization the electron spin is polarized in opposite directions in the X^{-1} and X^{+1} trions [13]. This is consistent with the respective electron spin systems that were identified in Sect. 5.3.1.

Generally, the expectation value of the Overhauser field in a weak magnetic field B along the z-axis and in the presence of nuclear dipolar interactions can be expressed as [3, 35, 36, 43]:

$$B_{\text{nuc}} = f \frac{B^2 \langle S_z \rangle}{B^2 + \xi B_{\text{loc}}^2}, \qquad (5.13)$$

where $B = B_{\text{el}} + B_{\text{ext}}$ is the total effective magnetic field seen by the nuclei. $\langle S_z \rangle$ is the expectation value of the electron spin along the z-axis, ξ is a numerical factor of order unity and f is a proportionality constant determined by (5.6) and (5.8). The fact that we observe DNSP even if $B_{\text{ext}} = 0$ suggests that the Knight field of a single spin-polarized electron is strong enough to ensure $B_{\text{el}} > B_{\text{loc}}$.[6]

---

[6] We measured the stray field in our set-up to be $0.05 \pm 0.01\,\text{mT}$ at an angle of $\sim 25°$ to the optical axis.

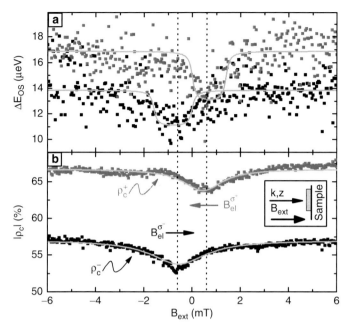

Fig. 5.3. Overhauser shift ΔE_{OS} (**a**) and PL polarization (**b**) as a function of applied external magnetic field B_{ext}. Here, the measured spin splitting is determined by a weighted average of the X^{-1} spectral lines measured by the spectrometer. The *light gray curves* in (**a**) and (**b**) are fits according to the model described in the text. Observation of correlated dips in spin splitting and in polarization as a function of B_{ext} suggests an average Knight field $B_{el} \approx 0.6\,\text{mT}$ seen by the nuclei. Under σ^+ (σ^-) excitation, the corresponding Knight field $B_{el}^{\sigma^{+(-)}}$ is parallel (antiparallel) to the wave-vector **k** of laser excitation. The schematic in the *inset* of (**b**) sketches the orientations of the laser wave-vector and a positive external magnetic field

Based on (5.13), it could be concluded that application of an external field that cancels the Knight field should result in the complete disappearance of DNSP. Figure 5.3a shows the dependence of the observed Overhauser shift ΔE_{OS} of X^{-1} on $|\mathbf{B}_{ext}|$. A dip in the Overhauser shift at $B_{ext} = -B_{el} \approx \pm 0.6\,\text{mT}$ is observed under excitation with $\sigma^\pm$ polarized light, which gives a direct measurement of the average Knight field B_{el} experienced by the QD electrons. The direction of B_{el} is determined by the direction of the spin of the QD electron: Excitation with σ^+ polarized light leaves a residual electron with spin up in the QD. Since $A > 0$, (5.7) implies that the Knight field indeed has to point in the negative direction.

Even when $B_{ext} = -B_{el}$, ΔE_{OS} is only reduced from ~ 16 to $\sim 12\,\mu\text{eV}$, indicating that the cancelation of the Knight field B_{el} by the external field is not complete. The Knight fields we measured with this method range from ± 0.6 to $\sim \pm 3\,\text{mT}$, depending on the degree of PL polarization, the excitation light intensity and the QD that was studied. The measured values indicate a time-averaged electron spin polarization between 3% and 30%. A fully polarized electron spin would have given rise to a Knight field on the order of 10–20 mT (cf. Sect. 5.3.3). Estimating the maximal value of B_{el} in a self assembled QD is difficult due to large uncertainties of the exact confinement length-scale and in the composition of the QD.

The principal reason for the incomplete reduction in DNSP at $B_{ext} = -B_{el}$ is the inhomogeneity of the Knight field. Since B_{ext} is homogeneous, the condition $B = 0$ is satisfied only for a small class of nuclei at any given B_{ext}. The rest of the nuclei still experience a sizable total magnetic field and as a result, the Overhauser field is only slightly modified when $B_{ext} = -B_{el}$. To demonstrate the role of the inhomogeneous nature of the Knight field, we extended (5.13) to account for the inhomogeneity: We assume an in-plane Gaussian electron wavefunction $|\psi(\mathbf{R}_i)|^2 \propto \exp[-(x_i^2 + y_i^2)/l^2]$ which we convolve with (5.13) to estimate the total contribution of the different classes of QD nuclei. The choice of a maximum Knight field of 1.5 mT in the center of the dot, $B_{loc} = 0.11$ mT and $l = 20$ nm gives a reasonable description of the experimental data (solid curves in Fig. 5.3a), even though we only assumed a single nuclear species.

Remarkably, a minimum in the degree of PL polarization is also observed for the same B_{ext} where ΔE_{OS} has a minimum (Fig. 5.3b): this is at first surprising since polarization of the X^{-1} trion line is solely determined by the hole-spin which has a negligible coupling to the nuclear spins. A possible explanation is based on AEI: after the resonant excitation of the QD, the electron excited into a p-shell state is expected to tunnel out into the n-doped GaAs layer in $\lesssim 10$ ps [26]. After tunneling, the QD is neutral and the remaining electron–hole pair is subject to AEI which rotates the electron–hole spin [44]. This coherent rotation is then interrupted by reinjection of another electron from the n-doped GaAs layer into the QD s-shell to form a ground-state electron-singlet in $\tau_{el} \approx 20$ ps, as required by the charging condition. Because tunneling is a random process, the time the QD spends in the neutral state is random and the posttunneling hole-spin state is partially randomized, which leads to a finite ρ_c.

The Overhauser-field competes with the exchange interaction; a reduction in DNSP will therefore lead to a reduction in ρ_c as depicted in Fig. 5.3b. The PL polarization in the presence of an Overhauser shift ΔE_{OS} and an exchange-splitting ΔE_{ex} due to AEI can be approximated as [45]:

$$\rho_c = \frac{1 + \Delta E_{OS}^2 \tau_{el}^2/\hbar^2}{1 + (\Delta E_{OS}^2 + \Delta E_{ex}^2)\tau_{el}^2/\hbar^2}, \quad (5.14)$$

provided other spin relaxation processes are neglected. Fitting the polarization $\rho_c(X^{-1})$ with the measured spin splitting in Fig. 5.3a, $\tau_{el} = 30$ ps is obtained.[7]

---

[7] As shown in Fig. 5.1b, the PL from this QD exhibits an asymmetric polarization under $\sigma^{\pm}$ pumping: the origin of this asymmetry is not clear and was not observed on all QDs studied.

Below saturation, a reduction in the excitation power results in a decrease in both spin-splitting and ρ_c: this observation corroborates the model described by (5.14).

The electron (spin) exchange with the n-doped GaAs layer also explains how QD electron-spin pumping is achieved in a negatively charged QD: irrespective of the preexcitation electron state, the sequential tunneling ensures that the QD ends up in a trion state where the electrons form an s-shell singlet. Preservation of hole-spin in these QDs then implies that the post-recombination electron is always projected into the same spin-state.

5.5 Bistability of the Electron Nuclear Spin System

DNSP behaves qualitatively different in low and high external magnetic fields. While the low field case was studied in the previous section, we now study a situation where the coupled electron–nuclear spin system is exposed to a magnetic field which is on the order of the Overhauser field B_{nuc}. There, the external magnetic field can fully compensate the nuclear field B_{nuc}, thereby greatly enhancing the electron nuclear spin coupling which depends on the effective Zeeman splitting of the electron spin.

In external magnetic fields, nuclear spin polarization manifests itself in a difference in emission energies between excitation with circularly and linearly polarized light as was established in Sect. 5.3.3. Figure 5.4 shows the X^{-1} emission energies of a single QD under excitation with circularly polarized light as a function of external magnetic field. Gray (black) denotes excitation with σ^+ (σ^-) light, while squares (triangles) stand for co- (cross-) circular detection. The polarizations for excitation and detection are denoted as $(\sigma^\alpha, \sigma^\beta)$ where σ^α and σ^β correspond to excitation and detection, respectively. The data shown in Fig. 5.4 was obtained in a single sweep from $B = -2\,\text{T}$ to $B = +2\,\text{T}$, varying excitation and detection polarization for each B-field value in the order $(\sigma^+, \sigma^-) \Rightarrow (\sigma^+, \sigma^+) \Rightarrow (\sigma^-, \sigma^+) \Rightarrow (\sigma^-, \sigma^-)$ such that any memory of the nuclear spin system is erased during the sweep. The data for $|B| < 500\,\text{mT}$ was taken with smaller magnetic field steps in order to highlight the detailed behavior of DNSP at low fields. Every data point represents the center of mass of the emission peak of X^{-1} taken from a single spectrum with 1 s integration time and a signal to noise ratio (SNR) of $\sim$100:1 for co-circular detection (Fig. 5.4b). The effects of nuclear polarization can be seen in the range of $|B_{\text{ext}}| \lesssim 1.2\,\text{T}$ where emission energies for a given detection polarization depend strongly on the helicity of the laser light. Excitation with σ^+ light creates a residual electron with its spin pointing in the positive z-direction (see Fig. 5.4a). According to (5.5) and (5.8), this creates a nuclear spin polarization in the same direction and, due to the negative sign of the g^*_{el}, a nuclear field pointing in the negative z-direction. This scenario is consistent with the polarization sequences and lineshifts observed in Fig. 5.4c. Above $1.2\,\text{T}$, the emission energies of the QD are almost independent of excitation

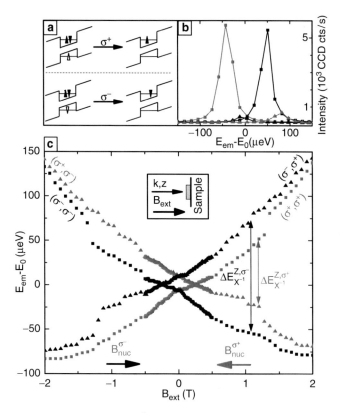

Fig. 5.4. (a) Configurations of X^{-1} before and after the emission of a $\sigma^\pm$ polarized photon. *Open (filled) triangles* denote the spin of the hole (electron). (b) Raw spectra at $B_{\text{ext}} = -0.96$ T for the four excitation/detection configurations in the circular basis: *gray (black)* denotes excitation with σ^+ (σ^-) polarized light. Detection is co- or cross-circular (*squares* and *triangles*, respectively). (c) Energy dispersion of X^{-1} under circularly polarized excitation: The emission energies (E_{em}) are different for σ^+- and σ^--polarized excitation due to DNSP and the resulting effective nuclear magnetic field $B_{\text{nuc}}^{\sigma\pm}$ under $\sigma^\pm$ excitation (orientation indicated by the arrows in the figure). An energy $E_0 \approx 1.3155$ eV was subtracted from E_{em}. The *inset* shows the relative orientation of **k**-vector, quantization axis z and positive magnetic field

light polarization, indicating that nuclear effects become very small. Another striking feature in this data is the symmetry under simultaneous reversal of the excitation light helicity and the sign of the magnetic field. However, the data is not symmetric under the reversal of only one of these parameters. This asymmetry indicates that the system distinguishes between nuclear fields pointing along or against the external magnetic field – we will see in the following that it is more efficient for the system to create a nuclear field pointing against B_{ext} than one that points along this field.

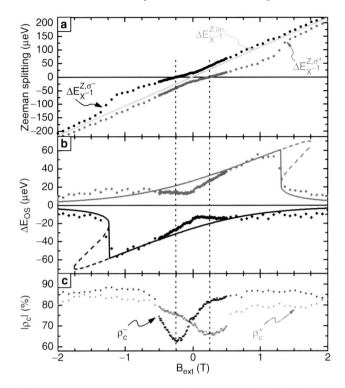

Fig. 5.5. Nuclear Polarization in external magnetic fields: (**a**) Spin splitting of X^{-1} under circularly polarized excitation. *Gray and black symbols* correspond to excitation with σ^+- and σ^--polarized light, respectively. The *solid line* is a linear fit to the data as described in the text. (**b**) Deviation of the spin splitting between circular and linear excitation: Overhauser shift for σ^+- and σ^--excitation (*gray and black diamonds*, respectively). The *solid and dashed lines* are the results of the fits according to the model discussed in the text. (**c**) QD spin polarization $\rho_c^{\pm}$ under $\sigma^{\pm}$ excitation extracted from PL intensities as described in the text. The polarization shows a minimum at the magnetic field where the exciton Zeeman splitting is zero, consistent with our model of carrier relaxation (Sect. 5.4)

In order to obtain a more quantitative picture of the magnetic field dependent DNSP, we performed the following analysis steps on the data (see Fig. 5.5): We first extract the Zeeman splittings for excitation with σ^+ and σ^- light from the raw data shown in Fig. 5.4c. To this data, we fit a linear Zeeman splitting such that the fit coincides with the data at magnetic fields $B_{\text{ext}} > 1.8\,\text{T}$ where nuclear polarization is very small (Fig. 5.5a). The excitonic g-factor, $g_{\text{ex}}^* \equiv g_{\text{el}}^* + g_{\text{h}}^* = -1.87$ that we find with this fitting procedure matches within a few percent to an independent measurement of g_{ex}^* that we performed with linearly polarized excitation (not shown here). The Overhauser shift can now be extracted from this fit with the help of (5.11); the result is plotted in Fig. 5.5b. There is a striking difference when polarizing the nuclei along or against the external field. Nuclear polarization with B_{nuc} pointing along the applied field is rather inefficient and shows a slight decrease with increasing magnitude of the applied field. Polarization with B_{nuc} pointing against the external magnetic field on the other hand shows a much richer behavior: The nuclear polarization first increases almost linearly as the magnitude of the external field increases and then shows a sudden drop when $|B| \approx 1.2\,\text{T}$. At $B = 0$, DNSP abruptly changes its dependance on magnetic field and $dB_{\text{nuc}}/dB_{\text{ext}}$ shows a sudden jump.

From the spectral data we can also extract information about the hole spin polarization before- and the residual electron spin polarization after recombination of X^{-1}. For this, we define a degree of QD spin polarization as $\rho_c^\pm \equiv (I_{(\sigma^\pm,\sigma^+)} - I_{(\sigma^\pm,\sigma^-)})/(I_{(\sigma^\pm,\sigma^+)} + I_{(\sigma^\pm,\sigma^-)})$ under $\sigma^\pm$ excitation. $I_{(\sigma^\alpha,\sigma^\beta)}$ are the intensities of the dominant PL-peaks in the corresponding analyzer/polarizer configurations. At $B_{\text{ext}} = 0$, $\rho_c^\pm$ is equivalent to the PL circular polarization. At finite B_{ext} however, $\rho_c^\pm$ measures the relative weight of the two (perfectly circularly polarized) Zeeman split emission lines and is therefore a measure of hole spin preservation during relaxation from the p- to the s-shell. The measured quantity $\rho_c^\pm$ is plotted in Fig. 5.5c as a function of external magnetic field. It is roughly constant and on the order of 85% over a wide range of magnetic fields. Only for the fields where the trion Zeeman splitting vanishes due to a cancelation of the exciton Zeeman splitting with the Overhauser field, $\rho_c^\pm$ dips to roughly 65%. This behavior is consistent with the rotation of the exciton spin during relaxation of the optically created electron from the excited p-shell state to the s-shell via the electron reservoir (see Sect. 5.4).

As in the data presented in Sect. 5.4, there is a certain asymmetry in the data shown in Fig. 5.5c that remains unexplained: ρ_c^- is larger than ρ_c^+ at high magnetic fields and the dip in ρ_c^+ at lower fields is less pronounced than for ρ_c^-. A possible reason for this asymmetry could be the different excitation efficiencies in the QD for σ^+ and σ^- excitation.

5.5.1 Modeling of the Data

Most of the above-mentioned nuclear effects in the presence of an external magnetic field can be described by a simple rate equation model proposed earlier [13, 32, 42] and originally based on work by Abragam and D'yakonov [27, 46]. The rate equation is based on the condition for dynamical equilibrium (5.6) between the electron and the nuclear spin system in the absence of any coupling to the environment. This equilibrium is reached on a typical timescale given by the nuclear spin relaxation time T_{1e}, which can be estimated to be [35]

$$\frac{1}{T_{1e}} = \frac{1}{T_{1e}^0} \frac{1}{1 + \Omega_{\text{el}}^2 \tau_{\text{el}}^2}. \tag{5.15}$$

Here, τ_{el} is the electron spin correlation time which broadens the electronic spin states. $\Omega_{\text{el}} = \Delta E_{\text{el}}^Z/\hbar$ is the electron Larmor frequency which depends on the degree of nuclear polarization through (5.8) and (5.9). For a given nuclear species, the nuclear spin relaxation time at zero electron Zeeman splitting is given by $1/T_{1e}^0 = f_{el}\tau_{el}(A_i/N\hbar)^2$ with N the number of relevant nuclei and f_{el} the fraction of time the QD is occupied with a single electron. The quantity $A_i/N\hbar$ corresponds to the precession frequency of a nuclear spin in the Knight field of the QD electron. The expression for T_{1e}^0 is valid for $A_i/N\hbar \ll \tau_{el}^{-1}$ and if we assume a homogenous electron wave function $\psi(\mathbf{r}) \propto \sqrt{8/v_0 N}$ which is constant within the QD volume and zero outside.

By adding a nuclear spin decay channel which is dominated by nuclear spin diffusion out of the QD on a timescale T_d, we end up with a rate equation of the form

$$\frac{d\langle I_z^i \rangle}{dt} = -\frac{1}{T_{1e}}(\langle I_z^i \rangle - I^i B_{I^i}(x)) - \frac{1}{T_d}\langle I_z^i \rangle, \tag{5.16}$$

$$\simeq -\frac{1}{T_{1e}}(\langle I_z^i \rangle - I^i) - \frac{1}{T_d}\langle I_z^i \rangle. \tag{5.17}$$

Due to the high degree of electron spin polarization deduced from the high value $\rho_c^\pm$, we expanded (5.6) to first order around $S_z = 1/2$ to arrive at (5.17).

This equation was obtained for the coupling of a single electron to a single nuclear spin. It can be approximately generalized to the case of an ensemble of different nuclei in the QD by considering the mean nuclear spin polarization $\langle I_z \rangle = \frac{1}{N}\sum_i \langle I_z^i \rangle$. For this, we replace the hyperfine constant A_i in (5.15) by the sum A, and the nuclear spin I^i in (5.17) by the average $\overline{I^i}$ over the two dominant nuclear spin species. With the values for A_i and I^i noted in Sect. 5.3.3, this results in $A = A_\text{In} + A_\text{As} = 102\,\mu\text{eV}$ and $\overline{I^i} = \frac{6}{2}$. We take these numbers to be fixed in the following, even though the In content varies drastically within a QD and accurate estimates of $\overline{A_i}$ and $\overline{I^i}$ are difficult to obtain.

We note that this model was previously applied to situations where DNSP was induced by neutral excitons [32,42]. There, anisotropic exchange interaction plays a crucial role and has to be included in the detuning factor in (5.15). In this work however, DNSP is induced by a single, spin-polarized electron for which exchange interaction plays no role.[8]

Since the electron-mediated nuclear spin relaxation time T_{1e} itself depends on nuclear spin polarization, (5.17) leads to the following self-consistent nonlinear steady state solution $\langle I_z^\text{ss} \rangle$ for the mean nuclear spin polarization:

$$\langle I_z^\text{ss} \rangle = \frac{\overline{I^i}}{1 + \frac{T_{1e}^0}{T_d}(1 + (\frac{\tau_{el}}{\hbar})^2(g_{el}^*\mu_B B_\text{ext} + A\langle I_z^\text{ss} \rangle)^2)}. \tag{5.18}$$

---

[8] We can rule out the possibility that DNSP is induced by the intermediate state where the QD is neutral during electron relaxation through the reservoir. In this case the direction of the nuclear field would be opposite to the one observed in the experiment.

We note that averaging (5.17) over the different nuclear species, our choice of a homogenous electron wavefunction and the fact that we neglected the magnetic field dependence of T_d can all limit the validity of this model.

In order to fit our experimental data, we numerically solved the implicit equation (5.18). The result of such a fit is shown in Fig. 5.5b. The model qualitatively reproduces the data. Still, some features, like the before-mentioned kink of DNSP around zero external field as well as the high residual spin polarization at high external magnetic fields, could not be explained within the model or any reasonable extension [42] to the rate equation picture employed here. In the region $1.2\,\text{T} < B_{\text{ext}} < 1.8\,\text{T}$ the model predicts three solutions: two stable states, one with a low and one with a high degree of DNSP and an unstable solution of intermediate nuclear spin polarization (the last two solutions correspond to the dashed lines in Fig. 5.5). Since in this experiment we changed excitation polarization from σ^+ to σ^- for each magnetic field value, the system always followed the solution with minimal nuclear spin polarization. The fact that the drop in DNSP in this measurement was rather smooth compared to the model prediction was probably due to the long timescale of the buildup of DNSP right before its disappearance: since in the experiment every point was taken with an integration time of 1 s, the nuclear system did not have time to reach its steady state polarization before the excitation light polarization was switched. The parameters used for the fitting curve in Fig. 5.5b were $T_{1e}^0/T_d = 3.4$, $\tau_{\text{el}} = 35\,\text{ps}$, $g_{\text{el}}^* = -0.69$, which are all realistic values for our QD.

The electron spin correlation time found in the fit can be explained with the resonant excitation scheme that we used in our experiments. The QD is excited from its ground state into the p-shell and after relaxation through an intermediate neutral excitonic state PL emission is observed from carriers recombining from the s-shell. Since this system is pumped close to saturation, the lifetime and thus the correlation time of the residual electron are limited by the relaxation time from the p-shell to the n^{++}-GaAs layer by tunneling. This timescale is expected to be on the order of 10 ps and the value of 35 ps we find here is in good agreement with the value found independently in Sect. 5.4.

The parameters obtained in this fit also allow us to estimate the nuclear spin relaxation time T_{1e}^0. Using the value $\tau_{\text{el}} = 35\,\text{ps}$, the corresponding value for $f_{\text{el}} = 0.035$ (assuming an exciton lifetime of 1 ns) and $N = 10^4$–10^5, we obtain $T_{1e}^0 = 1$–$100\,\text{ms}$.

5.5.2 Hysteresis in the Magnetic Field Sweeps

In this section, we focus on the bistable behavior of the coupled electron–nuclear spin system in the magnetic field range close to the "breakdown" of DNSP. Figure 5.6a shows a graphical representation of the solutions of the nonlinear equation (5.18). The result suggests that the maximal achievable degree of DNSP in our system leads to a maximal OS given by $\text{OS}_{\text{max}} = A\overline{I^i}(1 + T_{1e}^0/T_d)^{-1}$. This value is reached when nuclear spin relaxation is maximized, i.e., when the total electron Zeeman splitting is zero

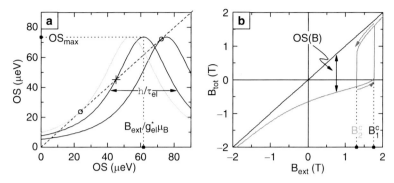

Fig. 5.6. (a) Graphical solution of (5.18): The *right* (*left*) *hand side* is represented by the *solid curves* (*dashed line*). These terms correspond to gain and loss of DNSP, respectively. *Circles* (*cross*) indicate the stable (unstable) solutions for nuclear spin polarization. The center of the Lorentzian shifts proportionally to the external magnetic field, explaining the magnetic field dependence of DNSP. The *dark* (*light*) *gray curve* show the situation at the critical field B_1^c (B_2^c). The figure illustrates that: (1) Bistability can only be observed if the maximal slope of the Lorentzian is bigger than 1 and (2) the difference between B_1^c and B_2^c is on the order of the width of the electron spin states in units of magnetic fields $\hbar/\tau_{\text{el}}g_{\text{el}}^*\mu_{\text{B}}$. (b) Total magnetic field B_{tot} seen by the QD electron under optical orientation of nuclear spins with σ^+-polarized light (*gray curve*). The curve is calculated from (5.17) with the parameters found from the fit presented in Fig. 5.5. The model shows that when the nuclear magnetic field B_{nuc} opposes the external field (i.e., for $0 < B_{\text{ext}} < 1.74\,\text{T}$), the nuclei overcompensate B_{ext} and the electron sees a total magnetic field $B_{\text{tot}} < 0$. When the nuclear field saturates to its highest value, B_{tot} is very close to zero and the nuclear spin polarization becomes unstable

(cf. (5.15)). It can also be seen from the figure that there is a regime of external magnetic fields where two stable solutions for DNSP coexist. One solution leads to a high degree of nuclear polarization, reaching OS_{max} at its maximum, while the other one shows a low degree of nuclear polarization. The graphical solution also shows that bistability is an inherent property of the solutions of (5.17) for systems where OS_{max} is at least on the order of the width of the density of states of the electronic spins ($\hbar/\tau_{\text{el}}$), which is typically the case for localized carriers such as in QDs, but not for bulk systems. The two stable solutions can be understood as follows: When increasing an external field while creating a nuclear field in the opposite direction, the electron Zeeman splitting is reduced compared to the case where nuclear spin polarization is absent. Therefore, the nuclear spin relaxation rate T_{1e}^{-1} remains at a high value such that DNSP can be maintained. As soon as OS_{max} is reached, however, the system can no further compensate for an increasing external magnetic field. DNSP will start to drop, which eventually leads to an abrupt jump of DNSP to a low value at an external field B_1^c. This jump is due to the negative feedback of DNSP on T_{1e}^{-1}. When ramping the external field down again, now in the absence of nuclear polarization, the system will initially remain in a state of low DNSP since T_{1e}^{-1} is still low. At the same time, DNSP will slightly increase due to the increasing rate T_{1e}^{-1} of nuclear polarization with decreasing magnetic field strength. At a field B_2^c, the positive feedback of increasing DNSP on T_{1e}^{-1} will take over and an abrupt jump to a state of high nuclear polarization will occur. As can be seen from Fig. 5.6a, the difference between the fields B_1^c and B_2^c is on the order of the width of the electronic spin states in units of magnetic fields.

In order to observe the hysteretic behavior of DNSP we performed a magnetic field dependent PL experiment as described above, now by exciting the QD with light of constant helicity and by ramping the magnetic field from low to high values and back again. Hysteretic behavior can be expected if the nuclear fields created in that way are pointing against the external magnetic field. In our system such a situation is realized when exciting the QD with σ^+ light and applying an external field in the positive z-direction.

Figure 5.7 shows data obtained in this regime: Going from low to high field strength, DNSP is significant up to a magnetic field of $B_1^c = 1.74\,\mathrm{T}$, where

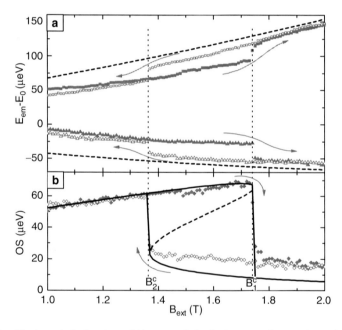

Fig. 5.7. Hysteresis behavior of the coupled electron–nuclear spin-system: Magnetic field sweeps under excitation with constant light polarization (σ^+). (**a**) Energy dispersion of X^{-1}, sweeping magnetic field up or down (as indicated by *arrows*). *Squares* (*triangles*) denote co- (cross-)circular detection with respect to excitation polarization. The *dashed line* is a fit to the case of linearly polarized excitation. (**b**) Overhauser shifts extracted from the data shown in (**a**) for the magnetic field sweeping up and down (*solid and open diamonds*, respectively). The *black line* shows the simulations described in the text

it suddenly drops. Sweeping the magnetic field back to low field amplitudes, DNSP reappears at a field $B_2^c = 1.36\,\text{T}$, a value different from B_1^c. The difference of $380\,\text{mT}$ between these two field is on the order of $\hbar/(\tau_{\text{el}} g_{\text{el}}^* \mu_\text{B})$ as predicted by the model.

A fit of (5.18) to the data is also shown in Fig. 5.7. The parameters used for this fit were $T_{1e}^0/T_\text{d} = 3.5$, $\tau_{\text{el}} = 33\,\text{ps}$, $g_{\text{el}}^* = -0.69$, consistent with the parameters used in the fit shown in Fig. 5.5. As in the previous fit, the residual nuclear polarization at high fields observed in this experiment is slightly higher than what is predicted by the model.

5.5.3 Discussion

We note that at the point of maximal Overhauser shift ($B_\text{ext} = B_1^c = 1.74\,\text{T}$), our model predicts that B_nuc is almost completely canceled by B_ext (see Fig. 5.6b). This point is therefore of particular interest because it enables a direct measure of the maximal nuclear field $B_\text{nuc}^\text{max} = \text{OS}_\text{max}/g\mu_\text{B} = B_1 = 1.74\,\text{T}$. Remarkably, the remaining exciton Zeeman splitting at this point is solely due to the Zeeman interaction of the hole with B_ext. We can therefore directly obtain the hole g-factor and find $g_h^* = -1.2$. This observation has found applications in the precise and systematic study of electron and hole g-factors in semiconductor QDs [47].

Our experiment along with the model also shows that the maximal nuclear polarization of $\sim$47% achieved in our system is limited by the fraction T_d/T_{1e}^0, i.e., the ratio between nuclear spin decay time and electron mediated nuclear spin relaxation time.[9] While T_d is a parameter given by the nature of the QD, T_{1e}^0 could potentially be modified by varying the pump power or the details of the excitation process [33].

We extended the rate equation (5.16) and included the dynamics of the mean electron spin $\langle S_z \rangle$ by expanding (5.6) to first order in $\langle S_z \rangle$. The evolution of $\langle S_z \rangle$ is described by a rate equation similar to (5.17). The main differences between the electron and the nuclear spin dynamics are that the electron spin system, in the absence of other relaxation mechanisms, reaches the thermal equilibrium state (5.6) at a rate N/T_{1e}. Compared to the nuclear spin relaxation rate, the electron spin relaxation is faster by the number of nuclei N in the system. In addition, the electron spin is repumped into its initial state $S_z^0 = \rho_c^\pm/2$ at the exciton recombination rate on the order of 1 ns. This extension, however, did not lead to any new insights on the behavior of the nuclear spin system. A numerical study of this extended model suggested though that the mean value of the electron spin decreases linearly with increasing nuclear spin polarization. The electron spin thus seems to follow the intricate dynamics of the nuclear spin system. This observation motivates further studies on the positively charged exciton where PL light polarization gives a direct measure of the mean electron spin [48].

---

[9] We estimate the degree of polarization by comparing $\text{OS}_\text{max} = 70\,\mu\text{eV}$ to the OS corresponding to full nuclear spin polarization, which was given in Sect. 5.3.3.

The qualitative disagreement of the model with our data in the low field regime where the measured DNSP shows a clear "kink" as a function of magnetic field indicates that our simple approach does not give a full description of the nonlinear processes that lead to an equilibrium value of DNSP in a QD. A further extension of the model could include light-induced nuclear spin relaxation due to the nuclear quadrupolar interaction [49] which could induce an additional loss of nuclear spin polarization at low external magnetic fields. Another possible nuclear depolarization mechanism relevant at low fields is the coupling of the nuclear Zeeman reservoir to the nuclear dipolar reservoir [28]. Since the heat conductivity for dipolar spin temperature is larger than for Zeeman spin temperature [50], the rate of nuclear spin depolarization in the QD will increase as soon as the two reservoirs couple. This coupling of nuclear Zeeman and dipolar reservoirs might therefore also explain the observed "kink" of DNSP at low fields. While our rate equation approach was purely classical, it could also be conceived that the quantum mechanical nature of the electron spin system would alter the behavior of DNSP at low fields and explain the unpredicted features in our measurement. In order to confirm this hypothesis further theoretical and experimental studies are required.

5.6 Buildup and Decay of Nuclear Spin Polarization

A key ingredient for the understanding of the coupled electron–nuclear spin system is the knowledge of the relevant timescales of the dynamics of nuclear spin polarization. This has already become apparent in Sect. 5.5, where (5.18) shows that the maximal nuclear spin polarization in a QD is limited by the ration of buildup and decay times of the nuclei. Many other aspects like the respective roles of nuclear spin diffusion, quadrupolar relaxation and trapped excess QD charges influence the dynamics of DNSP and remain essentially unexplored up to now. While the buildup time of DNSP (τ_{buildup}) is likely to depend on the way the nuclear spin system is addressed, the DNSP decay time (τ_{decay}) is an inherent property of the isolated nuclear spin system of a QD. Furthermore, experimental determination of τ_{decay}, which directly yields the correlation time of the fluctuations of the Overhauser field along the axis in which the nuclei are polarized, can be crucial for understanding the limits of electron spin coherence in QDs [17].

In order to study the buildup and decay of DNSP, we extended our standard µPL setup (Sect. 5.2) by the ability to perform "pump-probe" measurements. An acousto-optical modulator (AOM) served as a fast switch of excitation light intensity, producing light pulses of variable lengths, with rise- and fall-times $\ll 1\,\mu$s. We differentiate between "pump" pulses of duration τ_{pump}, used to polarize the nuclear spins, followed by "probe" pulses of length τ_{probe}, used to measure the resulting degree of DNSP. The intensity of each pulse corresponds to the saturation intensity of the observed emission line, maximizing both the resulting OS and the SNR of the measurement. A mechanical shutter placed in the PL collection path is used to block the pump pulses while allowing the probe pulses to reach the spectrometer. Pump and probe pulses are separated by a waiting time τ_{wait} with a minimal length of 0.5 ms, limited by the jitter of the shutter opening time. In order to measure the buildup (decay) time of DNSP, τ_{pump} (τ_{wait}) are varied, respectively, while keeping all other parameters fixed. The timing and synchronization of the pulses is computer controlled via a digital acquisition card operating at a clock period of 2 μs, which sets the time resolution of the pulse sequences. Individual pump-probe sequences are repeated while the signal is accumulated on the spectrometer CCD in order to obtain a reasonable SNR. We verify a posteriori that individual pump-probe pairs are separated by much more than the measured DNSP decay time.

Figures 5.8b,c show the results for buildup and decay curves of DNSP obtained with this technique. The resulting curves fit surprisingly well to a

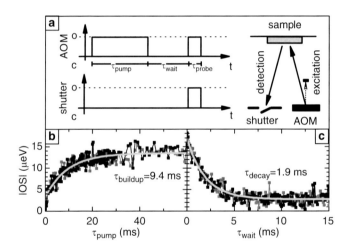

Fig. 5.8. (a) Schematic of the pulse sequences used in the buildup and decay time measurements of DNSP. An acousto optical modulator (AOM) deflects the excitation beam on and off the sample, serving as a fast switch (o (c) denote the open (closed) state, respectively). The AOM creates pump (probe) pulses of respective lengths τ_{pump} (τ_{probe}), separated by a waiting time τ_{wait}. A mechanical shutter blocks the pump pulse from reaching the spectrometer, while letting the probe pulse pass. (b) DNSP buildup curves obtained by varying τ_{pump} at fixed τ_{wait} (0.5 ms) and τ_{probe} (0.2 ms). The *gray* (*black*) data points correspond to QD excitation with light of positive (negative) helicity. The *light gray line* is an exponential fit, yielding a buildup time of $\tau_{\text{buildup}} = 9.4$ ms. (c) DNSP decay curves obtained by varying τ_{wait} at fixed τ_{pump} (50 ms) and τ_{probe} (0.5 ms). The color coding is identical to (a). The exponential fit reveals a decay time of $\tau_{\text{decay}} = 1.9$ ms

simple exponential, yielding $\tau_{\text{buildup}} = 9.4\,\text{ms}$ and $\tau_{\text{decay}} = 1.9\,\text{ms}$.[10] The small residual OS observed for $\tau_{\text{pump}} = 0$ ($\tau_{\text{wait}} \gg \tau_{\text{decay}}$) in the buildup (decay) time measurement is due to the nuclear polarization created by the probe pulse. Comparing our experimental findings to previous experiments is not straightforward, since, to the best of our knowledge, the dynamics of DNSP at zero external magnetic field has not been studied up to now. However, in experiments performed at external magnetic fields of $\sim 1\,\text{T}$, the buildup time of DNSP in QDs was estimated to be on the order of a few seconds [19, 42]. A further shortening of τ_{buildup} could arise from the strong localization of carriers in our QDs, which has been shown to be an important ingredient for efficient nuclear spin polarization [31]. Most strikingly, previous experimental results in similar systems revealed DNSP decay times on the order of minutes [30]. It is thus at first sight surprising that we find a DNSP decay time as short as a few milliseconds.

5.6.1 Electron Mediated Nuclear Spin Decay

A possible cause for the fast decay of DNSP is the presence of the residual QD electron even in the absence of optical pumping. We study its influence on τ_{decay} with the following experiment: While the nuclear spin polarization is left to decay, we apply a voltage pulse to the QD gate electrodes, ejecting the residual electron from the QD into the nearby electron reservoir. This is achieved by switching the QD gate voltage to a value where the dominant spectral feature observed in PL stems from the recombination of the neutral exciton. Using transient voltage pulses, we are able to perform this "gate voltage switching" on a timescale of $30\,\mu\text{s}$. Before sending the probe pulse onto the QD, the gate voltage is switched back to its initial value in order to collect PL from X^{-1} recombination. The dramatic effect of this gate voltage pulsing on DNSP lifetime is shown in Fig. 5.9b. On the timescale of the previous measurements, almost no DNSP decay can be observed. By prolonging τ_{wait} up to a few seconds (Fig. 5.9c), we estimate the spin decay time of the unperturbed nuclear system to be $\tau_{\text{decay}} \approx 2.3\,\text{s}$. We note that the increase of τ_{wait} necessary for this experiment results in a reduced SNR, which makes an exact determination of τ_{decay} difficult.

The role of the residual electron in depolarizing the nuclear spins was further confirmed in two independent measurements (not shown here). First, we perform a modified version of the gate voltage switching experiment: During the interval τ_{wait}, the gate voltage is switched to a regime where the QD ground state consists of two electrons in a spin singlet state [51]. This state doesn't couple to the nuclear spins and the measured τ_{decay} is again on the

---

[10] The rate equation model presented in Sect. 5.5 predicts deviations from an exponential dependance due to the feedback of DNSP on the nuclear spin cooling rate. However, the limited SNR of our experiment and the finite length of the probe pulses do not allow us to observe these deviations at zero magnetic field.

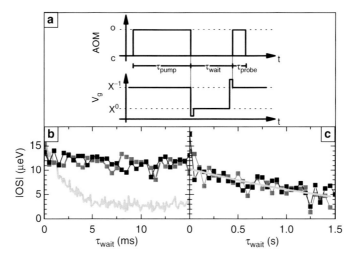

Fig. 5.9. (**a**) Timing diagram for the gate voltage switching experiment: during the period τ_{wait}, the QD gate voltage is switched to a value where the neutral exciton is the stable QD charge complex. Using transient pulses, the switching time is 30 μs. Ejecting the residual QD electron removes its effect on DNSP depolarization. This is demonstrated in (**b**), which shows DNSP decay measurements in the absence of the residual QD electron. The *gray (black) data points* represent DNSP decay under σ^+ (σ^-) excitation. For comparison, the *light gray curve* shows the mean of the data presented in Fig. 5.8b. (**c**) Same measurement as in (**b**), but over a longer timescale. The exponential fit (*light gray*) indicates a decay time constant of $\tau_{\text{decay}} \approx 2.3\,\text{s}$.

order of seconds. The second control experiment consists in measuring DNSP dynamics at a constant gate voltage where X^{+1} is the stable QD charge complex. As we showed in Sect. 5.4, optically pumping the X^{+1} exciton can also lead to DNSP. However, in this case, the optically created electron polarizes the nuclear spins and no electron is left in the QD after exciton recombination. The corresponding DNSP decay channel is therefore not present. As expected, τ_{decay} is also on the order of seconds for this case. We argue that two mechanisms could lead to the efficient decay of DNSP due to the residual electron. The first mechanism is caused by the randomization of the residual QD electron spin through cotunneling to the close-by electron reservoir. Cotunneling happens on a timescale of $\tau_{\text{cot}} \approx 3\,\text{ns}$ for the structure studied in this work [26]. The resulting electron spin depolarization is mapped onto the nuclear spin system via hyperfine flip-flop events. Taking into account the detuning ΔE_{el}^Z of the two electron spin levels and using (5.15), the nuclear spin depolarization rate can be estimated to be $T_{1e}^{-1} \simeq (A/N\Delta E_{\text{el}}^Z)^2/\tau_{\text{cot}}$ [35]. In order to get a rough estimate of the resulting timescale, we take ΔE_{el}^Z to be constant and equal to half the maximum measured OS. With these values, we obtain a nuclear spin depolarization time on the order of 100 ms, roughly consistent with our measurement.

A second possible mechanism is the indirect coupling of nuclear spins due the presence of a QD (conduction band) electron [27]. While this process conserves the total angular momentum of the nuclear spin system, it can lead to a decay of the OS by redistributing the nuclear spin polarization within the QD and by increasing the nuclear spin diffusion rate out of the QD. The resulting decay rate for the nuclear field has been estimated to be on the order of $T_{\text{ind}}^{-1} \simeq A^2/N^{3/2}\Delta E_{\text{el}}^Z$ as discussed in Sect. 5.3.4. The corresponding estimate for τ_{decay} of $\sim$1 ms was obtained in the limit of $\Delta E_{\text{el}}^Z \gg A$, which is the only regime where theoretical predictions for the timescale of indirect nuclear spin interactions in QDs are available.

Our study of DNSP timescales was complemented by adding a permanent magnet to our sample. The resulting magnetic field is antiparallel to the excitation beam direction and has a magnitude of $B_{\text{ext}} = -220$ mT at the site of the QD. The buildup and decay time measurements in the presence of B_{ext} are shown in Fig. 5.10. In accordance with the discussion of Sect. 5.5, an asymmetry between the cases of σ^+ and σ^- excitation is observed. The situation where B_{nuc} opposes B_{ext} is more efficient than the one where B_{nuc} aligns with B_{ext}; equilibrium is therefore reached faster and at a higher nuclear spin polarization in the first case, which corresponds to σ^--excitation for the present magnetic field direction. The measurements presented in Fig. 5.10a,b confirm this picture: we find that τ_{buildup} and τ_{decay} are both increased by a factor of $\sim$2–3 when the polarization of the excitation light is changed from σ^- to σ^+.

We again performed the "gate voltage switching" experiment in the presence of B_{ext} (Fig. 5.10c). Since in this case DNSP decay is not mediated by the residual QD electron, no dependance of τ_{decay} on excitation light helicity was found and only the average between the two data sets (σ^+ and σ^- excitation) is shown. Compared to the case of zero external magnetic field, the decay of nuclear spin polarization is further suppressed. Even though extracting exact numbers is difficult in this case due to the required long waiting times, we estimate τ_{decay} to be on the order of a minute. This further suppression of DNSP decay rate can be induced with a magnetic field as small as $\sim$1 mT as shown in the inset of Fig. 5.10c: Keeping $\tau_{\text{wait}} = 1$ s fixed, we sweep an external magnetic field while measuring the remaining OS. The resulting dip around $B_{\text{ext}} = 0$ has a half-width of $\sim$1 mT. This indicates that nuclear spin depolarization at zero magnetic field is governed by the nonsecular terms of the nuclear dipole–dipole interactions (5.4) which can be suppressed by applying an external magnetic field that exceeds the local dipolar field $B_{\text{loc}} \approx 0.1$ mT [35]. The exact nature of this zero field decay of DNSP, however, is still unclear since nuclear dipole interactions should depolarize the nuclear spins in a much shorter time on the order of $T_2 \approx 10$–$100\,\mu$s. We suggest that the interplay of dipolar interactions and quadrupolar shifts could explain this experimental results at low magnetic fields.

We also investigated the possible role of nuclear spin diffusion and the resulting DNSP of the bulk nuclei surrounding the QD. For this purpose, we studied the dependance of τ_{decay} on the nuclear spin pumping time τ_{pump} for

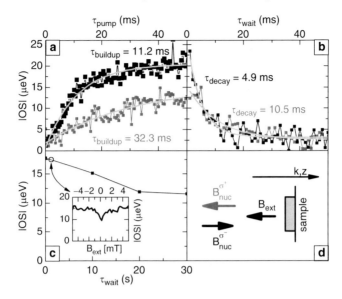

Fig. 5.10. Measurements of buildup and decay of DNSP in an external magnetic field $B_{\mathrm{ext}} \approx -220$ mT: (**a**) Buildup of DNSP. In the presence of B_{ext}, it is more efficient and thus faster to produce a nuclear magnetic field compensating the latter (*black*, σ^- excitation) than one that enforces it (*gray*, σ^+ excitation). (**b**) If DNSP decay is mediated through the residual QD electron, it is again more efficient to depolarize the nuclei if the total effective magnetic field seen by the electron is minimized. The color coding is the same as in (**a**). *Solid curves* in (**a**) and (**b**) show exponential fits to the data, the resulting buildup- and decay times are given in the figures. (**c**) Decay of DNSP in the absence of the QD electron. Compared to the zero-field case (Fig. 5.9c), DNSP decay time is prolonged to $\tau_{\mathrm{decay}} \approx 60$ s. The *inset* shows OS after a waiting time of 1 s as a function of external magnetic field. DNSP decay is suppressed on a magnetic field scale of ~1 mT, indicative of DNSP decay mediated by nuclear dipole–dipole interactions. (**d**) shows the respective directions of the external magnetic field and the nuclear fields $B_{\mathrm{nuc}}^{\sigma^+}$ ($B_{\mathrm{nuc}}^{\sigma^-}$) induced by QD excitation with σ^+ (σ^-) polarized light

$\tau_{\mathrm{pump}} \gg \tau_{\mathrm{buildup}}$ in the absence of the QD electron. A nuclear spin polarization in the surrounding of the QD would lead to an increase of τ_{decay} with increasing τ_{pump} [30]. However, within the experimental parameters accessible with our experiment, we were unable to see such a prolongation and hence any effect of polarization of the surrounding bulk nuclei. We interpret this fact as a strong indication that we indeed create and observe a very isolated system of spin polarized nuclei. The QD boundaries constitute a barrier for spin diffusion and the small remaining flux of nuclear spin polarization is too low to saturate the bulk material surrounding the dot.

Finally, we note that we have tested the dependance of our experimental results on the polarization of the probe pulse. We repeated all experiments presented in this section using probe pulses both with linear polarization or with orthogonal polarization with respect to the pump pulse. This test revealed that our experimental findings are independent of the polarization of the pump pulse. While for experiments performed in the presence of an external magnetic field this shows that our probe pulses indeed does not alter the nuclear spin polarization, this observation is surprising for experiments in zero magnetic field. As we discussed in Sect. 5.3.2, in the absence of external magnetic fields, nuclear spin polarization is destroyed on a timescale $T_2 \approx 10$–$100\,\mu s$ by nuclear dipole–dipole interactions. At the same time, nuclear spin temperature has a lifetime $T_1 \gg T_2$, even at zero magnetic field (see [35, Chap. 5]). In Sect. 5.4, we explained our experimental observation of DNSP at zero field with the presence of a relatively strong Knight field during optical pumping of QD nuclear spins. Since this Knight field is zero in the absence of laser excitation, one would expect DNSP to decay within a time T_2 after switching off the pump pulse. The probe pulse arriving after τ_{wait} (with $T_2 < \tau_{\text{wait}} < T_1$) would then lead to the reappearance of DNSP oriented along the Knight field created by the probe pulse, which is opposite to the Knight field of the pump pulse if the helicity was switched between the pump and the probe pulse. The fact that we do not see the corresponding reversal of the sign of DNSP indicates that the nuclear T_2 time is on the order of T_1 for the QD nuclear spin system. A possible reason for this could be the strong nuclear quadrupolar interactions in QDs [52] which can partly suppress the effect of nuclear dipole–dipole interactions (V. Kalevich, K. Kavokin, I. Merkulov 2008, private communication).

5.6.2 High Field Nuclear Spin Dynamics

In view of the nonlinear coupling between the electron and the nuclear spin system that was demonstrated in Sect. 5.5, the purely exponential buildup and decay curves measured in Sect. 5.6 might come as a surprise. Since the nuclear spin relaxation rate T_{1e} due to the QD electron depends on electron spin detuning, the buildup and decay rates of DNSP should depend on the degree of nuclear spin polarization and therefore change during the time traces presented in Fig. 5.8. These nonlinear effects are most prominent at the moment where the external and nuclear magnetic fields cancel. Since at low external magnetic fields this corresponds to the regime of almost zero nuclear spin polarization (i.e., to the beginning of the buildup- resp. to the end of the decay-curves), the experimental signature of the nonlinear character of the electron–nuclear spin system is not very pronounced there.

When increasing the external magnetic field, the nuclear spin dynamics slow down due to the increasing electron Zeeman splitting. However, in a configuration where B_{nuc} opposes B_{ext}, the total magnetic field felt by the QD electron crosses through zero at some point during the buildup and the decay of DNSP (cf. Fig. 5.6). At this point, the nuclear spin dynamics speed up again and $T_{1e} = T_{1e}^0$.

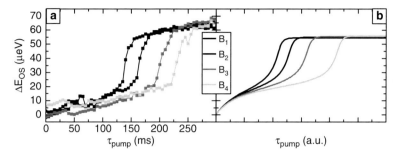

Fig. 5.11. (a) Buildup of DNSP in external magnetic fields on the order of the Overhauser field. The experiment was performed with the procedure and external parameters described in the main text at magnetic fields $B_1 = 1.1\,\text{T}$, $B_2 = 1.2\,\text{T}$, $B_3 = 1.3\,\text{T}$, and $B_4 = 1.4\,\text{T}$. (b) Simulations according to the classical nonlinear rate equation (5.17) with the parameters found in the fit for Fig. 5.5. The magnetic fields used for the simulation are $B_1 = 1.22\,\text{T}$, $B_2 = 1.24\,\text{T}$, $B_3 = 1.26\,\text{T}$, and $B_4 = 1.28\,\text{T}$

In order to observe the nonlinear buildup and decay curves discussed above, we performed the pump-probe measurement of DNSP described earlier in a regime of positive magnetic field and with excitation light of positive helicity (σ^+). Additionally, the nuclear spin polarization is reset between every pump-probe pulse pair by illumination with linearly polarized light for 100 ms. This ensures that individual pump-probe sequences are independent of each other.

Figure 5.11 shows the buildup curves of DNSP measured at various external magnetic fields. The nonlinear effects discussed before are clearly visible in this measurement. Also shown in the figure is a numerical simulation of the dynamics described by the nonlinear equation of motion (5.17) at the corresponding magnetic fields. The parameters for these curves are directly taken from the fit to the data presented in Fig. 5.5 without any further adjustments.[11] We note that the magnetic fields applied in the measurement do not completely agree with the magnetic fields used in the simulation. This is most probably due to slightly different excitation conditions (in terms of excitation power and energy) between this experiment and the one presented in Sect. 5.5. Changing these parameters can significantly alter the critical magnetic fields $B_{1,2}^c$ defined in Sect. 5.5 and therefore the magnetic field dependance of the buildup of DNSP.

A much more interesting situation arises for the decay of DNSP in sizable external magnetic fields. Since the nuclear spin decay rate depends strongly on the electronic environment of the nuclei, the dependance on ΔE_{el}^Z of the electron-mediated DNSP decay rate can have various forms, depending on the relative importance of the different possible mechanisms discussed in Sect. 5.6.1.

---

[11] Since these fits only gave the relative time-constants of buildup and decay but not their absolute values, the time axis in Fig. 5.11b is given in arbitrary units.

A good picture of the different decay characteristics at various QD gate voltages in high magnetic fields can be obtained by measuring DNSP simultaneously as a function of gate voltage and time. The nuclei are initialized in a state of maximal DNSP at a gate voltage V_1 corresponding to the center of the X^{-1} plateau. The gate voltage is then switched to a value V_2 and DNSP is measured after a waiting time τ_{wait}. In this measurement we scan τ_{wait} first and step to the next value in V_2 after a full time trace is recorded. The measurement result as a function of V_2 and τ_{wait} is shown in Fig. 5.12a, where the final degree of DNSP is encoded in gray-scale. The voltages corresponding to the crossover between the (n)- and $(n+1)$-electron regimes as determined from a gate voltage dependant PL experiment (at the lowest

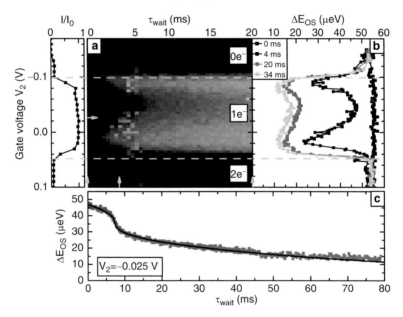

Fig. 5.12. Decay of DNSP in an external magnetic field of $B_{\text{ext}} = 1\,\text{T}$. The nuclear spin polarization was initialized with a 100 ms, σ^+-polarized pump pulse at a gate voltage V_1 corresponding to the middle of the $1e^-$-plateau, resulting in an initial Overhauser shift $\Delta E_{\text{OS}}(0) \approx 55\,\mu\text{eV}$. Immediately after this nuclear spin initialization, the gate voltage was switched to a value V_2. (**a**) Measurement of ΔE_{OS} as a function of waiting time t_{wait} and gate voltage V_2. *Gray lines* indicate the transition between the QD charging states identified by the low power PL experiment shown in the *left panel*. (**b**) and (**c**) are line-cuts through figure (**a**) at fixed t_{wait} and V_2, respectively, as indicated by the *gray arrows* in (**a**). (**a**, **b**, and **c** have all been obtained by independent measurements.) The *black line* in (**c**) is a fit according to (5.17)

possible excitation power) are marked in the figure.[12] In agreement with the discussion in Sect. 5.6, the measurement shows three clearly distinct regions of DNSP decay: When the QD is occupied by a single electron, DNSP decays in tens of milliseconds; when zero or two electrons are present, there is no decay of DNSP on the timescale of the presented measurements. Even when increasing the measurement time to 5 min, DNSP shows virtually no decay in these regions. We speculate that the corresponding nuclear T_1 time is on the order of hours.

In the region where the QD is occupied with one electron, the fast DNSP decay shows a much richer behavior. The decay rate shows a marked increase when V_2 approaches the edge of the 1e$^-$-plateau, where cotunneling rates increase substantially [26]. This illustrates the importance of cotunneling in electron-mediated DNSP decay, which is twofold: Cotunneling ensures that the mean electron spin polarization is zero due to the coupling to the (unpolarized) electron reservoir; this sets the equilibrium nuclear spin polarization through (5.6). Furthermore, cotunneling limits the electron spin correlation time τ_{el}, which broadens the electron spin states and allows for electron–nuclear spin flips to happen at first place.

When approaching the 1e$^-$–2e$^-$ transition point however (0.02 V < V_2 < 0.04 V), nuclear spin lifetime increases again, even though the stable configuration of the QD is still singly charged. We believe that this is a signature of motional narrowing: While a finite τ_{el} is necessary to overcome the energy mismatch of the initial and final states of an electron–nuclear spin flip-flop, the nuclei cannot undergo such a transition if the electron spin fluctuations become too fast. This becomes apparent by inspecting (5.15) which shows that T_{1e} has a maximum for $\tau_{\text{el}} = 1/\Omega_{\text{el}}$. We observe the maximal electron–nuclear spin relaxation rate at a gate voltage $V_2 = 0.02$ V. Since at $\tau_{\text{wait}} = 0$ the total electron Zeeman splitting ($\Delta E_{\text{OS}} + g_{\text{el}}^* \mu_B B_{\text{ext}}$) amounts to $\sim$20 µeV, the corresponding electron cotunneling rate at this gate voltage is on the order of 30 GHz, consistent with independent calculations of this quantity [26]. Motional narrowing is not observed on the 0e$^-$–1e$^-$ transition, where one would at first sight expect a similar behavior as in the 1e$^-$–2e$^-$ transition since cotunneling processes are equivalent for these two regimes. However, this is not strictly true since the tunneling rate also has a dependance on gate voltage and increases exponentially with increasing voltage. Therefore, cotunneling is slower on the 0e$^-$–1e$^-$ transition than on the 1e$^-$–2e$^-$ transition and τ_{el} never reaches the value $1/\Omega_{\text{el}}$ on the low-voltage side of the 1e$^-$-plateau.

The center of the 1e$^-$-plateau shows a lifetime of DNSP of roughly 10 ms, limited by the interactions of the QD nuclei with the residual QD electron which is randomized by cotunneling with the reservoir. Due to the nonlinear

---

[12] We note that these voltages do not correspond to the voltages indicated in Fig. 5.1. The reason for this is light-induced accumulation of space-charges in our gated structures during laser excitation. These charges screen the applied gate voltage, thereby shifting it to lower values.

electron–nuclear spin coupling, the decay in this region shows a highly nonlinear behavior as presented in Fig. 5.12c. Nuclear spin depolarization initially happens at a rather slow rate but speeds up as soon as B_{nuc} and B_{ext} cancel after $t_{\mathrm{wait}} \approx 7\,\mathrm{ms}$ and at $\Delta E_{\mathrm{OS}} \approx 36\,\mathrm{\mu eV}$. Once $|B_{\mathrm{nuc}}| < |B_{\mathrm{ext}}|$, the electron Zeeman splitting increases again and DNSP decay slows down. We fitted the measured decay curve by a numerical solution of the rate equation (5.16) for $S_z = 0$ and an initial condition $\Delta E_{\mathrm{OS}}(0)$. The parameters found in this fit were $\Delta E_{\mathrm{OS}}(0) = 47\,\mathrm{\mu eV}$, $T_{\mathrm{1e}}^0 = 3.3\,\mathrm{ms}$, $T_{\mathrm{d}} = 110\,\mathrm{ms}$, $\tau_{\mathrm{el}} = 360\,\mathrm{ps}$ and $g_{\mathrm{el}}^* = -0.62$, in agreement with the corresponding numbers found in Sect. 5.5.2. However, the electron spin correlation time we found is surprisingly short given that the measurement was performed in the absence of optical excitation. Even though this measurement of DNSP decay was not performed exactly at the center of the 1e$^-$-stability plateau (cf. Fig. 5.12a), we would expect τ_{el} to be on the order of 1 ns. Another surprising feature is the remaining nuclear spin depolarization time of $T_{\mathrm{d}} = 110\,\mathrm{ms}$. We note that this nuclear spin relaxation cannot be due to spin diffusion mediated by nuclear dipole–dipole interactions, since this rate has to be much smaller given the long DNSP lifetime indicated by the measurement in the 0e$^-$- and 2e$^-$-regions of Fig. 5.12a. The mechanism causing nuclear spin relaxation at the rate $T_{\mathrm{d}}^{-1} \approx 0.1\,\mathrm{Hz}$ must therefore be caused by the presence of the residual QD electron. We believe that indirect nuclear spin interactions mediated by the QD electron cause this decay through the mechanism discussed in Sect. 5.3.4. This decay is not expected to be exponential and further study of the DNSP decay dynamics in this regime is required to fully understand the measurement presented in Fig. 5.12c.

A surprising feature arises in the transition between the 1e$^-$ and 0e$^-$ regimes ($-0.14\,\mathrm{V} < V_2 < -0.1\,\mathrm{V}$). Figure 5.12b shows that in this regime DNSP has an initial, fast decay (on a timescale $\ll 1\,\mathrm{ms}$, not resolved in this measurement) after which it settles at a finite value of DNSP and shows no further decay on the timescale of our experiments. This unexplained observation is very surprising since it occurs in a regime where the QD is in a neutral state and only virtual occupations by a single electron are allowed. Furthermore, DNSP decay seems to stop at a value where the total electron Zeeman splitting is smaller than its value at $\tau_{\mathrm{wait}} = 0$ and where one would therefore expect DNSP decay to be enhanced. Understanding the interesting dynamics in this QD charging regime requires further investigation.

We note that we have repeated the experiment discussed above for external magnetic fields of $B_{\mathrm{ext}} = 0$, 0.5, and 1.5 T. The basic features described above have been observed at all those fields. At $B_{\mathrm{ext}} = 0$, DNSP decay in the 1e$^-$-region is too fast to observe a variation of cotunneling rates over the 1e$^-$-stability plateau. We also checked the dependance of our results on the exact form of the gate voltage-overshoots used to switch V_{g} between V_1 and V_2 (cf. Fig. 5.9) and found that the gate voltage-switching does not influence our results. A further check for the validity of our results was to change the order in which τ_{wait} and V_2 were measured. Sweeping V_2 and stepping τ_{wait} gave results identical to the ones depicted in Fig. 5.12.

We conclude that we have good understanding of the decay dynamics of DNSP in the regimes where the QD electron occupancy number is well defined. In the cross-over regions where the QD charging state is changed, DNSP decay shows unexpected features that warrant further investigation.

5.7 Future Perspectives

One of the principal experimental findings presented in this work was the identification of the QD electron as an efficient source of nuclear spin decay. While in our approach the electron is indispensable for building up a nuclear spin polarization in the first place, it also acts against this buildup by inducing indirect nuclear spin interactions as well as cotunneling mediated depolarization. In order to increase the maximal attainable degree of nuclear spin polarization it would therefore be interesting to investigate possibilities of suppressing this electron-mediated DNSP decay. This could be achieved by increasing the tunnelling barrier between the QD and the electron reservoir in our structures. However, changing this parameter also has the consequence of changing the electron spin correlation time τ_{el} and the fraction of QD electron occupation, f_{el}. Both these factors crucially influence the QD nuclear spin dynamics in various ways. Systematically studying the dependance of DNSP on the QD tunneling barrier thickness could therefore yield valuable information on how to enhance DNSP in self-assembled QDs.

For the interpretation of all of our experimental results, considering the nuclear spins as an ensemble of classical magnetic moments was sufficient. Investigating their quantum mechanical nature would be interesting for both fundamental reasons and applications that aim at tailoring the fluctuations of the mean nuclear spin [7]. The back-action of a (quantum-mechanical) measurement of the nuclear spin polarization along a given axis would be an interesting experiment in this direction. In order to perform a projective measurement on the QD nuclear spin system, the accuracy of the detection of the Overhauser-shift has to be greatly improved as compared to our experimental technique. The necessary energy resolution for such an experiment has recently been estimated to be on the order of $A/N^{3/2}$ [7]. Using optimistic numbers, this corresponds to an energy resolution of ~ 0.1 neV or ~ 25 kHz. While this resolution is out of reach with our present spectroscopic techniques, more advanced methods like optically detected electron spin resonance [53] or EIT [54] are close to reaching the required sensitivity. A first step in this direction would be the use of differential transmission measurements [55] to prepare and detect DNSP.

Our measurements of the Overhauser-shift of a QD electron give information about the mean nuclear field that the electron is exposed to. Investigating the role that the different nuclear species play in the dynamics of the nuclear field requires a further extension of our experimental techniques. Applying an NMR field resonant with a given nuclear spin species could induce additional heating for those spins, thereby giving information about the contribution of the different spin species to the measured Overhauser-shift. Such experiments are complicated by the large inhomogeneous broadening of the corresponding NMR lines caused by strain-induced quadrupolar shifts. Using more refined NMR and optical detection techniques, however, NMR experiments could prove useful in investigating the dynamics of QD nuclear spins in greater detail.

Another exciting perspective is the experimental observation of the onset of nuclear order in a single QD. Different theoretical scenarios for these nuclear phase transitions have been proposed. Based on the experimental results presented in this work, realizing these proposals seems to be within experimental reach. Nuclear self polarization was predicted to occur for the case where nuclear spins couple to an electron spin system which is artificially maintained in a disordered spin state at sufficiently low electron temperatures [46]. Combining recent advances in electron spin resonance in self-assembled QDs [53] with the fast, noninvasive measurement of DNSP via pulsed PL could allow us to observe this spontaneous nuclear spin polarization which was predicted to occur at moderate temperatures of a few Kelvin. At much lower temperatures, a true ferromagnetic phase transition of the nuclear spins was recently predicted to occur in semiconductor nanostructures [56]. A possibility for reaching low nuclear spin temperatures in the micro-Kelvin range is adiabatic demagnetization of QD nuclear spins. Bringing the optically cooled nuclear spin system from a field on the order of 1 T adiabatically to zero field could result in a nuclear spin temperature being two to three orders of magnitude lower than what can be achieved through direct optical cooling.

Acknowledgments

We thank A. Badolato for crystal growth and J. Dreiser for help with sample preparation. P.M. would like to express his gratitude toward C.W. Lai who taught him the experimental techniques which made this work possible. Furthermore we acknowledge fruitful discussions with A. Högele, C. Latta and S.D. Huber. This work is supported by NCCR-Nanoscience.

References

1. A. Overhauser, Phys. Rev. **92**, 411 (1953)
2. G. Lampel, Phys. Rev. Lett. **20**, 491 (1968)
3. D. Paget, G. Lampel, B. Sapoval, V.I. Safarov, Phys. Rev. B **15**, 5780 (1977)
4. A.V. Khaetskii, D. Loss, L. Glazman, Phys. Rev. Lett. **88**, 186802 (2002)
5. A.C. Johnson, J.R. Petta, J.M. Taylor, A. Yacoby, M.D. Lukin, C.M. Marcus, M.P. Hanson, A.C. Gossard, Nature **435**, 925 (2005)
6. F.H.L. Koppens, J.A. Folk, J.M. Elzerman, R. Hanson, L.H.W. van Beveren, I.T. Vink, H.P. Tranitz, W. Wegscheider, L.P. Kouwenhoven, L.M.K. Vandersypen, Science **309**, 1346 (2005)

7. D. Klauser, W.A. Coish, D. Loss, Nuclear spin dynamics and zeno effect in quantum dots and defect centers (2008). arXiv:0802.2463v1
8. W.A. Coish, D. Loss, Phys. Rev. B **70**, 195340 (2004)
9. D. Gammon, S.W. Brown, E.S. Snow, T.A. Kennedy, D.S. Katzer, D. Park, Science **277**, 85 (1997)
10. S.W. Brown, T.A. Kennedy, D. Gammon, E.S. Snow, Phys. Rev. B **54**, R17339 (1996)
11. K. Ono, S. Tarucha, Phys. Rev. Lett. **92**, 256803 (2004)
12. A.S. Bracker, E.A. Stinaff, D. Gammon, M.E. Ware, J.G. Tischler, A. Shabaev, A.L. Efros, D. Park, D. Gershoni, V.L. Korenev, I.A. Merkulov, Phys. Rev. Lett. **94**, 047402 (2005)
13. B. Eble, O. Krebs, A. Lemaitre, K. Kowalik, A. Kudelski, P. Voisin, B. Urbaszek, X. Marie, T. Amand, Phys. Rev. B **74**, 081306 (2006)
14. C.W. Lai, P. Maletinsky, A. Badolato, A. Imamoglu, Phys. Rev. Lett. **96**, 167403 (2006)
15. A.I. Tartakovskii, T. Wright, A. Russell, V.I. Fal'ko, A.B. Van'kov, J. Skiba-Szymanska, I. Drouzas, R.S. Kolodka, M.S. Skolnick, P. Fry, A. Tahraoui, H.Y. Liu, M. Hopkinson, Phys. Rev. Lett. **98**, 26806 (2007)
16. I.A. Akimov, D.H. Feng, F. Henneberger, Phys. Rev. Lett. **97**, 056602 (2006)
17. I.A. Merkulov, A.L. Efros, M. Rosen, Phys. Rev. B **65**, 205309 (2002)
18. R.J. Warburton, C. Schaflein, D. Haft, F. Bickel, A. Lorke, K. Karrai, J.M. Garcia, W. Schoenfeld, P.M. Petroff, Nature **405**, 926 (2000)
19. P. Maletinsky, C.W. Lai, A. Badolato, A. Imamoglu, Phys. Rev. B **75**, 35409 (2007)
20. M. Bayer, G. Ortner, O. Stern, A. Kuther, A.A. Gorbunov, A. Forchel, P. Hawrylak, S. Fafard, K. Hinzer, T.L. Reinecke, S.N. Walck, J.P. Reithmaier, F. Klopf, F. Schäfer, Phys. Rev. B **65**, 195315 (2002)
21. M. Atatüre, J. Dreiser, A. Badolato, A. Högele, K. Karrai, A. Imamoglu, Science **312**, 551 (2006)
22. M.E. Ware, E.A. Stinaff, D. Gammon, M.F. Doty, A.S. Bracker, D. Gershoni, V.L. Korenev, S.C. Badescu, Y. Lyanda-Geller, T.L. Reinecke, Phys. Rev. Lett. **95**, 177403 (2005)
23. P.Y. Yu, M. Cardona, *Fundamentals of Semiconductors: Physics and Materials Properties*, 3rd edn. (Springer, Berlin, 2001)
24. M.J. Snelling, E. Blackwood, C.J. McDonagh, R.T. Harley, C.T.B. Foxon, Phys. Rev. B **45**, 3922 (1992)
25. S. Amasha, K. MacLean, I.P. Radu, D.M. Zumbühl, M.A. Kastner, M.P. Hanson, A.C. Gossard, Phys. Rev. Lett. **100**, 046803 (2008)
26. J.M. Smith, P.A. Dalgarno, R.J. Warburton, A.O. Govorov, K. Karrai, B.D. Gerardot, P.M. Petroff, Phys. Rev. Lett. **94**, 197402 (2005)
27. A. Abragam, *The Principles of Nuclear Magnetism* (Clarendon, Oxford, 1961)
28. M. Goldman, *Spin Temperature and Nuclear Magnetic Resonance in Solids* (Oxford University Press, Oxford, 1970)
29. J.A. McNeil, W.G. Clark, Phys. Rev. B **13**, 4705 (1976)
30. D. Paget, Phys. Rev. B **25**, 4444 (1982)
31. A. Malinowski, M.A. Brand, R.T. Harley, Physica E **10**, 13 (2001)
32. I. Merkulov, Phys. Usp. **45**, 1293 (2002)
33. P.F. Braun, B. Urbaszek, T. Amand, X. Marie, O. Krebs, B. Eble, A. Lemaitre, P. Voisin, Phys. Rev. B **74**, 245306 (2006)

34. E.I. Gryncharova, V.I. Perel, Sov. Phys. Semicond. **11**, 997 (1977)
35. F. Meier, *Optical Orientation* (North-Holland, Amsterdam, 1984)
36. M.I. Dyakonov, V.I. Perel, Sov. Phys. JETP **38**, 177 (1974)
37. J.R. Petta, A.C. Johnson, J.M. Taylor, E.A. Laird, A. Yacoby, M.D. Lukin, C.M. Marcus, M.P. Hanson, A.C. Gossard, Science **309**, 2180 (2005)
38. G. Giedke, J.M. Taylor, D. D'Alessandro, M.D. Lukin, A. Imamoglu, Phys. Rev. A **74**, 032316 (2006)
39. D. Klauser, W.A. Coish, D. Loss, Phys. Rev. B **73**, 205302 (2006)
40. R. Hanson, L.P. Kouwenhoven, J.R. Petta, S. Tarucha, L.M.K. Vandersypen, Rev. Mod. Phys. **79**, 1217 (2007)
41. W. Yao, R.B. Liu, L.J. Sham, Phys. Rev. B **74**, 195301 (2006)
42. D. Gammon, A.L. Efros, T.A. Kennedy, M. Rosen, D.S. Katzer, D. Park, S.W. Brown, V.L. Korenev, I.A. Merkulov, Phys. Rev. Lett. **86**, 5176 (2001)
43. V.L. Berkovits, C. Hermann, G. Lampel, A. Nakamura, Phys. Rev. B **18**, 1767 (1978)
44. S. Laurent, B. Eble, O. Krebs, A. Lemaitre, B. Urbaszek, X. Marie, T. Amand, P. Voisin, Phys. Rev. Lett. **94**, 147401 (2005)
45. E.L. Ivchenko, Pure Appl. Chem. **67**, 463 (1995)
46. M.I. Dyakonov, V.I. Perel, JETP Lett. **16**, 398 (1972)
47. R. Kaji, S. Adachi, H. Sasakura, S. Muto, Appl. Phys. Lett. **91**, 261904 (2007)
48. R. Kaji, S. Adachi, H. Sasakura, S. Muto, Hysteretic response of electron-nuclear spin system in single inalas quantum dots:excitation power and polarization dependences (2007). arXiv:0709.1382
49. D. Paget, T. Amand, J. Korb, Light-induced nuclear quadrupolar relaxation in semiconductors (2008). arXiv:0801.2894v1
50. C. Ramanathan, Dynamic nuclear polarization and spin-diffusion in non-conducting solids (2008). arXiv.org:0801.2170
51. B. Urbaszek, R.J. Warburton, K. Karrai, B.D. Gerardot, P.M. Petroff, J.M. Garcia, Phys. Rev. Lett. **90**, 247403 (2003)
52. R.I. Dzhioev, V.L. Korenev, Phys. Rev. Lett. **99**, 037401 (2007)
53. M. Kroner, K. Weiss, B. Biedermann, S. Seidl, S. Manus, A. Holleitner, A. Badolato, P. Petroff, B. Gerardot, R. Warburton, K. Karrai, Phys. Rev. B **78**, 075429 (2008)
54. A. Imamoglu, Phys. Stat. Sol. (b) **243**, 3725 (2006)
55. A. Hogele, M. Kroner, S. Seidl, K. Karrai, M. Atature, J. Dreiser, A. Imamoglu, R.J. Warburton, A. Badolato, B.D. Gerardot, P.M. Petroff, Appl. Phys. Lett. **86**, 221905 (2005)
56. P. Simon, D. Loss, Phys. Rev. Lett. **98**, 156401 (2007)

6

Quantum Dot Single-Photon Sources

Peter Michler

Summary. In this contribution, we briefly recall basic concepts of quantum optics and semiconductor quantum-dot physics which are necessary to understand the physics of single-photon generation with single quantum dots. The classification of light states and the photon statistics as well as the electronic and optical properties of the quantum dots are discussed. We then review the recent progress on extending the wavelength range and show how polarization control and high repetition rates have been realized. New generations of electrically driven single-photon LEDs lead to ultralow pump currents, high repetition rates, high collection efficiencies, and elevated temperature operation. Furthermore, new developments on coherent state preparation and single photon emission in the strong coupling regime are reviewed. The generation of indistinguishable photons and remaining challenges for practical single-photon sources are also discussed.

6.1 Introduction

A single photon source, which is able to generate photons on demand allows the ultimate quantum control of the photon generation process, i.e., single photons can be generated within short time intervals with a deterministic dwell time between successive photon generation events. Such a source has the potential of enabling many new applications in the field of photonics and quantum information technology [1]. This is particularly true for quantum cryptography, which exploits the fundamental principles of quantum mechanics to provide unconditional security for communication. Possible other applications are optical quantum computing, ultrasensitive metrology, random number generators, and quantum teleportation (see Chap. 11).

The requirements on the properties of single-photon sources depend on their specific application. Desirable properties for all sources are a high and constant internal quantum efficiency, low multiphoton emission probability, a high emission efficiency into a single mode, a low jitter, and high clock rates. Applications which rely on two-photon interferences on beam splitters, e.g., linear optics quantum computation, demand indistinguishable photons, i.e., long coherence times are indispensable. This property is not necessary for applications in quantum cryptography or for a standard of optical brightness. Therefore, more practical aspects like electrical pumping and high-temperature operation can be considered for these kind of applications. Furthermore, the emission wavelength of the source should be one which minimizes optical losses in the transmission, e.g., 1.3 and 1.5 µm for transmission in glass fibers, and maximize the photo-detection efficiency. Currently, single photon detectors, e.g., avalanche photodiodes possess their highest quantum efficiency in the visible spectrum.

The chapter is organized as follows: in Sect. 6.2 we give a brief review on the classification of light states and photon statistics and introduce the first and second order correlation functions $g^{(1)}(\tau)$ and $g^{(2)}(\tau)$. We discuss the coherence properties of single photon pulses and the interference of two photons on a beam splitter since a number of applications rely on these processes. The interference contrast depends on the indistinguishability of the photons which is also introduced. In Sect. 6.2.2 we describe the electronic shell structure of a quantum dot (QD) and explain the excitonic fine structure and their implications on bright and dark states. Furthermore, nonresonant excitation into the wetting layer or barrier, quasiresonant excitation into higher shells, and truly resonant excitation schemes into the s-shell are discussed. The positive influence of cavity effects on single photon emission is described in the last part of this section. In Sect. 6.3 the different measurement techniques of the first and second order correlation functions are introduced and a realization of the two-photon interference experiment is presented. Special emphasis is given on the comparison of experimentally accessible quantities with the corresponding theoretical ones. In Sect. 6.4 we give a discussion of the most recent developments on semiconductor single-photon sources. Especially, the progress on extension of the wavelength range, polarization control, high single-photon emission rates, high temperature operation, electrical pumping, photon indistinguishability, coherent state preparation, and single photon emission in the strong coupling regime are reviewed. Finally, Sect. 6.5 concludes the review and tries to identify the future challenges towards commercially useable single-photon sources.

6.2 Theoretical Background

The quantum theory of light is an indispensable tool in the description and understanding of basic measurement results on the single-photon level such as photon antibunching in the emission of a single emitter and single-photon interference. The goal of this chapter is to give an introduction into the basic theoretical concepts of the characterization of nonclassical light fields. Thermal light, coherent light, and photon number states with their corresponding photon statistics and correlation functions will be considered. Furthermore, the spatiotemporal modes of the photons, i.e., the wavepackets of the single photons have to be taken into account for the description, e.g., of the two-photon interference of single photons at a beam splitter. Complete presentations are given in textbooks on quantum optics [2–4].

The properties of the emitted light critically depend on the excitation mechanisms of the single emitter, i.e., the QD and its local environment. Therefore, a detailed knowledge of the physical properties of QDs is a precondition for their specific application in a single-photon source. Optical and electrical pumping as well as resonant and nonresonant pumping schemes will be discussed in the second part of the chapter.

6.2.1 Classification of Light States and Photon Statistics

In order to better understand the special features of a single-photon source it is instructive to compare first the statistical properties of thermal light and coherent light.

Thermal, Coherent and Photon Number States

Different light fields can be characterized by their specific photon number fluctuations. Thermal radiation results from a thermal equilibrium of emission and absorption between a radiation field and an ensemble of emitters (e.g., atoms in a discharge lamp). In such a system the photon number probability distribution can be written in terms of the mean photon number $\langle n \rangle$ as the geometric distribution [2]

$$P(n) = \frac{\langle n \rangle^n}{(1 + \langle n \rangle)^{1+n}}. \tag{6.1}$$

It can be seen from Fig. 6.1a that $n = 0$ has always the largest probability and the distribution falls off monotonically with increasing n. Please note, the photon probability distributions show no particular peak at $\langle n \rangle$. The photon number fluctuations are expressed in statistical terms by the variance of the distribution $(\Delta n)^2 = \langle n^2 \rangle - \langle n \rangle^2$. For a thermal mode, the variance is found to be $(\Delta n)^2 = \langle n \rangle^2 + \langle n \rangle$ [2]. Thus, the magnitude of the fluctuation Δn is approximately equal to $\langle n \rangle$ for $\langle n \rangle \gg 1$.

The coherent state (*Glauber* state) is characterized by the minimum uncertainty product [5] in accordance with the Heisenberg uncertainty relation for energy and time. An example for a coherent light source is a laser which is operated well above the threshold. The photon number distribution in this state is a Poisson distribution, i.e., the probability to find n photons in the coherent state with respect to the mean photon number is given by

$$P(n) = e^{-\langle n \rangle} \frac{\langle n \rangle^n}{n!}. \tag{6.2}$$

Poisson distributions are shown in Fig. 6.1b for two mean photon numbers $\langle n \rangle$ and they are peaked at the mean photon numbers, respectively.

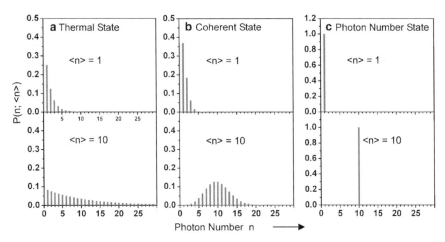

Fig. 6.1. Photon number probability distribution for single mode emission of (**a**) a thermal state, (**b**) a coherent state, and (**c**) a photon number state for two different mean photon numbers $\langle n \rangle = 1, 10$, respectively

The variance is calculated to $(\Delta n)^2 = \langle n \rangle$ [2]. Whereas for thermal radiation the fluctuations are always comparable with the mean photon number in the mode, the relative fluctuations $(\Delta n / \langle n \rangle = 1/\sqrt{\langle n \rangle})$ for coherent light approach zero with increasing photon numbers. Therefore, coherent light best approaches the pure classical picture of waves with fully determined amplitude and phase, i.e., revealing zero uncertainties. It is already obvious from Fig. 6.1, that single photons on demand cannot be generated by simply attenuating a light beam steaming from a thermal or a coherent light source. Such a nonclassical light state can only be generated by a single-photon source as will be shown below.

A photon number or a *Fock* state $|n\rangle$ is a truly nonclassical light state. It can be generated by a single-photon source. The Fock state is the eigenstate of the photon number operator: $\hat{n}|n\rangle = n|n\rangle$, i.e., a mode which is excited in this state is occupied by exactly n photons and the variance $\Delta n = 0$. Figure 6.1c illustrates the photon number distributions for the Fock states where nonzero probability is given exclusively for the mean photon number $\langle n \rangle = n$.

Photon Detection and Correlation Functions

This section introduces the prerequisite formalism for the theoretical description of the experimental results on the coherence and photon statistics of the light sources. We will restrict ourselves in this section on a *single mode* description and on *pure* light states. A more general, detailed and careful discussion also for multimode and mixed states can be found in [2–4].

Photons are typically detected by photoelectrons, e.g., photomultipliers or avalanche photodiodes. In this case the photon is absorbed and as a consequence the quantum state of the light field is altered. The precise nature of the photodetector is not of interest here and the properties like dead time, time-resolution and efficiency will be discussed in Sect. 6.3. The probability of photon detection w_1 which is associated with a transition from an initial field state Ψ to possible final states Ψ_f via absorption of a photon is given by

$$w_1(t) \sim \sum_f |\langle \Psi_f | \hat{a}(t) | \Psi \rangle|^2 \,, \tag{6.3}$$

where the sum goes over all possible final states and $\hat{a}(t)$ denotes the photon-destruction operator in the Heisenberg picture. We can rewrite the above equation with the identity operator $\sum_f |\Psi\rangle\langle\Psi| = 1$ to

$$w_1(t) \sim \langle \Psi | \hat{a}^\dagger(t) \hat{a}(t) | \Psi \rangle. \tag{6.4}$$

The *first-order coherence* function is defined as

$$g^{(1)}(t_1, t_2) = \frac{\langle \Psi | \hat{a}^\dagger(t_2) \hat{a}(t_1) | \Psi \rangle}{[\langle \Psi | \hat{a}^\dagger(t_1) \hat{a}(t_1) | \Psi \rangle \langle \Psi | \hat{a}^\dagger(t_2) \hat{a}(t_2) | \Psi \rangle]^{1/2}} \tag{6.5}$$

and the corresponding degree of *second-order coherence* is defined by

$$g^{(2)}(t_1, t_2) = \frac{\langle \Psi | \hat{a}^\dagger(t_1) \hat{a}^\dagger(t_2) \hat{a}(t_2) \hat{a}(t_1) | \Psi \rangle}{\langle \Psi | \hat{a}^\dagger(t_1) \hat{a}(t_1) | \Psi \rangle \langle \Psi | \hat{a}^\dagger(t_2) \hat{a}(t_2) | \Psi \rangle}. \tag{6.6}$$

The first and second order coherence functions are commonly measured by a *Michelson interferometer* and a *Hanbury-Brown and Twiss* setup, respectively (details see Sect. 6.3). When the light is stationary, only the relative time $\tau = t_2 - t_1$ is relevant. As a consequence both degree of coherence depend only on a single argument. Introducing a reduced notation[1] the above expressions reduce to a compact form

$$g^{(1)}(\tau) = \frac{\langle \hat{a}^\dagger(t) \hat{a}(t+\tau) \rangle}{\langle \hat{a}^\dagger(t) \hat{a}(t) \rangle} \tag{6.7}$$

and

$$g^{(2)}(\tau) = \frac{\langle \hat{a}^\dagger(t) \hat{a}^\dagger(t+\tau) \hat{a}(t+\tau) \hat{a}(t) \rangle}{\langle \hat{a}^\dagger(t) \hat{a}(t) \rangle^2}. \tag{6.8}$$

Comparing (6.7) with (6.4),[2] we see that the photon detection probability, i.e., the counting rate of the photodetector is in fact just the first-order correlation function except for a scale factor. Furthermore, the first-order correlation is insensitive to the photon statistics, since (6.7) only depends on

---

[1] The following shortened notation is used: In the case of a pure state Ψ the average value of an observable is: $\langle \hat{O} \rangle = \langle \Psi | \hat{O} | \Psi \rangle$ whereas in the general case of a mixed state: $\langle \hat{O} \rangle = \text{Tr}(\hat{\rho}\hat{O})$ where $\hat{\rho}$ is the corresponding density operator and $\hat{O}$ is a quantum mechanical operator.
[2] Here we assume $t_1 = t_2 = t$ which means we are dealing with one event at a particular time.

the average photon number $\langle n \rangle = \langle \hat{a}^\dagger \hat{a} \rangle$. In other words, spectrally-filtered thermal light and coherent light of the same spectral width exhibit the same degree of first-order coherence. In contrast, the second-order coherence allows to distinguish between the different type of light fields. It is instructive to compare $g^2(0)$ for thermal, coherent, and photon number states:

1. Coherent[3] light $|\alpha\rangle$

$$g^{(2)}(0) = \frac{\langle \alpha | \hat{a}^\dagger \hat{a}^\dagger \hat{a} \hat{a} | \alpha \rangle}{\langle \alpha | \hat{a}^\dagger \hat{a} | \alpha \rangle^2} = 1. \tag{6.10}$$

2. Photon number states[4] $\langle n |$

$$g^{(2)}(0) = \frac{\langle n | \hat{a}^\dagger \hat{a}^\dagger \hat{a} \hat{a} | n \rangle}{\langle n | \hat{a}^\dagger \hat{a} | n \rangle^2} = 1 - \frac{1}{n}. \tag{6.11}$$

This is a truly nonclassical result since $g^{(2)}(0) < 1$ for $n \geq 1$. The light is called to be in a *sub Poissonian* state. Especially, for a true single-photon source ($n = 1$) $g^{(2)}(0) = 0$.

3. Thermal light

$$g^{(2)}(0) = 1 + \frac{(\Delta n)^2 - \langle n \rangle}{\langle n \rangle^2} = 2. \tag{6.12}$$

This indicates that photons have the tendency to be detected simultaneously at the photoelectrons. The light is called to be in a *Super Poissonian* state.

The calculation of the photon correlation function $g^{(2)}(\tau)$ as a function of the delay time τ is somewhat more elaborate and can be found in [2]. Figure 6.2 shows the results for $g^{(2)}(\tau)$ for a thermal, a coherent, and a sub-Poissonian light source [2]. For a coherent source $g^{(2)}(\tau) = 1$ which means that the photons are completely uncorrelated. In contrast, a light source with a Super-Poissonian statistics shows a clear excess of coincidences ($1 < g^{(2)}(\tau) < 2$) for times shorter than the coherence time T_2 of the light source. This phenomenon ($g^{(2)}(\tau) \leq g^{(2)}(0)$) is called *photon bunching*. On the other hand a Sub-Poissonian light source satisfies $g^{(2)}(\tau) < 1$, this indicates that two photons are unlikely to be detected simultaneously by the detectors. A light source, e.g., emission from a single quantum emitter, whose degree of second-order coherence satisfies the inequality ($g^{(2)}(0) \leq g^{(2)}(\tau)$) is called antibunched light (*photon antibunching*).

---

[3] The coherent state is defined as

$$|\alpha\rangle = \sum_{n=0}^{\infty} \exp(-\frac{|\alpha|^2}{2}) \frac{\alpha^n}{\sqrt{n!}} |n\rangle \tag{6.9}$$

with the properties $\hat{a}|\alpha\rangle = \alpha|\alpha\rangle$ and $\langle \alpha | \hat{a}^\dagger = \langle \alpha | \alpha^*$.

[4] For the following we use the bosonic commutator relation $[\hat{a}, \hat{a}^\dagger] = 1$.

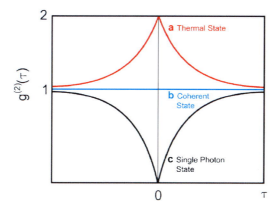

Fig. 6.2. Continuous-wave (cw) photon correlation functions $g^2(\tau)$ for (**a**) thermal, (**b**) coherent, and (**c**) a single photon state, respectively

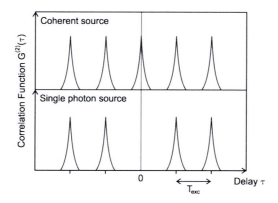

Fig. 6.3. Correlation trace from a pulsed coherent and a pulsed single photon source. The zero delay peak vanishes for a true single photon source

Under *pulsed* excitation conditions, the second-order correlation function $g^{(2)}$ consists of a series of correlation peaks separated by the repetition period T. In the case of a coherent source all peak areas are identical (see Fig. 6.3), i.e., showing a Poisson distributed statistics. Sub-Poisson statistics occur if the peak area of the central pulse at $\tau = 0$ is smaller than the peak areas of the other peaks. For perfect single photon emission the central peak is absent indicating the generation of only one photon per pulse (see Fig. 6.3).

Single-Photon Pulses and Indistinguishability of Photons

Linear optics quantum computing schemes [6] and quantum teleportation techniques are based on two-photon interference of two single-photon pulses on a beam splitter which require *indistinguishable photons*. This means that the two photons have to be Fourier-transform limited and possess the same

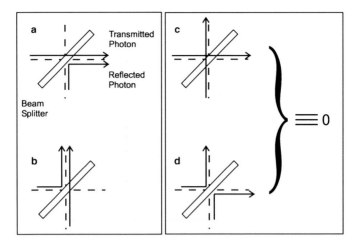

Fig. 6.4. When two indistinguishable single photons enter a 50/50 beam splitter, they may be transmitted or reflected in various ways: one may reflected and the other transmitted (**a** and **b**); both photons may be transmitted (**c**), or both may be reflected (**d**). It turns out that by quantum interference, the probabilities for "both transmitted" and "both reflected" cancel out. Thus the two photons always emerge from the same port

pulse width, bandwidth, carrier frequency, polarization, transverse mode profile, and arrival time at the beam splitter. Figure 6.4 shows schematically the interference of two indistinguishable photons at a beam splitter. When two indistinguishable photons enter the two different input arms of a 50/50 beam splitter at the same time, it turns out by quantum interference, that the probabilities for "both transmitted" and "both reflected" cancel out. Thus, the two photons emerge from the beam splitter along the same output port. Its origin lies in the bosonic nature of the photons and has been first predicted and observed by Hong, Ou, and Mandel [7]. However, if the two photons are not indistinguishable, they can behave independently, and the two-photon interference effect is reduced.

A single photon which propagates along an axis can be described by a wave packet, whose spectrum and time-dependent amplitude are determined by the parameters of the source [2,8]. Therefore, the physical properties of the emitter, the whole excitation and recombination processes have to be taken into account for a complete description of single photon pulses. If the radiative transition between two quantum states of the emitter is broadened solely by the spontaneous emission process, the single photon pulses are Fourier-transform limited, i.e., the characteristic temporal widths of the photons are $T_2 = 2/\Delta\omega$ with a natural linewidth of $\Delta\omega$. In the other case, if dephasing processes are present, the photon is in an incoherent mixed state as opposed to a pure state. First-order coherence measurements can be applied to characterize their coherence length (see Sect. 6.3). The coherence length l of a single photon, which is emitted from an individual quantum dot, is determined by the coherence time T_2 ($l = cT_2$, c = speed of light) of the excited state. T_2 can be considered as the dephasing time of the excited state and is defined by

$$\frac{1}{T_2} = \frac{1}{2T_1} + \frac{1}{T_2^*}, \tag{6.13}$$

where T_1 is the radiative lifetime of the emitter, and T_2^* is the pure dephasing time. For the case of the quantum dot, e.g., exciton–phonon interactions and carrier–carrier scattering are a source of pure dephasing.

Besides dephasing processes of the excited states, a time jitter, i.e., an uncertainty in the arrival time of the photons at the beam splitter will reduce the degree of the two-photon interference. In the case of an incoherent pumping scheme over an excited state, e.g., p-shell or wetting layer of a QD, the carriers must relax to the ground state through phonon emission with a relaxation rate r before the photon can be emitted. This relaxation process will cause a time jitter which affects the temporal overlap of the single-photon pulses [9, 10].

The interference dip, i.e., the probability $p_{34}(\delta t)$ that the two photons emerge from different output arms of a (50/50) beam splitter as a function of the temporal offset δt between them depends on the dephasing of the QD, on the time jitter, as well as on the bandwidth of the detectors [11]. The lineshape of the interference dip can be described by [2]

$$p_{34}(\delta t) = \frac{1}{2}(1 - |J|^2_{\delta t}), \tag{6.14}$$

where $|J|^2_{\delta t}$ is the *overlap integral* between the pairs of photons incident on the 50/50 beam splitter. Note, that δt can be adjusted by the delay line in an interferometric setup (see Sect. 6.3).

In general, under consideration of pure dephasing and a relaxation rate r from an excited to the ground state the single-photon wavefunctions can be described by [9]

$$|\Psi\rangle \propto \int_{t_0}^{\infty} dt e^{-\gamma(t-t_0)+i\phi(t)} \hat{a}^\dagger(t)|0\rangle, \tag{6.15}$$

where $\gamma = 1/2T_1$ is the spontaneous-emission amplitude decay rate, t_0 is a random variable with probability density re^{-rt_0} for $t_0 > 0$ and zero otherwise, and $\phi(t)$ is a random function whose probability distribution describes the pure dephasing process. For the case of a time-invariant pure dephasing process and a Lorentzian emission spectrum the spectral linewidth is given by $\Delta\omega = 2/T_2 = 2(\gamma + \alpha)$ with $\alpha = 1/T_2^*$ the pure dephasing rate. The overlap integral of the wavepackets is given by [9]

$$|J|^2_0 = \frac{\gamma}{\gamma + \alpha} \frac{r}{r + 2\gamma}. \tag{6.16}$$

The value of the overlap integral is unity for perfectly indistinguishable photons giving rise to a zero two-photon interference dip ($p_{34}(0) = 0$). Therefore, the overlap integral is appropriate to quantify the *degree of indistinguishability*. From the above results it becomes clear that a truly resonant pumping scheme of the emitter would be advantageous since it avoids the time jitter from the relaxation process. For $r \to \infty$ the overlap integral is equal to $T_2/2T_1$.

6.2.2 Emitter Properties: Semiconductor Quantum Dots

Excitonic States

High-quality self-assembled quantum dot structures can be grown by modern epitaxy techniques. Mainly, strain induced islands are used as a process for forming self-assembled quantum dots. In general, the dots are formed on random positions on the wafer. However, lithographic techniques allow to control the dot position by prestructuring the growth surface (see Chap. 2). This active positioning is a considerable progress with respect to the applications of single quantum dots, e.g., in single-photon sources. Quantum dots combine atom-like properties such as a discrete energy spectrum and sharp lines in photoluminescence with the advantage that their emission wavelengths can be tailored to a large extend. In order to understand their optical spectra, the electronic shell structure, the spin structure, and the many-body interactions between electrons and holes have to be considered (see also Chap. 1). For an even number of electron and holes in a QD, neutral complexes form, and depending on the population, recombination from, e.g., the exciton (one electron, one hole), the biexciton (two electrons, two holes) and multiexcitonic (N electrons, N holes) states can be observed. For an odd number of particles in the QD, charged exciton transitions take place. Due to the Coulomb correlations between the carriers all transitions possess distinct and somewhat different energies. Therefore, by spectral filtering, they can all be used in principle as a source for single photon emission provided some restrictions are fulfilled which will be discussed below. However, most demonstrations have used the biexcitonic or the excitonic transition. They show a distinct fine-structure in their emission spectra which can be understood if the spin structure of electrons and holes are considered.

The projection of the electronic spin on the z-axis (growth axis) of the dot is either $1/2$ or $-1/2$, whereas the heavy-hole spin projection is either $3/2$ or $-3/2$. This results in four distinct spin values for one electron–hole pair (exciton) in the quantum dot. The two bright states $|-1/2, 3/2\rangle$ and $|1/2, -3/2\rangle$ have total z-spin of ± 1 and are coupled to the light field whereas the other two states $|1/2, 3/2\rangle$ and $|-1/2, -3/2\rangle$ have total z-spin of ± 2 and are therefore optically decoupled, i.e., dark, due to the selection rules for dipole transitions (see Fig. 6.5). In the case where the dot possesses a perfect cylindrical symmetry around its growth axis, the $\pm|1\rangle$ states are degenerate. In practice, most of the dots are asymmetric. As a consequence, the degeneracy is lifted by the electron–hole exchange interaction resulting in a doublet structure shown in Fig. 6.5. This fine structure splitting is, e.g., for InGaAs/GaAs in the range of ~ 0–$100\,\mu eV$. Thus, the excitonic recombination line is split in energy and the

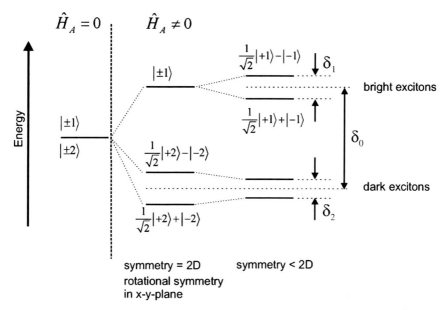

Fig. 6.5. Energy level scheme for the excitonic QD ground state without ($\hat{H}_A = 0$) and with ($\hat{H}_A \neq 0$) inclusion of exchange interaction between electron and hole spins. The initial fourfold exciton degeneracy for noninteracting spins is lifted thus forming optically allowed "bright" and forbidden "dark" configurations. The bright states are further split into a doublet under the conditions of lowered in-plane symmetry [12]

individual components possess orthogonal linear polarizations [12]. Typically, one of the linear polarizations is aligned with one of the substrates cleave directions (i.e., for GaAs (110) or (1-10)).

The biexciton is a spin-singlet state which does not reveal a fine structure itself but decays to one of the two optically bright excitonic states. Thus the polarization of the biexcitonic recombination lines is also determined by the excitonic states. The polarization properties of the biexciton–exciton radiative cascade are discussed in detail in the Chap. 7 in this volume.

The simplest charged excitonic configuration (trion $X^{\pm}$) is given by one s-shell exciton plus a single excess carrier (electron or hole). Exemplarily, for the negatively charged trion complex X^- the two s-electrons must have opposite spins due to the Pauli exclusion principle, whereas both spin orientations are allowed for the hole. As a consequence, there is no fine structure splitting. However, the polarization of the emitted photon is ruled by the spin of the excess carrier in the dot. Recombination lines from other charge configurations in the dot are rarely used for single photon sources. Therefore we will limit the discussion to the above mentioned configurations.

Important properties of a perfect single photon source are an internal quantum efficiency of unity, a predetermined photon emission time (no jitter), and the indistinguishability of the emitted photons. In practice, the properties e.g.,

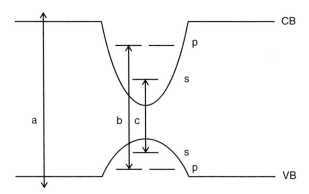

Fig. 6.6. (a) Non-resonant, (b) quasi-resonant, and (c) resonant optical excitation schemes

internal quantum efficiency, jitter, coherence time, polarization of the emitted single photons critically depend on the excitation process. Here, we distinguish between nonresonant excitation into the wetting layer or barrier, quasiresonant excitation into a higher shell, e.g., p-, d-shell, or resonant excitation into the s-shell (see Fig. 6.6).

Non-resonant Excitation

After pulsed excitation of the quantum dot above the barrier bandgap electron–hole pairs are mainly generated in the barrier and subsequently electron and holes are captured by the QDs and relax to the lowest energy levels within a short time scale ($\sim$1–100 ps). In practice, even or odd carrier numbers in the dot are possible and have been observed. The detailed occupation depends on the pump power and on the specific environment of the dot (e.g., donors, electric fields). Exemplarily, we discuss the scenario for an even number of carriers in the dot. The recombination of this multiexcitonic state occurs in a cascade process, i.e., in sequential optical transitions of the multiexciton states $NX, (N-1)X, \ldots$ the biexciton $2X$, and the exciton $1X$. All transitions possess typical radiative lifetimes in the range of a few 100 ps up to $\sim$1 ns. The energy of the photons emitted during relaxation depends significantly on the number of charge carriers that exist in the QD, due to Coulomb interactions enhanced by strong carrier confinement. If the recombination time of the multiexcitonic states are longer than the recombination time of the free electron–hole pairs in the barriers, each excitation pulse can lead to at most one photon emission event at the corresponding NX transition. Therefore, regulation of photon emission process can be achieved due to a combination of Coulomb interactions creating an anharmonic multiexciton spectrum and slow relaxation of highly-excited QDs leading to vanishing reexcitation probability following the corresponding photon emission event at the NX-transition.

Thus, specific photons from the cascade process, e.g., the $2X$ and $1X$, can be spectrally filtered out and can be used to generate single photons.

Due to the nonresonant excitation process the number of captured carriers in the dot is given by a Poisson distribution. To achieve a quantum efficiency close to unity for a certain transition, the pump power has to be adjusted well above the average occupation number of the corresponding transition. This has serious implications for the time jitter of the emission process. It is determined by the carrier capture times, the relaxation times within the dot, and the spontaneous emission time of the corresponding transition. A typical jitter is in the range between several 100 ps up to a few nanoseconds. Due to the typically shorter biexcitonic lifetime, the biexcitonic state produces single photons with significantly less jitter in their emission time than the single exciton state [13].

Another drawback of the nonresonant excitation process is that charge carriers can be captured by adjacent traps or defect centers in the vicinity of the dot. This might lead to fluctuations in the emission wavelength between different pulses and is known as spectral diffusion, a major line broadening effect for quantum dot transitions. In addition pure dephasing occurs due to elastic scattering with other carriers. This will further reduce the indistinguishability of the photons.

Furthermore, the time averaged biexciton/exciton emission is fully unpolarized since both fine structure components contribute equally to the emitted light. Thus, a polarization filter is necessary to select, e.g., linearly polarized light which reduces the efficiency by a factor of 2.

Quasi-resonant Excitation

The quasi-resonant excitation into a higher shell, e.g., the p-shell (see Fig. 6.6) has some important advantages against the non-resonant pumping scheme discussed above. While the above-barrier excitation might be approximately considered as an incoherent excitation process, the p-shell pumping can be arranged as a coherent process. This opens the possibility of a controlled occupation of one electron–hole pair in the p-shell by applying a π-pulse. After relaxation into the s-shell, each generated electron–hole pair delivers a single photon. Thus, a high quantum efficiency (close to one) can be possible. Another important aspect is a nearly complete suppression of background light for quasi-resonant pumping. This leads to an almost complete suppression of multiple ($n \leq 2$) emission events and therefore nearly perfect triggered singephoton emission is achieved ($g^{(2)}(0) \sim 0$) [14]. In addition, pure dephasing processes should be drastically reduced since the charge carriers are exclusively generated within the desired dot.

Resonant Excitation

The ultimate pumping process would be a direct excitation into the s-shell of the dot (see Fig. 6.6). Using a π-pulse, exactly one of the two optically bright exciton states could be excited. This scheme possess the same advantages as the quasi-resonant excitation scheme, but in addition, no additional relaxation process would be necessary before the photon emission process. That would reduce the time jitter to solely the radiative lifetime. Furthermore, the polarization of the photon used in the excitation process is carried to the emitted photon if spin relaxation could be neglected. This scheme has not been used so far for implementation of single photon sources due to serious problems with laser stray-light. However, very recently, Mueller and coworkers [15] demonstrated resonance fluorescence from a coherently driven semiconductor quantum dot in a cavity. The dot was embedded in a planar optical microcavity and excited in a wave-guide mode so as to discriminate its emission from residual laser scattering (see Sect. 6.4). This promises both improved photon indistinguishability and high efficiency. Resonant electrical pumping could may be achieved with double-heterojunction resonant-tunneling structures as proposed in [16].

Other excitation schemes like stimulated Raman scattering involving adiabatic passage (STRAP), or variants of them, have been discussed for QDs [10] and also demonstrated in atomic systems [17] for the controlled generation of single photons. These schemes allow an active control of both the excitation and decay processes in a *coherent* way. Therefore, they transcend the temporal separation of the excitation and decay process and promise a high degree of indistinguishability of the single photons. However, these techniques rely on two tunable lasers or one laser and a high finesse cavity exactly tuned to the desired transition which requires a rather elaborate execution.

In summary, the implementation of the nonresonant pumping process is easy and robust since no specific excitation wavelength is necessary and the corresponding electrical design is accomplished with the standard semiconductor technology like p-i-n structures [18]. This scheme should be applicable for applications in quantum cryptography and standard of optical brightness. The more sophisticated schemes of quasiresonant, resonant pumping, and STIRAP techniques are inevitable if a high degree of photon indistinguishability is necessary like for applications in quantum computing or for quantum teleportation experiments.

Influence of the Dark Excitonic States

The dark exciton state (DS) can influence the optical and quantum optical properties and thus the single photon emission of a quantum dot. While the dark state does not usually play a major role in the optical properties of epitaxially grown QDs, e.g., InGaAs, it is much more evident in those of colloidal QDs as the splitting between the exciton bright state (BS) and the DS is usually considerably larger (2–20 meV [19,20]). However, the DS can also distinctly effect the properties of epitaxially grown QDs [21]. Exciton relaxation to the lowest QD state (DS) leads to a quenching of the BS photoluminescence and thus reduces the single-photon quantum efficiency. On the other hand, due to thermalization back into the BS, the uncertainty in the time of photon emission is increased leading to a larger time jitter.

6.2.3 Cavity Effects on Single Photon Emission

Coupling the emitter to a single cavity mode is very desirable with respect to high repetition rates, high quantum efficiencies, and a large degree of indistinguishability of the emitted photons.

The spontaneous emission of photons is randomly distributed over the full solid angle of 4π of space and is therefore subject to photon losses by total internal reflections on the semiconductor-vacuum surface, this reduces the photon-capture efficiency of all optically coupled devices (e.g., fibers for collection of the emission). It is therefore desirable to channel a large fraction β of the emitted photons to a single spatial mode of a micro- or nano-cavity. Examples of modern semiconductor cavity structures are micropillar cavities, microdisk cavities (see Fig. 6.7), and photonic crystal cavities (see Chap. 9). These resonator structures are characterized by well defined spectral and spatial mode profiles as a consequence of a strong lateral and vertical confinement of the light. Moreover, they possess high quality factors Q and small mode volumes V_m. These properties increase β even more by an enhancement of the spontaneous decay rate which is given by the Purcell effect (see Chaps. 8 and 9). In addition, the shorter radiative recombination times allow for higher repetition rates for the single photon emission. If a quantum emitter is spectrally matched with a single cavity mode, located at a maximum of the electric field, and its dipole is aligned with the local electric field, the Purcell factor is given by $F_p = (3/4\pi^2)(\lambda/n)^3 Q/V$ [22], where λ is the emission wavelength and n the refractive index of the semiconductor. The β factor can be estimated to $\beta = F/(F + \gamma)$, where $\gamma/\tau_{\text{free}}$ is the spontaneous emission rate into leaky modes and F/τ_{free} its spontaneous emission rate into the cavity mode [23]. Since γ is in the order of one, very large values of β are obtained for moderate

Fig. 6.7. SEM image of a micropillar cavity (*left*) and a microdisk cavity (*right*)

Purcell factors (e.g., $\beta > 0.8$ for $F > 5$). Thus, for achieving a high β value the spatial position of the QD and the energy of its transition has to be adjusted with respect to the maximum of the mode profile and the energy of the cavity mode, respectively. The former task is still a challenge although considerable progress has been made on prepatterning of samples grown by MBE to create nucleation spots on which QD growth takes place (see Chap. 2). An alternative way is the positioning of the cavity around a QD. This idea has allowed a deterministic coupling of a single QD to a single mode of a photonic crystal cavity [24] (see also Chap. 9).

The energetic resonance condition can be achieved by temperature tuning, by electric field tuning via the quantum confined Stark effect, or by local thermal annealing (see also Chap. 2). In most practical cases, temperature tuning is very limited due to the phonon induced broadening of the linewidth. Typically, this tuning mechanism is used in the 0.1 to few meV range. A comparable tuning range is accessible by the quantum confined Stark effect. By local thermal annealing values up to 15 meV have been reported [25]. Up-to-date the QD transition energies are only defined within a relatively broad inhomogeneous linewidth ($\sim$10–30 meV) of the corresponding QD ensemble. That implies that several QDs have to be placed inside a cavity with prepatterned QDs so that one can be definitely tuned into resonance. However, such a procedure has some serious drawbacks since even off resonance QDs may contribute considerably to the mode emission [26,27]. The physical mechanism is not yet fully understood but it degrades the purity of the single-photon emission. Therefore, a better control of QD exciton energies during or after growth by, e.g., an annealing technique would certainly constitute a breakthrough for numerous applications in the future.

The *quantum efficiency* η of the single photon source can be viewed as the product of the coupling efficiency of the emitter to the cavity mode β and the extraction efficiency of the single photon into a single-mode traveling wavepacket. In order to determine η one has to distinguish between the strong and weak coupling regime (see Chap. 8).

For a micropillar cavity in the *weak coupling regime* η is given by

$$\eta = \beta \frac{Q}{Q_{\text{int}}}, \qquad (6.17)$$

where $1/Q_{\text{int}}$ is the intrinsic loss due to the finite DBR reflectivity and $1/Q$ is the total cavity loss. A detailed discussion can be found in [23]. Quantum efficiencies up to 10% have been reported in photolithographically defined pillar microcavities [28]. In the case of *strong coupling* η can be expressed by

$$\eta = \beta \frac{\kappa}{\kappa + \gamma}, \qquad (6.18)$$

where κ is the intracavity field decay rate, and γ the emitter decay rate [29]. Recently, large values as high as 97% have been demonstrated in a single quantum dot microcavity system in the strong coupling regime [30] which is very promising for quantum information applications.

It is important to note that a high degree of indistinguishability and quantum efficiency cannot be realized in the case of an incoherent excitation scheme via an excited state in the quantum dot, simultaneously. This is caused by the time-jitter which is induced by the relaxation from the excited state (e.g., p-shell) to the ground state (e.g., s-shell) of the emitter. A detailed study of this effect has been performed by Kiraz and coworkers [10] and one main result of their study is displayed in Fig. 6.8. With increasing Purcell factor an increase of the β-factor is observed. In contrast, the degree of indistinguishability $(1 - p_{34})$ decreases monotonically to 44% for $F_p = 100$. In order to achieve both, a high degree of indistinguishability and quantum efficiency they propose a coherent pumping scheme, such as one involving cavity-assisted spin flip Raman transitions.

For some implementations, e.g., in quantum cryptography linear *polarized* single photons are required. It has been shown, that the polarization of the emission of quantum dots embedded in elliptical microcavities possess a high degree of linear polarization. Pillars with *elliptical* cross section lift the degeneracy of the lowest order mode. These modes will exhibit Purcell enhancements for just one polarized state of the single dot. Therefore, losses with

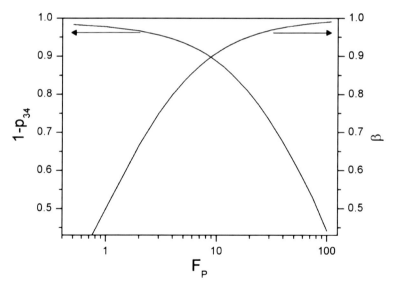

Fig. 6.8. Dependence of indistinguishability and collection efficiency on the cavity-induced decay rate of a quantum dot. Parameter values are $\Gamma_{\text{spon}} = 10^9 \text{ s}^{-1}$, $\Gamma_{\text{relax}} = 10^{11} \text{ s}^{-1}$, $\gamma_{\text{deph}} = 0$, and the excitation laser is a Gaussian beam with a pulse width of 10^{-11} s. Peak laser Rabi frequency is changed between 1.1×10^{-11} and 0.93×10^{-11}. After [10]

external polarizers could be avoided. Recently, linear polarization degrees of more than 90% have been reported from high-quality micropillar cavities [31].

6.3 Experimental

6.3.1 Measurement of the First-Order Coherence Function

First order coherence measurements can be applied to characterize the coherence length l and coherence time $(T_2 = l/c)$ of a single photon source. In a quantum optics picture, T_2 represents the duration of a wave train that is associated with the emitted photon. The length of the wave train as well as the uncertainty of the photon energy $(\Delta E \times T_2 \geq h/2\pi)$ cannot be measured for a single photon. The operators of phase and the number of photons do not commute. Thus, the definition of a single photon $(n = 1)$ as an oscillatory wave with a definite phase is meaningless. The phase of a number state is completely undefined. However, in experiments where the single quantum system is repeatedly excited, the frequency distribution of the emitted wave packet, ΔE, and its phase information become experimentally accessible [32]. Typically, a Michelson interferometer (MI) is applied to measure the first-order coherence function in order to determine the coherence time T_2 (see Fig. 6.9). It is realized with a 50/50 beam splitter, a fixed mirror in one arm, and a

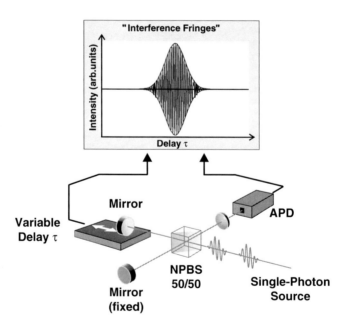

Fig. 6.9. Schematic sketch of a Michelson interferometer (MI) and a typical measurement result for the first order coherence function of a single photon source

movable second mirror in the other arm. The path length of the second arm is typically scanned by a computer-controlled translation stage within a mm to cm range with a spatial resolution of $\sim\lambda/10$. A charge-coupled device (CCD) or an APD is used to record the intensity at the MI output. A CCD offers the advantage that several dots could be studied in parallel [33].

If we assume a quasimonochromatic incoming light field of the form $E_{in} = E_0(t)\exp(i\omega_0 t)$ with the central frequency ω_0, a linewidth $\Delta\omega$, and the time averaging is performed on a larger time scale than the coherence time T_2 the balanced MI output intensity is given by [34]:

$$I = 2I_0\left[1 + |g^{(1)}(\tau)|\cos\varphi(\tau)\right], \tag{6.19}$$

where I_0 is the intensity delivered to the MI. The delay time τ is given by the optical path difference of the interferometer, $\tau = 2(L_1 - L_2)/c$, with c being the speed of light, and $\varphi(\tau) = \arg(g^{(1)}(\tau))$, i.e., the phase difference. Interference fringes with a period of ω_0 are observed. The visibility of the interference fringes, $V(\tau)$, is defined as

$$V(\tau) = \frac{I_{\max}(\tau) - I_{\min}(\tau)}{I_{\max}(\tau) + I_{\min}(\tau)}, \tag{6.20}$$

where $I_{\max}$ and $I_{\min}$ represent the maximum and minimum intensity of the interference pattern, respectively. The dependence of the visibility on τ for a balanced MI can be written in terms of the first-order coherence function $g^{(1)}(\tau)$

$$V(\tau) = |g^{(1)}(\tau)|. \tag{6.21}$$

Thus, the magnitude of the degree of temporal coherence may therefore be measured by monitoring the visibility of the interference pattern as a function of delay time. In many experimental situations, the decrease of $|g^{(1)}(\tau)|$ is described by an exponential ($\sim\exp[-|\tau|/T_2]$) or Gaussian ($\sim\exp[-(\pi\tau^2)/(2T_2^2)]$) function and T_2 can be determined from the corresponding fit [33, 35–37].

It is important to note that the temporal fluctuations of the light, the first-order coherence, and the frequency spectrum are manifestations of the same physical properties of the emitters that constitute the light source. The relation between the spectrum of the light $F(\omega)$ and its degree of first-order coherence is given by the *Wiener–Kintchine theorem* [2]

$$F(\omega) = \frac{1}{\pi}\mathrm{Re}\left[\int_0^\infty d\tau g^{(1)}(\tau)e^{i\omega\tau}\right]. \tag{6.22}$$

This theorem has practical relevance since sometimes, it is experimentally more convenient to measure $g^{(1)}(\tau)$ instead of the spectrum. This technique is known as Fourier-transform spectroscopy. A consequence of (6.22) is an inverse relation of linewidth (FWHM) $\Delta\omega$ and coherence time T_2. The exact relation between them depends on the spectral profile of the light source. For a Lorentzian (Gaussian) spectral lineshape the Fourier transform is an exponential (Gaussian) curve with a linewidth of $\Delta\omega = 2/T_2$ ($\Delta\omega = \sqrt{8\pi\ln 2}/T_2$).

6.3.2 Measurement of the Second-Order Coherence Function

Different experimental techniques are applied to measure the second-order coherence function $g^{(2)}(\tau)$. The probably most direct method would be to measure the times of a single-photon detector's counting events and to calculate the correlation function according to (6.8). However, this method would limit the measurements to time scales longer than the dead time of the detector which amounts to several 10 ns. To overcome this problem, detection schemes using two independent detectors in a Hanbury-Brown–Twiss type setup are usually used (see Fig. 6.10). The setup consists of two orthogonally arranged pathways centered around a nonpolarizing (50/50) beam splitter. Each arm of the HBT interferometer is equipped with a high sensitive single-photon detector. In such an arrangement, the second detector can detect an event while the first one is still dead. Now, the time-resolution is typically determined by the response time of the detector. Currently, used avalanche photodetector modules (APDs), which offer the highest detection efficiencies (∼40–70%) in the visible and near infrared spectrum have response times of 400–700 ps. APDs with slightly lower detector efficiencies (∼5–35%) possess response times in the 30–50 ps range.

Technically, two operation methods can be distinguished. In the first approach, a time-tagged method for recording of individual photon events with their arrival time on both channels is used. This allows the most flexible off-line analysis of the photon dynamics. Drawbacks are large data sets and the fact that the computation of the correlation function is very time consuming. In the second method, only the time differences between the detection

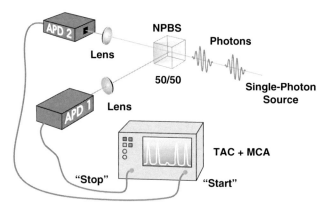

Fig. 6.10. Photon autocorrelation setup after Hanbury-Brown and Twiss. The collected photon stream from a single photon source is prefiltered by a spectrometer (alternatively: narrow band filters) and sent to fast avalanche photo detectors (APDS) trough a 50/50 nonpolarizing beam splitter (NPBS). The photon coincidence statistics $n(\tau)$ are measured by combined time-amplitude conversion (TAC) and multichannel analysis (MCA)

events (Start and Stop) are registered and in a subsequent process a time-to-amplitude conversion followed by a multichannel analysis is performed in order to generate a histogram of coincidence events $n(\tau)$. One has to consider that the measured coincidence function $n(\tau)$ differs from the original second-order coherence function $g^{(2)}(\tau)$. The probability to measure a time difference at time τ is given by [38]: $n(\tau)$ = (probability to measure a stop event at time τ after a start event at time 0) × (probability that no stop detection has occurred before)

$$n(\tau) = (G^{(2)}(\tau) + R_{\text{dark}})(1 - \int_0^\tau n(\tau')\mathrm{d}\tau'), \quad (6.23)$$

where $G^{(2)}(\tau)$ is the unnormalized second-order coherence function and R_{dark} describes the detector dark counts. The measured histogram of coincidence counts $n(\tau)$ approaches $G^{(2)}(\tau)$ in the limit when R_{dark} is much smaller than the signal count rate R, and the average arrival time of the photons $1/R$ is much smaller than the observed delay time τ. This means that the probability that no stop detection has occurred before, is approximately 1. Losses, like undetected photons lead only to a global decrease of $G^{(2)}(\tau)$ which can be compensated, e.g., with a longer measuring time. An exact solution of the above integral equation shows that $n(\tau)$ exhibits an exponential decrease on a time scale given by the detector count rate [38].

Normalization of $G^{(2)}(\tau)$ can be achieved by considering a Poisson light source ($g^{(2)}(\tau) = 1$) with the same average detector count rates. For such a source, the experimental expectation values of correlation counts $n(\tau)$ under either pulsed or cw excitation can be estimated directly from

$$C_{\text{Poisson}}^{\text{pulsed}} = R_{\text{Start}} R_{\text{Stop}} \Delta t_{\text{Laser}} t_{\text{int}},$$

$$C_{\text{Poisson}}^{\text{cw}} = R_{\text{Start}} R_{\text{Stop}} \Delta t_{\text{MCA}} t_{\text{int}},$$

with R_{Start}, R_{Stop} the detector count rates, t_{int} the total integration time, and either the laser pulse repetition period Δt_{Laser} or the Multichannel Analyzer (MCA) time bin width Δt_{MCA}. The Poisson-normalization of $G^{(2)}(\tau)$ is then simply achieved by division of each individual correlation pulse peak integral (detected at integer multiples of the laser repetition period $n \times \Delta t_{\text{Laser}}$) by $C_{\text{Poisson}}^{\text{pulsed}}$ or as the ratio of the full cw-correlation trace $g^{(2)}(\tau) = G^{(2)}(\tau)/C_{\text{Poisson}}^{\text{cw}}$, respectively.

6.3.3 Measurement of the Indistinguishability

The indistinguishability of photons from a single-photon source can be measured by colliding two individual photon wavepackets at a beam splitter in a Hong–Ou–Mandel-type experiment [36, 39]. The statistics of the outcome of the photons from the beam splitter is detected by single-photon detectors. If the duration of the single photon wavepackets exceed the response time of the detectors interference effects occur and can be studied in a time-resolved manner [11]. However, the photon wavepackets emitted from single semiconductor QDs and QDs in microcavities are typically short (10 to several 100 ps) with respect to the typical response time of the detectors. So far, this was the case for all experiments with quantum dots and temporal effects are therefore neglected in the following.

Furthermore, it is important to note that two photon interference can always be observed if the signal is spectrally filtered by a sufficiently narrow interference filter in front of the detection system. Such an approach would lower the temporal resolution of the detection system and therefore reduce the maximum possible repetition rate of the single-photon source to prevent photons from overlapping [8].

Different types of Hong–Ou–Mandel (HOM) interferometers are commonly used. In a first type, a free-space optics setup has been implemented in a Michelson-type interferometer shown in Fig. 6.11 [36, 40]. A widely used method is to excite the active emitter (single QD) of the single-photon source by a pair of laser pulses separated by ΔT (~2–3 ns). This cycle is repeated with

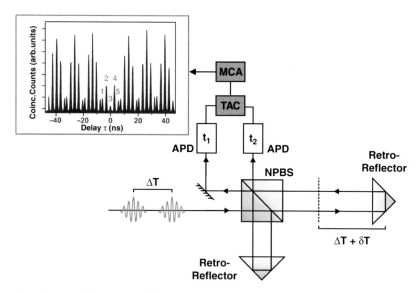

Fig. 6.11. Two-photon interference experiment. Two single photons with a delay of ΔT enter a nonpolarizing beam splitter. The pulses are interfered with each other using a Michelson type interferometer with a path length difference of $\Delta T + \delta t$. The photons at the interferometer outputs are detected by APDs. The photon coincidence statistics $n(\tau)$ are measured by combined time-amplitude conversion (TAC) and multichannel analysis (MCA). The typical histogram shows clusters of five peaks separated by the pump laser repetition period. The five different peaks are due to different combinations of photon paths through the interferometer (see text). The small area of peak 3 demonstrates two-photon interference

the pump laser repetition period (~12–13 ns). Upon each excitation event, ideally the QD emits one single photon. After a polarization selection, the emitted photons are sent to the two arms of the MI interferometer which introduces a propagation delay between the short and long arms of $\Delta T + \delta t$. The two output ports of the beam splitter are fed to single photon counting modules where the time differences between the detection events (Start (t_1) and Stop (t_2)) are registered. In a subsequent step a time-to-amplitude conversion followed by a multichannel analysis is performed to generate a histogram of coincidence events of the time intervals $\tau = t_2 - t_1$. Figure 6.11 presents such a histogram for $\delta t = 0$. The histogram shows clusters of five peaks separated by the pump laser repetition period. The five different peaks are due to different combinations of photon paths through the interferometer. The peaks at $\tau = \pm 2\Delta T$ arise from the first photon taking the short arm and the second taking the long arm. For the peaks at $\tau = \pm \Delta T$ both photons pass through the same arm. The central peak $\tau = 0$ corresponds to the situation where the first photon takes the long arm and the second photon takes the short arm. Therefore, both photons arrive at the beam splitter at the same time. The reduced coincidence signal at $\tau = 0$ is the signature of the two-photon interference. The probability of two photons colliding in the beam splitter and leaving in opposite directions can be defined by the quantity

$$p_{34}(\delta t) = \frac{A(0)}{A(T) + A(-T)}, \tag{6.24}$$

where $A(\tau)$ is the area of the peak at time interval τ on the histogram where the delay of the MI interferometer is set to $\Delta T + \delta t$. The coincidence dip $p_{34}(\delta t)$ is then measured by varying the interferometer path length offset δt.

The advantage of the free space realization of the HOM interferometer is a high throughput. Special care has to be taken to satisfy the mode matching condition using free space optics. Alternatively, a polarization maintaining optical fibre Mach–Zehnder interferometer (MZI) with fibre optic beam splitters may be used [41]. There, the critical mode matching conditions are automatically fulfilled. However, losses from fiber in- and out-coupling may occur and reduce the signal collection efficiency.

6.4 Single-Photon Sources

In this section we give a discussion of recent novel developments on semiconductor single-photon sources. The different approaches report progress on one or more of the following properties, e.g., extension of the demonstrated wavelength range, polarization control, high single-photon emission rates, high temperature operation, electrical pumping, photon indistinguishability, coherent state preparation, and single photon emission in the strong coupling regime.

6.4.1 Available Wavelengths

For practical applications the emission wavelength of the single photon source should be one which minimize optical losses in the transmission and maximize the photon detection efficiency. Currently, Si-Avalanche photodiodes are the most efficient single photon detectors. They possess their highest quantum efficiency ($\sim$70–75%) in the wavelength range 630–750 nm. The emission wavelength of the source should be in the absorption minimum of the transmitted medium. For free space applications this is fulfilled in the visible spectral region, for a plastic optical fiber within an absorption minima of the fiber (470, 560, 650 nm), and for a glass fiber at 1.3 µm (dispersion minimum) and around 1.55 µm (absorption minimum).

Since the first demonstration of single-photon emission on the basis of a single quantum dot [42] studies on single-photon sources have been performed on a wide variety of semiconductor alloys such as $In_xGa_{1-x}As/GaAs$ ($\sim$870–970 nm) [42–44], $In_xGa_{1-x}As/Al_yGa_{1-y}As$ ($\sim$800–815 nm) [45–47], $In_{0.75}Al_{0.25}As/In_{0.7}Ga_{0.3}As$ (770 nm) [48], $GaAs/Al_xGa_{1-x}As$ ($\sim$735–790 nm) [49], InP/GaInP ($\sim$660–690 nm, 750 nm) [50–52], CdSe/ZnSe ($\sim$520 nm) [53], and GaN/AlN ($\sim$350 nm) [54].

Recently, single photon emission at the telecom wavelength $\sim$1.3 µm has been also realized [55, 56]. By carefully controlling the critical strain of InAs/GaAs self-assembled quantum dots Ward et al. [55] produced a microcavity sample with a low density of large dots. The larger dots emit around 1.3 µm. Figure 6.12 displays a PL spectrum from the micropillar cavity at two different pump powers of a picosecond-pulsed laser at 785 nm. Both, exciton (X) and biexciton (X_2) lines can be identified in the spectrum. However, it was not possible to decide whether the exciton and biexciton lines arise from the same dot nor whether the other lines observed correspond to emission from different dots. Performing photon statistics measurement at telecom-wavelength is a challenge since the detectors exhibit both lower efficiencies

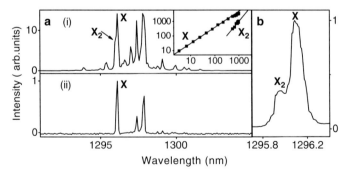

Fig. 6.12. (a) PL spectra from a micropillar cavity for two different laser powers at 5.2 K. *Inset:* Pump power dependence of the exciton and biexciton lines. (b) Detailed PL spectrum at a pump power of 10 µW. After [55]

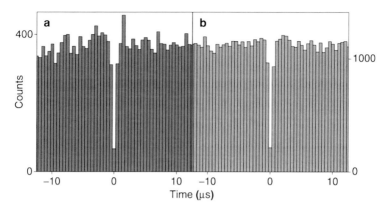

Fig. 6.13. (a) Photon correlation function of the X emission line at 5.2 K. (b) The same line at 32 K. After [55]

and larger dark count rates. The best signal-to-noise ratios can be achieved by using narrow time gate widths of the order of a few nanoseconds. Thus, the resulting correlation function is given by a discrete histogram rather than a full representation of $g^{(2)}(\tau)$. Figure 6.13 shows the results of a correlation measurement on the exciton line at 5.2 K [57]. The height of the zero delay peak is clearly below that of the neighboring channels. The measured suppression of the multiphoton pulses is 18.3% of the calculated Poisson-normalized level. Accounting for the dark counts in the detectors and unwanted light leakage through the used filters the residual multiphoton rate was estimated to be below 10%. A more comprehensive summary of telecom-wavelength single-photon sources based on single quantum dots can be found in [57]. In summary, the currently accessible spectral range spans from the near infrared (1.3 μm) up to the UV spectral region (350 nm).

6.4.2 Polarization Control and High Single-Photon Emission Rates

For many applications in quantum information technology the single-photon source should provide a single polarization of the emitted photons together with high single-photon emission rates. Both properties can be realized by embedding a quantum dot in an adequate cavity structure. We will focus the discussion here on micropillar structures due to their high practical relevance.

The fundamental cavity mode of a circular shaped micropillar cavity is polarization degenerate. This degeneracy is lifted in structures with an elliptical cross section [31, 58–60]. Due to the shape anisotropy the fundamental mode is split into two opposite linear polarized modes. The polarization is aligned along the major and minor axes of the pillar cavity and the energy splitting depends on the ellipticity, i.e., on the ratio of the axes lengths. Splittings as large as 15 meV have been realized in micropillar cavities [58].

The polarization of the emission of a quantum dot, e.g., from an excitonic transition depends on the spin state of the exciton (see Sect. 6.2.2). For QDs outside a cavity there is no preferred polarization since both spin states are equal likely. For QDs inside the cavity the mode structure in which the QDs can emit is modified. The origin of the usually observed strong polarization degree in the emission is twofold. First, photons are preferentially emitted into the polarized mode due to the Purcell effect which prepares the photon into a given quantum state. Second, the polarized emission is more efficiently collected by a collection lens due to the directionality of the mode profile. Polarization degrees higher than 90% have been found in the emitted light in one linear polarization state [31,59].

A more sophisticated device would also allow for an active selection of the polarization states on the chip, e.g., with an applied external field. Recently, Strauf and co-workers have realized such a single photon source where both high single-photon rates and control over the polarization state have been achieved [61]. Figure 6.14 shows a schematic of their device design. The QDs were embedded into the center of a one-wavelength thick cavity which is sandwiched between two distributed Bragg reflectors. The two gates allow for controlling a QD loading process and a local current heating within the cavity. Trenches are fabricated to define oxide apertures and therefore optical

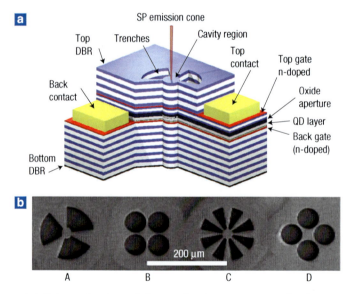

Fig. 6.14. (a) Schematic view of the single-photon source. The device consists of a λ-cavity region which is embedded between two distributed Bragg reflectors. The cavity contains a single layer of InAs QDs (*grey*) and a mode-confining tapered AlO_x region (*dark blue*). The top and back contacts (*yellow*) are formed adjacent to the QDs. (b) Scanning electron micrograph pictures of different trench designs. The inner lateral cavity area has a diameter of $\sim$20 nm. After [61]

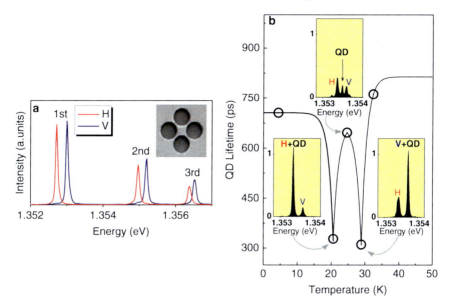

Fig. 6.15. (a) Mode spectra of the single photon source showing the first three nondegenerate cavity modes. H (V) indicates horizontal (vertical) polarization. (b) Temperature tuning of a single QD exciton transition trough the H and V modes. After [61]

mode-confinement. The design provides Q-factors up to 50,000 together with the possibility of controlling the mode degeneracy. Figure 6.15 shows a cavity mode spectrum showing a linear polarization splitting of 270 µeV for the fundamental mode. A contrast ratio of 57:1 has been observed between the H- and V-polarized modes. From the lifetime reduction when the QD is either in resonance with the vertical or horizontal mode a Purcell factor of 2.7 has been determined. The switching speed using temperature tuning is very limited. However, the authors mention that Stark-shift tuning will also be possible with their device design which should allow high modulation frequencies. Another important feature of the device is the record high single-photon emission rate. Under pulsed (c.w.) optical excitation a single photon rate of 4 MHz (10.1 MHz) have been detected. There, the excitation conditions have been chosen to a working point where $g^{(2)}(0) = 0.4$, that is, when the single photon source still performs 2.5 times better than an attenuated Poissonian source of the same average intensity. If corrected with the 13% detection efficiency of the setup the device emits into the first lens with a rate of 31 MHz. This ultrahigh rate is a combination of several effects. Besides the already discussed effects of the Purcell enhancement and the large geometrical extraction efficiency of the cavity mode two other reasons are responsible. To avoid signal loses by exciton relaxation to the dark states (see Sect. 6.2.2) the trion transition has been used as the optically active transition. This procedure increased the signal rate by a factor of three. Furthermore, a positive bias voltage created a fivefold intensity enhancement, most probably due to a field-enhanced carrier capture process.

In a very recent work, Rossbach et al. [52] demonstrated saturation count rates of 1.5 MHz (8.1 MHz into the first lens, with an extraction efficiency of 4.1%) at 659 nm under continuous wave excitation. The result has been achieved by using self-assembled quantum dots embedded in a planar microcavity realized by monolithically grown AlGaAs DBRs. This is a promising result since in pillar cavities the signal rate should be even higher due to a better mode confinement. The operation wavelength is in a transmission minimum of the plastic optical fiber and well suited for free space applications. Moreover, current Si-based single-photon detectors have their highest photon detection efficiency in the red spectral range. In summary, ultrahigh single photon rates are already available at different wavelengths and in carefully designed semiconductor devices.

6.4.3 High Temperature Operation

From the viewpoint of applications, e.g., in quantum cryptography, it would be important to be able to operate the single-photon sources at elevated temperatures, achievable by either liquid nitrogen cooling, thermoelectric cooling, or ultimately without any cooling. For high-temperature operation several requirements have to be fulfilled. First, large electronic band offsets and strong quantum confinement effects for both electrons and holes are necessary to prevent carrier thermalization and thermal emission of carriers into the wetting layer and/or barriers. Second, a large biexciton binding energy is important to prevent spectral overlapping of exciton and biexciton recombination lines at elevated temperatures. Wide-bandgap semiconductors like II–VI- and group-III nitride semiconductors offer large biexciton binding energies. For example, biexciton binding energies of $\sim$20 meV [62] and 30 meV [63] have been reported for CdSe/ZnSSe and GaN/AlN quantum dots, respectively. The highest temperature operation so far (200 K) has been reported on the basis of these two material systems.

K. Sebald et al. reported on the generation of triggered single photons from self-assembled CdSe quantum dots for temperatures up to 200 K [53]. For temperatures above 40 K an increasing multiphoton emission probability due to spectrally overlapping acoustic phonon sidebands of neighboring quantum dots was observed. The authors reported that the multiphoton probability of the bare quantum dot (background subtracted) was strongly suppressed even at 200 K. These results demonstrate the large potential of self-assembled CdSe/ZnSSe quantum dots for high temperature operation if the QD density can be further reduced. Very recently, R. Arians and co-workers reported the luminescence from a single CdSe quantum dot at room temperature [64]. In this case the self-assembled dots were embedded in ZnSSe/MgS barrier layers which provide a higher electronic confinement than bare ZnSSe layers.

Very promising, the intensity dropped by less than a factor of 3 between 4 K and room temperature. The room temperature linewidth of the single dot emission was 25 meV which the authors attribute to the interaction of excitons with optical phonons. So far, there has been no report in terms of photon-correlation studies at 300 K.

Recently, Kako et al. [65] reported triggered single photon emission from GaN/AlN quantum dots at temperatures up to 200 K. Figure 6.16 shows PL spectra under pulsed excitation at 3.5 and 200 K together with the corresponding photon correlation functions. The selected exciton line at 3.5 K exhibits a relatively broad linewidth which is most probably due to spectral diffusion. The biexciton at slightly higher energy is clearly separated from the exciton line. A value of $g^{(2)}(0)$ of 0.42 has been obtained at 3.5 K. At 200 K the exciton and biexciton lines are no longer clearly separated due to phonon broadening.

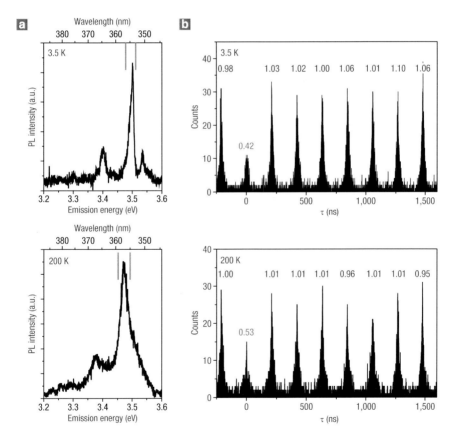

Fig. 6.16. (a) PL spectra under pulsed excitation at 3.5 K (*top*) and 200 K (*bottom*). (b) Photon correlation measurement of the exciton lines, respectively. The selected part of the spectrum which has been used for the autocorrelation measurement is indicated by vertical bars. After [65]

The two-photon probability increases slightly to 0.53 most probably due to a partial overlap with the biexciton line. In addition a broadband background from the AlN matrix contributes to the detected signal. For future improvements it will be crucial to improve the material quality and to reduce the built-in piezoelectric fields for the group-III nitride materials. One possible solution could be the growth on semi-polar or nonpolar surfaces.

6.4.4 Electrical Pumping: Single-Photon LED

First attempts on electrical pumping has been performed on a semiconductor heterostructure sample utilizing the Coulomb blockade effect in quantum wells at ultralow temperatures (some tens of millikelvin) [66]. The photon collection efficiency from the sample was weak preventing the study of the photon correlation function. More efficient schemes at helium temperatures (5–10 K) have been reported by Andrew Shields' group. Their first device was an electrically-driven p-i-n diode where the electroluminescence of the single quantum dot was isolated through a micrometer-diameter emission aperture in the opaque top contact [18]. An improved device design which is based on a cavity structure has been demonstrated by the same group [67,68]. Figure 6.17a shows a schematic of the single photon LED. The quantum dots are embedded in a 3 λ-cavity which is sandwiched between a high-reflectivity bottom DBR and the semiconductor–air interface in the aperture. This low-Q cavity enhances the collection efficiency (tenfold) due to a modified emission pattern [69]. A photon collection efficiency into a NA = 0.5 lens of $4.7 \pm 0.5\%$ has been achieved. Single-photon emission has been demonstrated from both exciton and biexciton transitions (see Fig. 6.17b). The finite peak at delay time $\tau = 0$ is most probably caused by background emission from layers other than the dot.

Electrical pumping schemes also opens the possibility to tailor the time jitter, i.e., the uncertainty in the time of single photon emission events. Typically, short voltage pulses which are used to inject carriers into the optically

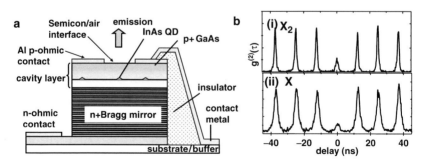

Fig. 6.17. (a) Schematic of the single-photon LED. (b) Autocorrelation functions recorded for the exciton and biexciton emission lines, respectively. After [68]

active region are superimposed on a fixed DC voltage. Bennett et al. [70] used 300 ps long voltage pulses of height V_pulse on top of a fixed bias V_dc chosen between 1.45 V and −0.16 V. By decreasing the applied DC bias they observed a decreasing exciton lifetime. The reason for the lifetime reduction occurs because when the voltage pulse ends a large electric field across the QDs causes a tunneling of electrons out of the QDs. Thus, the neutral exciton and biexcitons states are quenched which effectively caused their lifetime reduction. In this way they were able to demonstrate single-photon emission rates up to 1.07 GHz around 900 nm from both the exciton and biexciton states. Obviously, the operation in the low time-jitter regime comes with the price of a lower internal quantum efficiency since not all excitons are given time for a radiative recombination process.

Recently, Ward et al. [71] demonstrated an electrically driven telecommunication wavelength (∼1.29 μm) single-photon source at low temperature (∼12 K). Here, negative-going 0.5 V pulses (∼10 ns) have been applied to the device shortly after each positive-going excitation pulse to reduce the time jitter. In this case the mechanism for jitter reduction was different. The large InAs/GaAs QDs possess a higher electronic confinement potential which allows the emission to be Stark shifted out of the passband of the used spectral filter (FWHM ∼ 0.5 nm) before all carriers are removed from the dot. In addition, the authors proposed that the time-varying Stark shift also allows to eliminate multiphoton emission processes due to reexcitation during the relatively long (220 ps) electrical pump pulses. Another benefit of the combination of Stark shifting and spectral filtering is the generation of indistinguishable photons. Bennett et al. [72] used this technique to populate and control a single-photon state on timescales below the dephasing time of the state. In this way, they were able to observe two-photon interference with a visibility of 64%.

Another interesting approach is to use an aperture in an oxide layer to restrict carrier injection into a single QD [73–75]. This technique has been already successfully used in the VCSEL technology for current and mode confinement [76]. A thin AlAs layer is grown some 10 nm above the QD layer within the intrinsic region of the p-i-n layer of the device. After a mesa structure has been fabricated a wet oxidation process is performed to convert the outer edge of the AlAs layer to insulating AlO_x. Apertures down to 100 nm diameter can be produced. The advantages of such a design are lower injection currents, lower background emission since ideally only one QD is excited, a good lateral mode confinement together with mechanical stability. The electroluminescence of such a device showed a monochromatic emission (∼950 nm) at a continuous wave excitation (870 pA, 1,65 V, 10 K) over a spectral range of 550–1,050 nm [77]. Furthermore, the observation of Purcell enhancement in the decay rate of a single quantum dot has been reported in an electrically driven oxide confined LED structure [78]. This allowed single photon electroluminescence up to repetition rates of 0.5 GHz.

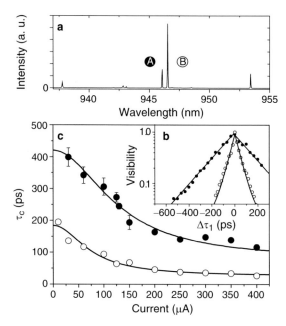

Fig. 6.18. (a) Electroluminescence spectra exhibiting two charged exciton recombination lines. (b) Visibility curves for c.w. injections currents of 200 μA. *Solid (empty) circles* correspond to line A (B). (c) Coherence time τ_c as a function of the current. The *black lines* are theoretical fits. After [79]

Very recently, the dependence of the coherence time T_2 on injection current has been reported by R.B. Patel [79]. Figure 6.18a exhibits the electroluminescence spectra of the used device showing two emission lines which are both attributed to charged exciton recombination. Figure 6.18b shows two visibility curves which have been measured by a Michelson interferometer setup (see also Sect. 6.3). The visibility curve is described by an exponential ($\sim \exp[-|\tau|/T_2]$) and the coherence time has been determined from the corresponding fit. Figure 6.18c shows the extracted coherence times of the two states as a function of injection current. Very long coherence times of up to 400 ps at low pump currents are found and a reduction of the coherence time with increasing current is observed. This dephasing has been attributed to charge fluctuations in the vicinity of the dot. The charge fluctuations lead to a time varying quantum confined Stark effect and consequently to variation of the emission wavelength. The experimental pump current dependence of the coherence time can be well described by a model of spectral diffusion of the transition line [79]. In addition, two photon interference experiments have been performed under DC operation using a fiber-coupled Mach–Zehnder interferometer setup. Injection currents of 100 μA which results in coherence times of ~325 ps have been used for the experiment. The authors observed two-photon interference visibility of 0.33 ± 0.06. This value was consistent with the assumption that interference was entirely limited by the resolution of their detection system and that there was a 100% overlap of the wavefunctions. The observation of higher visibility is envisioned for improved timing resolution.

It is very important for quantum communication applications to develop electrically driven sources at elevated temperatures. This may be achieved by using wide-bandgap semiconductors (see discussion above) and other alloys providing high quantum confinement. Very recently, an electrically driven single-photon emitter in the visible spectrum range, working up to 80 K, has been realized by a p-i-n diode structure based on InP quantum dots as active layers [80]. The InP dots were embedded between two 30 nm thick $Al_{0.2}Ga_{0.8}InP$ layers which were sandwiched between two 100 nm thick $Al_{55}Ga_{45}InP$ layers. This design ensured both good carrier confinement and high internal quantum efficiency. Figure 6.19a shows an electroluminscence spectrum of the sample at 80 K. Two well separated emission lines are visible above a certain background. The emission line at higher energy possesses a linewidth of 0.8 meV and has been identified as the biexciton recombination. Figure 6.19b shows the autocorrelation measurement of the biexciton line. Clear photon antibunching ($g^{(2)}(0) = 0.43$) is observed. The deviation from the ideal source ($g^{(2)}(0) = 0$) is caused by two effects, the limited temporal resolution of the system (500 ps) and background contributions to the signal. After deconvolution with the experimental response function a value of $g^{(2)}(0) = 0.25$ was determined. Subtracting the background gave a value for $g^{(2)}(0)$ of 0.03. This result demonstrates that even at 80 K, these QDs are suitable to provide electrically driven single-photon emission. At that point it is not clear if even higher temperatures can be reached with this dot system due to a possible spectral overlap between exciton and biexciton lines which would degrade the purity of single photon emission.

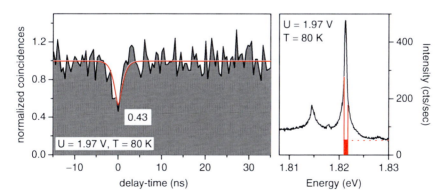

Fig. 6.19. (a) Electroluminescence spectrum of a single quantum dot. (b) Autocorrelation function recorded for the biexciton line. The *vertical lines* represent the part of the spectrum which was detected. The *horizontal line* indicates the signal height of the background. After [80]

6.4.5 Indistinguishable Photons

As we discussed in Sect. 6.2.1 photon indistinguishability is a necessary precondition for linear optics quantum computing schemes and quantum teleportation techniques. The work which is discussed in the following has been performed with QDs embedded in microcavities and the QDs have been quasi-resonantly pumped via an excited state of the QD. Single cavities have been used were the QD is excited twice by a pair of laser pulses separated by a small time window (2–3 ns) (see Sect. 6.3). The microcavities have been employed for utilization of the Purcell effect in order to reduce the radiative lifetime (see Sect. 6.2.1). Excitation via the excited state reduce pure dephasing effects (see Sect. 6.2.2). Both tasks together enlarge the overlap integral $|J|_0^2$ which is the crucial quantity for the observation of strong two-photon interference.

The first result on two photon interference of two single photons emitted from a quantum-dot micropillar cavity was reported from Santori et al. [36]. Their measurements have been performed with a free space HOM interferometer (see Sect. 6.3). The maximum measured factor of reduction in the two photon coincidence count rates which corresponds to the overlap integral $|J|_0^2$ was 0.69. After correction of experimental imperfections a value of 0.81 has been estimated. Furthermore, they made an interesting observation that the mean overlap between two consecutively emitted photons from the same microcavity could be larger than that would be expected from (6.16). This is possible if some spectral broadening results from a spectral-diffusion process which is slower than the separation of the two consecutively emitted photons ($\sim$2 ns) from the quantum dot.

Varoutsis et al. [40] reported measured reduction factors of the two-photon interference of 0.75 which is the highest raw value up to now. Contrary to the above mentioned work, their results fit perfectly to the theoretical expectations, estimated by $1 - T_2/2T_1$ implicating the absence of slow spectral diffusion processes and supposes an ultrafast relaxation rate r (compare (6.16)). A similar result has been achieved by the same group from a single-quantum dot in a two-dimensional photonic crystal cavity [81].

Bennett and co-workers [82] studied the influence of the exciton dynamics on the two-photon interference by using a fiber-based interferometer setup. They found that the two-photon interference visibility is affected by a long lived population of the biexciton state in the dot. Optimum reduction factors of the two-photon interference of 0.66 have been found at low excitation powers. Furthermore, they also reported that the observed mean overlap was larger than expected from (6.16), also indicating slow spectral-diffusion processes.

In summary, the present results on photon indistinguishability are encouraging. However, the coherence and the overlap integral must be improved further if the single-photon sources are to be useful for many quantum information applications. It becomes clear from our discussion in Sect. 6.2.1 and 6.2.2 that a truly resonant pumping scheme into the s-shell of the QD would be advantageous since it avoids the time jitter from the relaxation process and it should cause longer coherence times T_2 due to reduced carrier–carrier scattering processes.

6.4.6 Coherent State Preparation and Single-Photon Emission in the Strong Coupling Regime

Recently, Ester et al. reported on deterministic single photon emission after coherent optical state preparation in the *p*-shell of a single InGaAs/GaAs quantum dot [83]. Under the condition of π-pulse excitation into the *p*-shell they demonstrated nearly perfect single photon emission ($g^{(2)}(0) = 0.02$). However, as discussed in Sect. 6.2.2 a high degree of indistinguishability and quantum efficiency cannot be realized in case of an excitation scheme via an excited state in the quantum dot, simultaneously. Therefore, a truly resonant pumping scheme would be advantageous. Very recently, first approaches on coherent state preparation in the *s*-shell have been reported by Mueller et al. [15]. Both, continuous wave and pulsed optical excitation (8 ps long pulses) have been realized by the group and a full Rabi cycle has been achieved in the single QD resonance fluorescence under the latter. The QD was embedded in a planar optical microcavity and excited in a waveguide mode (see Fig. 6.20). This technique allowed the efficient discrimination of the QD emission from the residual laser scattering. Continuous wave second-order autocorrelation measurements revealed a pronounced antibunching dip, confirming the successful generation of single photons.

Single-photon emission in the *strong-coupling regime* has been observed from a QD-micropillar cavity [30] and from a QD-photonic crystal cavity [27]. The strong-coupling regime is of great interest for many quantum information schemes (see chapters by S. Reitzenstein and A. Forchel [Chap. 8] and

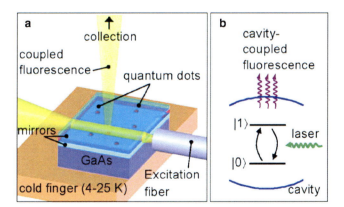

Fig. 6.20. (a) Scheme for orthogonal excitation and detection of the planar microcavity. (b) Energy level diagram and scheme of the cavity in which the quantum dots are embedded. After [15]

M. Scholz et al. [Chap. 11]). In the case of a single photon source it has been anticipated that extremely high efficiency and photon indistinguishability could be achieved [10]. The strong coupling regime requires $g^2 > (\gamma_c - \gamma_x)^2/16$ where g is the exciton-mode coupling frequency, and γ_c and γ_x are the full-width half maxima (FWHM) of the cavity and exciton modes, respectively. Typically, γ_x is in the range of a few microelectronvolts whereas γ_c is considerably larger for semiconductor microcavities (50–100 µeV). In this case the condition for strong coupling reduces to $g > \gamma_c/4$. In the strong coupling regime the decay rates of the coupled QD cavity system are typically very short. In the above mentioned studies values in the 10–20 ps range have been reported. To avoid multiple capture and emission processes from the emitter after a single laser excitation pulse, resonant or quasiresonant excitation schemes are essential to achieve high-purity single photon emission $g^{(2)}(0) \sim 0$. Otherwise, in the case of above barrier pumping process, long lived excitons may be captured by the QD after the emission of a first single photon pulse which leads to multiple photon emission processes.

In the QD-micropillar cavity system, a quasiresonant excitation via an excited state has been performed which also mostly prevented background emitters from being excited. A coupling constant of $g = 35\,\mu\text{eV}$ and a mode linewidth of $\gamma_c = 85\,\mu\text{eV}$ have been reported [30]. This results in a ratio of $g/\gamma_c = 0.41 > 1/4$ which satisfies the strong coupling condition. A value of $g^{(2)}(0) = 0.18$ could be achieved which proves that the dominant emission stems from a single QD.

In the case of the QD photonic-crystal cavity, nonresonant excitation into the barrier has been realized. The research group realized a positioning of an individual QD relative to the photonic crystal cavity with a 30 nm accuracy [27]. Therefore, the QD was located typically at ~90% of the electric-field maximum. A coupling constant of $g = 76\,\mu\text{eV}$ and a mode linewidth of $\gamma_c = 100\,\mu\text{eV}$ have been achieved, clearly entering the strong coupling regime. The autocorrelation exhibits a value of $g^{(2)}(0) = 0.54$ indicating a lower purity of the single-photon emission than in the micropillar case discussed above. The authors discuss two effects for the nonideal behavior. First, multiple carrier capture events per excitation pulse due to the nonresonant pump process. Second, the detected light is also collected from a third peak which is observed between the two anticrossing polariton peaks whose physical nature is not fully understood.

From the discussion above it becomes clear that both resonant-pumping and a positioning of a single quantum dot relative to the microcavity would be highly advantageous.

6.5 Summary and Outlook

During the last few years encouraging progress in understanding the physics of the single-photon generation process and in device fabrication of semiconductor quantum-dot based single-photon sources has been obtained.

Important progress has been achieved in the understanding of the QD excitation process and its role on the coherence time of the photons, on the limitations of photon indistinguishability and collection efficiency, on cavity effects on single photon emission, and on the influence of the excitonic fine structure (dark states) on emission efficiency.

Spectacular achievements are the extension of the wavelength range from the near infrared (1.3 µm) up to the UV spectral region (350 nm), polarization control and the demonstration of ultrahigh measured single-photon emission rates (4 MHz), high temperature operation (200 K) with wide-bandgap semiconductors, electrical pumping with a cavity design for enhanced photon collection efficiency, oxide apertures for ultralow electrical pump currents (870 pA), electrical pumping up to 80 K with an improved heterostructure design, ultrahigh repetition rates up to 1.07 GHz, high-purity single photon emission ($g^{(2)}(0) \sim 0.02$), the generation of high degrees of photon indistinguishability (75%), coherent state preparation in the p- and s-shell of an individual quantum dot, single-photon emission from positioned quantum dots in cavities, and single-photon emission in the strong coupling regime.

However, despite this remarkable progress many challenges remain for the realization of a commercial device. One important issue is that most of the above mentioned achievements have been individually realized and not in combination. Meaningful combinations strongly depend on the specific application of the single-photon source.

For example, for applications in quantum cryptography one would wish a low multiphoton probability, a high emission efficiency into a single mode, a low jitter, high clock rates, polarization control, electrical pumping at elevated temperatures, e.g., achievable by thermoelectric cooling, and an emission wavelength which minimizes transmission losses in combination with a high photon-detection efficiencies.

For application in linear optics quantum computation, low-temperature operation (~ 5 K) seems to be indispensable for long coherence times of the photons and high degrees of indistinguishability. Resonant pumping schemes, preferentially in the strong coupling regime, should be applied which should lead to high quantum efficiencies in combination with high-purity and large degrees of photon indistinguishability.

A precondition for mass production of devices is a scalable deterministic technology for the spatial and spectral matching of the QD with respect to the device architecture. This includes, e.g., the positioning of the QD with respect to a cavity structure. This task is still a challenge although considerable progress has been made on prepatterning of substrates or by forming cavities around selected QDs. The spectral resonance condition can be achieved by local thermal annealing, by temperature tuning, by utilizing the quantum-confined Stark effect, or by tailoring the cavity resonance (e.g., via the pillar diameter) to the QD transition energy.

Finally, the development of semiconductor quantum-dot based single-photon sources will further provide us with interesting physics and pave the way for new applications in the exciting fields of quantum optics, nanophotonics, and quantum information science.

Acknowledgments

The author thanks Alper Kiraz for critical reading the manuscript. The demonstrations of the electrically driven single photon source at 80 K and count rates of 1.5 MHz at 659 nm have been made by our group in Stuttgart. The essential contributions of G. Beirne, M. Jetter, M. Reischle, R. Rossbach, W.M. Schulz are gratefully acknowledged.

I thank Y. Arakawa, E. Hu, A. Kiraz, S. Reitzenstein, A.J. Shields, C.K. Shih, and S. Strauf for providing me with several figures.

The work was supported by the DFG via the Forschergruppe FOR 485 Quantum Optics in Semiconductor Nanostructures and FOR 730 Positioning of Single Nanostructures – Single Quantum Devices.

References

1. B. Lounis, M. Orrit, Pep. Prog. Phys. **68**, 1129 (2005)
2. R. Loudon, *The Quantum Theory of Light*, 3rd edn. (Oxford Science, Oxford, 2000)
3. R. Meystre, M. Sargent III, *Elements of Quantum Optics*, 3rd edn. (Springer, Berlin, 1999)
4. R. Lambropoulos, D. Petrosyan, *Fundamentals of Quantum Optics and Quantum Information*, (Springer, Berlin, 2007)
5. R.J. Glauber, Phys. Rev. Lett. **10**, 84 (1963)
6. E. Knill, R. Laflamme, G.J. Milburn, Nature **409**, 46 (2001)
7. C.K. Hong, Z.Y. Ou, L. Mandel, Phys. Rev. Lett. **59** 2044 (1987)
8. M. Oxborrow, A.G. Sinclair, Contemp. Phys. **46**, 173 (2005)
9. C. Santori, D. Fattal, J. Vučković, G.S. Solomon, Y. Yamamoto, New J. Phys. **6**, 89 (2004)
10. A. Kiraz, M. Atatüre, A. Imamoglu, Phys. Rev. A **69**, 032305 (2004)
11. T. Legero, T. Wilk, A. Kuhn, G. Rempe, Appl. Phys. B **77**, 797 (2003)
12. V.D. Kulakowski, B. Bacher, R. Weigand, T. Kümmel, A. Forchel, E. Borovitskaya, K. Leonardi, D. Hommel, Phys. Rev. Lett. **82**, 1780 (1999)
13. R.M. Thompson, R.M. Stevenson, A.J. Shields, I. Farrer, C.J. Lobo, D.A. Ritchie, M.L. Leadbeater, M. Pepper, Phys. Rev. B **64**, 201302(R) (2001)
14. S.M. Ulrich, M. Benyoucef, P. Michler, N. Baer, P. Gartner, F. Jahnke, M. Schwab, H. Kurtze, M. Bayer, S. Farad, Z. Wasilewski, A. Forchel, Phys. Rev. B **71**, 235328 (2005)
15. A. Mueller, E.B. Flagg, P. Bianucci, X. Wang, D.G. Deppe, W. Ma, J. Zhang, G.J. Salamo, M. Xiao, C.K. Shih Phys. Rev. Lett. **99**, 187402 (2007)
16. M. Pelton, Y. Yamomoto, Phys. Rev. A **59**, 2418 (1999)
17. A. Kuhn, M. Hennrich, G. Rempe, Phys. Rev. Lett **89**, 067901 (2002)
18. Z. Yuan, B.E. Kardynal, R.M. Stevenson, A.J. Shields, C.J. Lobo, K. Cooper, N.S. Beattie, D.A. Ritchie, M. Pepper, Science **295**, 102 (2002)

19. M. Nirmal, D.J. Norris, M. Kuno, M.G. Bawendi, Al. L. Efros, M. Rosen, Phys. Rev. Lett. **75**, 3728 (1995)
20. M. Furis, H. Htoon, M.A. Petruska, V.I. Klimov, T. Barrick, S.A. Crooker, Phys. Rev. B **73**, 241313 (2006)
21. M. Reischle, G.J. Beirne, R. Roßbach, M. Jetter, P. Michler, Phys. Rev. Lett. **101**, 146402 (2008)
22. E.M. Purcell, Phys. Rev. **69**, 681 (1946)
23. J.M. Gerard, in *Single Quantum Dots, Series: Topics of Applied Physics*, vol. 90, ed. by P. Michler (Springer, Berlin, 2003), pp. 298
24. A. Badolato, K. Hennessy, M. Atatüre, J. Dreiser, E. Hu, P.M. Petroff, A. Imamoglu, Science **308**, 1158 (2005)
25. A. Rastelli, A. Ulhaq, S. Kiravittaya, L. Wang, A. Zrenner, O.G. Schmidt, Appl. Phys. Lett. **90**, 073120 (2007)
26. S. Strauf, K. Hennessy, M.T. Rakher, Y. S. Choi, A. Badolato, L.C. Andreani, E.L. Hu, P.M. Petroff, D. Bouwmeester, Phys. Rev. Lett. **96**, 127404 (2006).
27. K. Hennessy, A. Badolato, M. Winger, D. Gerace, M. Atatüre, S. Gulde, S. Fält, E.L. Hu, A. Imamoglu, Nature **445**, 896 (2007)
28. A.J. Bennett, D.C. Unitt, P. Atkinson, D.A. Ritchie, A.J. Shields, Opt. Express **13**, 50 (2005)
29. G. Cui, M.G. Raymer, Opt. Express **13**, 9660 (2005)
30. D. Press, S. Gtzinger, S. Reitzenstein, C. Hofmann, A. Lffler, M. Kamp, A. Forchel, Y. Yamamoto, Phys. Rev. Lett. **98**, 117402 (2007)
31. A. Daraei, A. Tahraoui, D. Sanvitto, J. A. Timpson, P.W. Fry, M. Hopkinson, P.S.S. Guimaraes, H. Vinck, D.M. Whittaker, M.S. Skolnick, A.M. Fox, Appl. Phys. Lett. **88**, 051113 (2006)
32. F. Jelezko, A. Volkmer, I. Popa, K.K. Rebane, J. Wrachtrup, Phys. Rev. A **67**, 041802(R) (2003)
33. V. Zwiller, T. Aichele, O. Benson, Phys. Rev. B **69**, 165307 (2004)
34. B.E.A. Saleh, M.C. Teich, *Fundamentals of Photonics*, 2nd edn. (Wiley, New Jersey, 2007), pp. 420–421
35. C. Kammerer, G. Cassabois, C. Voisin, M. Perrin, C. Delalande, Ph. Roussignol, J.M. Gerard, Appl. Phys. Lett. **81**, 2737 (2002)
36. C. Santori, D. Fattal, J. Vučković, G.S. Solomon, Y. Yamamoto, Nature **419**, 594 (2002)
37. A. Berthelot, I. Favero, G. Cassabois, C. Voisin, C. Delalande, Ph. Roussignol, R. Ferreira, J.M. Gerard, Nat. Phys. **2**, 759 (2006)
38. T. Aichele, Dissertation, Humboldt-Universitt zu Berlin (2005)
39. C.K. Hong, Z.Y. Ou, L. Mandel, Phys. Rev. Lett. **59**, 2044 (1987)
40. S. Varoutsis, S. Laurent, P. Kramper, A. Lemaitre, I. Sagnes, I. Robert-Philip, I. Abram, Phys. Rev. B **72**, 041303 (2005)
41. D.C. Unitt, A.J. Bennett, P. Atkinson, K. Cooper, P. See, D. Gevaux, M.B. Ward, R.M. Stevenson, D.A. Ritchie, A.J. Shields, J. Opt. B: Quantum Semiclass. Opt. **7**, 129 (2005)
42. P. Michler, A. Kiraz, C. Becher, W.V. Schoenfeld, P.M. Petroff, Lidong Zhang, E. Hu, A. Imamoglu, Science **290**, 2282 (2000)
43. C. Santori, M. Pelton, G. Solomon, Y. Dale, Y. Yamamoto, Phys. Rev. Lett. **86**, 1502 (2001)
44. V. Zwiller, H. Blom, P. Jonsson, N. Panev, S. Jeppesen, T. Tsegaye, E. Goobar, M.E. Pistol, L. Samuelson, G. Bjrk, Appl. Phys. Lett. **78**, 2476 (2001)

45. M.H. Baier, E. Pelucchi, E. Kapon, S. Varoutsis, M. Gallert, I. Robert-Philip, I. Abram, Appl. Phys. Lett. **84**, 648 (2004)
46. A. Malko, D.Y. Oberli, M.H. Baier, E. Pelucchi, F. Michelini, K.F. Karlson, M.-A. Dupertuis, E. Kapon, Phys. Rev. B **72**, 195332 (2005)
47. A. Malko, M.H. Baier, K.F. Karlson, E. Pelucchi, D.Y. Oberli, E. Kapon, Appl. Phys. Lett. **88**, 081905 (2006)
48. S. Kimura, H. Kumano, M. Endo, I. Suemune, T. Yokoi, H. Sasakura, S. Adachi, S. Muto, H.Z. Song, S. Hirose, T. Usuki, Phys. Stat. Sol. (c) **2**, 3833 (2005)
49. S. Kiravittaya, M. Benyoucef, R. Zapf-Gottwick, A. Rastelli, O.G. Schmidt, Appl. Phys. Lett. **89**, 233102 (2006)
50. V. Zwiller, T. Aichele, W. Seifert, J. Person, O. Benson, Appl. Phys. Lett. **82**, 1509 (2003)
51. G. Beirne, M. Jetter, H. Schweizer, P. Michler, J. Appl. Phys. **98**, 093522 (2005)
52. R. Rossbach, M. Reischle, G. Beirne, M. Jetter, P. Michler Appl. Phys. Lett **92**, 071105 (2008)
53. K. Sebald, P. Michler, T. Passow, D. Hommel, G. Bacher, A. Forchel, Appl. Phys. Lett. **81**, 2920 (2002)
54. C. Santori, S. Götzinger, Y. Yamamoto, S. Kako, K. Hoshino, Y. Arakawa, Appl. Phys. Lett. **87**, 051916 (2005)
55. M.B. Ward, O.Z. Karimov, D.C. Unitt, Z.L. Yuan, P. See, D.G. Gevaux, A.J. Shields, P. Atkinson, D.A. Ritchie, Appl. Phys. Lett. **86**, 201111 (2005)
56. T. Yamaguchi, T. Tawara, H. Kamada, H. Gotoh, H. Okamoto, H. Nakano, O. Mikami, Appl. Phys. Lett. **92**, 081906 (2008).
57. M.B. Ward, A.J. Shields, in *Semiconductor Quantum Bits*, ed. by F. Henneberger and O. Benson (Pan Stanford Publishing, 2009), pp. 389
58. B. Gayral, J.M. Gerard, Appl. Phys. Lett. **72**, 1421 (1998)
59. E. Moreau, I. Robert, J.M. Gerard, I. Abram, L. Manin, V. Thierry-Mieg, Appl. Phys. Lett. **79**, 2865 (2001)
60. D.C. Unitt, A.J. Bennett, P. Atkinson, D.A. Ritchie, A.J. Shields, Phys. Rev. B **72**, 033318 (2005)
61. S. Strauf, N.G. Stoltz, M.T. Rakher, L.A. Coldren, P.M. Petroff, D. Boumeester, Nat. Photonics **1**, 704 (2007)
62. G. Bacher, in *Single Quantum Dots*, ed. by P. Michler (Springer, Berlin, 2003), pp. 147
63. S. Kako, K. Hoshino, S. Iwamoto, S. Ishida, Y. Arakawa, Appl. Phys. Lett. **85**, 64 (2004)
64. R. Arians, T. Kmmell, G. Bacher, A. Gust, C. Kruse, D. Hommel, Appl. Phys. Lett. **90**, 101114 (2007)
65. S. Kako, C. Santori, K. Hoshino, S. Götzinger, Y. Yamamoto, Y. Arakawa, Nat. Mater. **5**, 887 (2006)
66. J. Kim, O. Benson, H. Kan, Y. Yamamoto, Nature **397** 500 (1999)
67. A.J. Bennett, D.C. Unitt, P. See, A.J. Shields, P. Atkinson, K. Cooper, D.A. Ritchie, Appl. Phys. Lett. **86**, 181102 (2005)
68. A.J. Shields, Nat. Photonics **1**, 215 (2007)
69. A.J. Bennett, P. Atkinson, P. See, M.B. Ward, R.M. Stevenson, Z.L. Yuan, D.C. Unitt, D.J.P. Ellis, K. Cooper, D.A. Ritchie, A.J. Shields, Phys. Stat. Sol. **243**, 3730 (2006)
70. A.J. Bennett, D.C. Unitt, P. See, A.J. Shields, P. Atkinson, K. Cooper, D.A. Ritchie, Phys. Rev. B **72**, 033316 (2005)

71. M.B. Ward, T. Farrow, P. See, Z.L. Yuan, O.Z. Karimov, P. Atkinson, K. Cooper, D.A. Ritchie, Appl. Phys. Lett. **90**, 063512 (2007)
72. A.J. Bennett, R.B. Patel, A.J. Shields, K. Cooper, P. Atkinson, C.A. Nicoll, D.A. Ritchie, Appl. Phys. Lett. **92**, 193503 (2008)
73. D.J. Ellis, A.J. Bennett, A.J. Shields, P. Atkinson, D.A. Ritchie, Appl. Phys. Lett. **88**, 133509 (2006)
74. A. Lochmann, E. Stock, O. Schulz, F. Hopfer, D. Bimberg, V.A. Haisler, A.I. Toropov, A.K. Bakarov, A.K. Kalagin, Electron. Lett. **42**, 774 (2006)
75. M. Scholz, S. Büttner, O. Benson, A.I. Toropov, A.K. Bakarov, A.K. Kalagin, A. Lochmann, E. Stock, O. Schulz, F. Hopfer, V.A. Haisler, D. Bimberg, Opt. Express **15**, 9107 (2007)
76. D.L. Huffaker, D.G. Deppe, T.J. Rogers, Appl. Phys. Lett. **65**, 1611 (1994).
77. A. Lochmann, E. Stock, O. Schulz, F. Hopfer, D. Bimberg, V.A. Haisler, A.I. Toropov, A.K. Bakarov, A.K. Kalagin, M. Scholz, S. Büttner, O. Benson, Phys. Stat. Sol. (c) **4**, 547 (2007)
78. D.J.P. Ellis, A.J. Bennett, S.J. Dewhurst, C.A. Nicoll, D.A. Ritchie, A.J. Shields, New J. Phys. **10**, 043035 (2008)
79. R.B. Patel, A.J. Bennett, K. Cooper, P. Atkinson, C.A. Nicoll, D.A. Ritchie, A.J. Shields, Phys. Rev. Lett. **100**, 207505 (2008)
80. M. Reischle, G.J. Beirne, W.-M. Schulz, M. Eichfelder, R. Roßbach, M. Jetter, P. Michler, Opt. Express **12**, 12771 (2008)
81. S. Laurent, S. Varoutsis, L. Le Gratiet, A. Lemaitre, F. Raineri, A. Levenson, I. Robert-Philip, I. Abram, Appl. Phys. Lett. **87**, 163107 (2005)
82. A.J. Bennett, D.C. Unitt, A.J. Shields, P. Atkinson, D.A. Ritchie, Opt. Express **13**, 7772 (2005)
83. P. Ester, L. Lackmann, S. Michaelis de Vasconcellos, M.C. Hübner, A. Zrenner, M. Bichler, Appl. Phys. Lett. **91**, 111110 (2007)

7

Entangled Photon Generation by Quantum Dots

Andrew J. Shields, R. Mark Stevenson, and Robert J. Young

Summary. We describe recent progress in generating pairs of polarisation entangled photons from the biexciton cascade of a single self-assembled quantum dot. Entanglement between the emitted photons is demonstrated by the presence of strong correlations for different orthogonal measurement bases, the invariance of the correlation to the linear polarisation basis angle and the form of the two-photon density matrix. Through cancellation of the exciton spin splitting and reduction in background emission, entanglement fidelities to the expected symmetric Bell state Ψ^+ exceeding 90% are achieved, along with a strong violation of Bell's inequality. A model to describe the factors limiting the fidelity is presented. We analyse the biexciton emission for the general case of a finite exciton spin splitting in the intermediate exciton state of the cascade. This reveals that the emitted entangled two-photon state contains a phase term which is dependent on the time delay between the two emitted photons. Only for the case of zero exciton splitting are all pairs created with the same phase, leading to the highest fidelities to Ψ^+. Resolving the time delay between the two photons allows a strong enhancement in the entanglement fidelity for dots with a finite exciton splitting, as well as the preparation of the Ψ^- Bell state.

7.1 Introduction

Quantum entanglement is the phenomenon linking the properties of seemingly disparate objects. It occurs when the wavefunction of the whole system cannot be separated into the product of the constituents. In such a scenario, a measurement upon one part of the system will alter the state of the other parts. Remarkably quantum theory predicts this to occur irrespective of the spatial separation of the entangled objects, even if they are thousands of kilometres apart!

Einstein, Podolsky and Rosen (EPR) highlighted this "paradox" in their 1935 paper and subsequent writings, to illustrate that quantum theory must be incomplete [1]. Famously Einstein referred to entanglement as "spukhafte Fernwirkung" or "spooky action at a distance". It was not until 1964, that

John Bell suggested that the correlations of spatially separated entangled photons or spins could violate an inequality based on EPR's assumptions of separability and locality [2]. Since then many experiments have verified the violation of Bell's inequality and the existence of entanglement.

Entanglement implies that not only do the objects possess correlated properties, but also that a measurement on any individual object produces an unpredictable result. In contrast classically correlated objects have well defined properties. John Bell gave a good example of classical correlations in the intriguingly titled "Bertlmann's Socks and the Nature of Reality" [3], in which he writes:

> ...the philosopher in the street, who has not suffered a course in quantum mechanics, is quite unimpressed by Einstein–Podolsky–Rosen correlations. He can point to many examples of similar correlations in everyday life. The case of Bertlmann's socks is often cited. Dr. Bertlmann likes to wear two socks of different colours. Which colour he will have on a given foot on a given day is quite unpredictable. But when you see that the first sock is pink you can be already sure that the second sock will not be pink. Observation of the first, and experience of Bertlmann, gives immediate information about the second. There is no accounting for tastes, but apart from that there is no mystery here. And is not the EPR business just the same? ...

The socks of Bell's colleague are classically correlated because their colours are fixed when he dresses in the morning. Quantum correlations are altogether much spookier. If his socks possessed quantum correlations they would not have defined colours until observed by someone. Furthermore it would be fundamentally impossible for anyone, including Dr. Bertlmann, to predict the colour of his socks before this observation. Despite the strange indeterminate nature of the colour of each individual sock, if the pair are entangled, the observation will always reveal that they are different.

In recent years it has been realised that quantum entanglement could have many applications in the nascent field of Quantum Information Technology [4]. For example, the correlations between entangled photon pairs may be used to distribute secret keys between remote parties [5]. Any attempt by an eavesdropper to measure one of the photons, causes collapse of the two-photon wavefunction, thereby destroying the entanglement. Measuring the entanglement fidelity of the received photon pairs thereby provides a way to test if the communication has been intercepted enroute and thus to test the secrecy of the key.

Many quantum communication protocols such as quantum teleportation, entanglement swapping and dense coding rely upon entangled photons [4]. In quantum teleportation a Bell-state measurement on an input photon with an unknown quantum state and one of the photons in an entangled pair may be used to teleport the state onto the second photon in the entangled pair. Here the result of the Bell measurement is used to transform the state of the target photon without direct measurement of the input. Remarkably this protocol works irrespective of the spatial separation of the two entangled photons and so may be used to teleport quantum information over very long distances.

Entanglement swapping can be used to extend the range of quantum communications by sharing entangled photons [6, 7]. This can be regarded as quantum teleportation of a photon in an entangled pair. In this protocol the entanglement between two pairs (X1, X2) and (Y1, Y2) is transferred to one photon from each pair (X1,Y2), through a Bell state measurement on X2 and Y1. It has been proposed that quantum communications on a global scale is possible using a quantum repeater [6]. Here, nested steps of entanglement swapping along a chain of quantum memories may be used to extend the range of quantum communication to arbitrary distances.

As photons are the natural choice for quantum communications, it is a natural progression that they are also applied to quantum computing. Indeed a quantum repeater may be regarded as a simple quantum computer. Photons have the advantages that it is straightforward to encode and manipulate their quantum state and that they can be transmitted over long distances with little decoherence. Their immunity to interaction becomes a disadvantage, however, when it comes to the design of logic gates in which two photons must interact. In linear optics quantum computing this problem is solved using projective measurements to induce an effective interaction between the photons [8]. Another possibility is to convert the photons to an electronic excitation in a solid state system such as a quantum dot to allow a stronger interaction.

Entangled photons are also useful in the field of quantum imaging and metrology. Images and patterns formed using entangled photons can contain finer detail than is possible using ordinary light of the same wavelength. The optical resolution of a far field imaging systems is determined by the wavelength (λ) of the light used and the numerical aperture (NA) according to $\lambda/2$ NA [9]. However, it has been shown that this resolution can be enhanced by a factor of two using pairs of entangled photons [10]. Entanglement may also be used to improve the sensitivity of various other techniques and measurements, such as timing and interferometry [11]. In the penultimate section an example of quantum interferometry using entangled photon pairs from a quantum dot is shown.

To date the highest fidelity entangled photon pairs have been produced by spontaneous parametric down conversion in a crystal with a χ_2 non-linearity, such as Lithium Niobate [12]. Here a photon in the incident laser field is converted into two lower energy photons, according to the conservation of energy and momentum. Under appropriate conditions the polarisations of the two photons are produced in an entangled state. Recently entangled pairs have also been generated using a χ_3 non-linearity in a semiconductor [13].

Parametric down-conversion can produce high fidelity entangled pairs and has been extremely useful for a wide range of experiments in quantum optics [14]. However, it has the inherent problem that the generation of an entangled pair is probabilistic and follows Poissonian statistics [15]. This means that there will be an appreciable probability of generating two or more pairs, especially for high source intensities. These multiple pairs cause errors in the quantum gate or application, degrading its performance. One way to limit the proportion of multiple pairs is to reduce the source intensity, although this has the undesirable effect of reducing the success probability of the gate. Thus, although parametric down-conversion has proven very useful for demonstrating the viability of small quantum circuits, it is difficult to scale to larger applications.

An ideal entangled photon source would produce exactly one pair in response to an external trigger. It has been shown that the radiative atomic cascade of Calcium may be used to generate pairs of entangled photons [16]. Due to the discrete nature of the chosen atomic transitions, and isolation of individual atoms, no more than a single photon pair is detected each excitation cycle. In principle, by increasing the collection efficiency the source can supply a single entangled pair with the high probability required for large-scale quantum circuits, and without introducing errors from multiple pair generation.

Even the earliest experiments using atomic sources produced very high fidelity entanglement, however harnessing single atoms remains a highly challenging technical feat, and one which is impossible to imagine implemented when quantum information technology becomes mainstream. For such large-scale, practical, applications the natural solution would be to implement an entangled light source using conventional semiconductor technology. This would have the additional advantage that other components could be integrated together with the light source – for example spin or charge based quantum logic gates.

7.2 Entangled Photon Pairs from the Biexciton Cascade

7.2.1 Ideal Quantum Dots

Semiconductor quantum dots are often described as "artificial atoms" due to the fact that like an atom, they possess discrete quantum states. In 2000, Oliver Benson, Yoshi Yamamoto and colleagues at Stanford suggested that an entangled photon pair may be generated by the biexciton cascade of a single quantum dot [17]. The biexciton state is formed when the quantum dot is occupied by two electrons and two holes. This state relaxes radiatively through the sequential recombination of two electron–hole pairs, releasing two photons, as shown schematically in Fig. 7.1a.

In the ground state , the two electrons of the biexciton have spin quantum numbers of $m_z = +1/2$ and $-1/2$, while the two heavy holes have spin $+3/2$ and $-3/2$. The light holes are not confined for small InAs quantum dots. Conservation of spin/angular momentum determines the polarisation of the photons emitted: recombination of the electron–hole pair with a total angular momentum of -1 produces a left-hand (L) circularly polarised photon, while the electron–hole pair with a total spin of $+1$ couples to a right-hand (R)

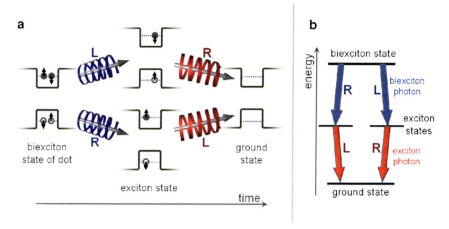

Fig. 7.1. (a) An illustration of the radiative decay process of the biexciton state from a single quantum dot. Recombination of electrons and holes results in the emission of a pair of photons via one of two intermediate exciton states. Ideally the emitted photons are right (R) and left (L) circularly polarised. (b) The same two-photon emission process in a simple energy level diagram. The biexciton photons are typically more energetic than the exciton photons

circularly polarised photon. Recombination of other configurations, with total angular momentum ± 2, are forbidden and they are normally dark in the optical spectrum.

As shown schematically in Fig. 7.1a there are two possible decay channels for the biexciton. The $+1/2$ electron and $-3/2$ hole could recombine first, producing a L polarised photon, and leaving an exciton consisting of a $-1/2$ electron and $+3/2$ hole trapped in the dot which subsequently recombine to emit a R photon. Alternatively the $-1/2$ electron and $+3/2$ hole could recombine first, resulting in a first R photon followed by the L. If there is no way to distinguish which path the system followed, other than by measuring the emitted polarisations, then the generated two-photon state should be described by the entangled state:

$$|\Psi^+\rangle = (|LR\rangle + |RL\rangle)/\sqrt{2}. \tag{7.1}$$

The unitary weighting of the two terms assumes that both decay channels are equally likely.

This state expressed in circular (c) polarisation basis can be re-written in a rectilinear polarisation basis (r) with horizontally and vertically polarised eigenvectors $|H\rangle = (|L\rangle + |R\rangle)/\sqrt{2}$ and $|V\rangle = i(|R\rangle - |L\rangle)/\sqrt{2}$, or alternatively in the diagonal polarisation basis (d) spanned by diagonally and antidiagonally polarised eigenvectors $|D\rangle = (|H\rangle + |V\rangle)/\sqrt{2}$ and $|A\rangle = (|H\rangle - |V\rangle)/\sqrt{2}$:

$$|\Psi^+\rangle = (|LR\rangle + |RL\rangle)/\sqrt{2} \equiv (|HH\rangle + |VV\rangle)/\sqrt{2} \equiv (|DD\rangle + |AA\rangle)/\sqrt{2}. \tag{7.2}$$

Thus it can be seen that the state appears entangled for each of the three measurement bases. Equation (7.2) predicts that the photons will have correlated polarisations if measured in rectilinear or diagonal bases, and anti-correlated if measured in circular polarisation. This striking prediction is one of the main tests for entanglement applied to the experimental data below.

Evidence for the existence of the biexciton cascade in a quantum dot was confirmed by cross-correlation measurements, which showed that the exciton photon was more likely to be detected after a biexciton photon had been registered [18]. However, polarisation sensitive correlation measurements, described below, showed that the polarisations of the biexciton and exciton photons were not entangled [19–21]. These measurements showed that like the case of Dr. Bertlmann's socks, the photons are generated in a well defined polarisation state and show only classical correlations.

The entanglement is destroyed by an energetic splitting of the intermediate exciton state, which allows the two decay channels of the cascade to be distinguished through the emitted photon energies. The splitting is driven by a difference in the electron–hole exchange interaction for the two intermediate levels, due to in-plane anisotropy including the shape and built-in strain of the dots [22, 23]. The exchange interaction hybridises the pure exciton spin states into a symmetric and anti-symmetric combinations aligned along the axes of the anisotropy in the layer plane, and split by an energy δ_1. These two exciton states, X_H and X_V, couple to biexciton and exciton photons polarised linearly along the horizontal and vertical directions.

As the H and V polarised photons produced by the cascade have different energies, they are unable to interfere and cannot be transformed into entangled states in the diagonal and circular bases, as is possible for the entangled state Ψ^+ in (7.2). As a consequence quantum dots with an exciton level splitting produce photon pairs with correlated polarisations aligned along the H and V directions. However, the correlations are lost if the measurements are made in the circular or diagonal bases.

7.2.2 Anisotropic Quantum Dots

The analysis of the biexciton cascade may be extended to the case of a finite splitting in the intermediate state by considering the phase evolution of the system during the decay. This is shown schematically in Fig. 7.2 depicting the energy stored in the quantum dot as a function of time. Recombination of the first electron–hole pair in the biexciton state, creates either a horizontally- or vertically-polarised XX photon (H_{XX} or V_{XX}), and leaves the quantum dot in the exciton state X_H or X_V. Thus the photon/exciton system is projected into the entangled state $(|H_{XX}X_H\rangle + |V_{XX}X_V\rangle)/\sqrt{2}$. As the two exciton states have different energies they will develop a phase difference with time, according to $e^{i\delta_1\tau/\hbar}$, where δ_1 is their energy separation. Thus if the exciton photon is emitted after a delay of τ, the two-photon state produced by the cascade is given by

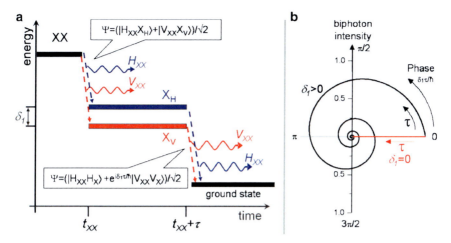

Fig. 7.2. (a) The energy of a quantum dot as a function of time following excitation to the initial biexciton (XX) state. The state Ψ is marked for times corresponding to emission of the first and second photons. (b) Represents the relationship between the biphoton intensity and the phase of the photon pair superposition for a dot with (*black*) and without (*red*) an energetic splitting δ_1 between the intermediate exciton states X_H and X_V

$$|\Psi\rangle = (|H_{XX}H_X\rangle + e^{i\delta_1\tau/\hbar}|V_{XX}V_X\rangle)/\sqrt{2}. \qquad (7.3)$$

Equation (7.3) is very similar in form to the state expected for an unsplit dot ($\delta_1 = 0$) given by (7.2), except for an additional phase term, which depends on the time delay τ between the two emitted photons.

To determine the total emission from the dot we must integrate over all possible time delays τ between the two photons. As the dot will emit pairs detected with a wide range of temporal spacings, up to the order of the biexciton and exciton decay times, the integration randomises the phase between the two terms in (7.3), destroying the entanglement. Thus only if the exciton splitting δ_1 is sufficiently small, of order of the lifetime broadened linewidth of the exciton, $\hbar/\tau_1$, is the phase relationship between the two terms well defined. This is illustrated in Fig. 7.2b which plots the intensity of the photon-pair wavefunction as a function of the phase $\delta_1\tau/\hbar$ for the case of zero and finite δ_1. For $\delta_1 = 0$ the phase is fixed and integration over all τ reveals a maximally entangled state. For finite δ_1 the phase revolves as τ progresses, integration over τ in this case results in a diminished superposition state.

Finally (7.3) suggests that even quantum dots with a finite exciton splitting, emit entangled photon pairs. If the time delay can be accurately measured for each photon pair emitted by the dot, we may be able to interpret and use each generated state. This is demonstrated experimentally in Sect. 7.7.

7.3 Biexciton Cascade of an InAs Quantum Dot

This chapter concentrates on self-organised quantum dots formed by growing a thin layer of InAs on GaAs at $\sim$500°C by Molecular Beam Epitaxy. A low areal density ($\sim 10^8 \, \text{cm}^{-2}$) of quantum dots can be prepared by stopping the In deposition just after the critical coverage required to form quantum dots. As these dots are relatively small, they have emission wavelengths that can be detected using Si CCD array detectors and Si avalanche photodiodes.

Figure 7.3 illustrates a typical device structure used to study entangled photon emission from a quantum dot. The quantum dot layer is grown in the centre of a λ-cavity designed to be resonant with the quantum dot emission energy and concentrate the emission in the vertical direction, into the collection optic. The cavity is formed between a 14–18 period GaAs/AlAs lower Bragg mirror and a 2–6 period upper mirror and is designed to have a relatively low quality factor, allowing enhancement of both the biexciton and exciton collection efficiencies. In order to isolate the emission from a single quantum dot, $\sim$2μm diameter apertures are formed in the opaque metal layer on the top surface.

7.3.1 Emission from Single Quantum Dots

Micro-photoluminescence (μPL) from individual apertures was recorded by focussing a picosecond pulsed laser, with a wavelength shorter than the GaAs

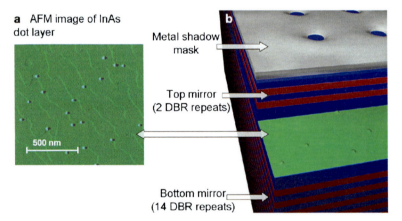

Fig. 7.3. (a) An image of InAs quantum dots grown on a GaAs surface taken using an Atomic Force Microscope (AFM). (b) An illustration of a typical device structure used to study individual quantum dots. The dot layer (*green*) is placed in an optical cavity to enhance the proportion of the emission collected with a microscope objective lens. The cavity is formed with alternating layers of GaAs (*blue*) and AlAs (*red*). A metal shadow mask containing apertures $\sim$2 μm in diameter allows individual dots to be studied. The diagram is not to scale

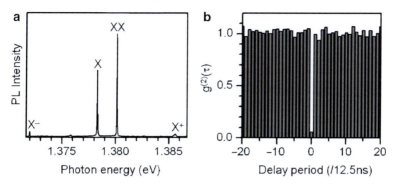

Fig. 7.4. (a) A photoluminescence energy spectrum from a single quantum dot. Emission lines corresponding to exciton (X), biexciton (XX), positively (X^+) and negatively (X^-) charged excitons are identified. (b) The second order correlation function measured for the exciton emission from a single quantum dot. A clear dip is observed at zero-delay indicating high quality single photon emission

bandgap, to a ~1 μm diameter spot using a microscope objective. The device was placed in a variable temperature He_4 cryostat and cooled to a temperature of ~10 K. PL was collected through the same microscope objective and analysed with a spectrometer and Si CCD array. A linear polariser was placed before the spectrometer and the transmitted linear polarisation was selected by a polarisation rotator.

Figure 7.4a plots a typical PL spectrum from an aperture containing a single InAs/GaAs quantum dot. The spectrum is dominated by two sharp lines due to the biexciton (XX) and exciton (X) transitions of the dot, as labelled. These lines are assigned by studying the dependence of their intensities on the laser power density. The X line shows a linear dependence on laser power before saturation, while for the biexciton the dependence is quadratic. The XX and X transition energies differ due to the difference in the Coulomb interaction between the electrons and holes trapped in the dot in their initial and final states. The sign of the biexciton binding energy depends on the growth conditions for the dots, and is frequently found to be negative for the small quantum dots studied here [24, 25].

It is often possible to observe a number of other lines in the spectrum which derive from charged excitons. In particular, the weak lines at 1.372 and 1.386 eV in Fig. 7.4a derive from the negatively and positively charged excitons, labelled X^- and X^+ respectively. These transitions may be assigned from their linear laser power dependence and absence of a polarisation splitting as described below.

The quantum nature of these transitions may be confirmed by studying the second-order correlation function, $g^{(2)}(\tau)$, using a standard Hanbury-Brown and Twiss set-up. Figure 7.4b plots the second order correlation function measured on the exciton transition. The strong suppression of $g^{(2)}(\tau)$ at $\tau = 0$ confirms that the dot emits one exciton photon per excitation pulse, inhibiting the occurrence of co-incidence counts for zero-delay. Indeed the height of the zero delay bin indicates that, tuned to the energy of the X transition, the source emits multi-photon pulses at 0.06 times the rate of a Poissonian source of the same intensity. The residual rate of coincidence counts derives from detector dark counts and weak background emission from the device due to layers other than the quantum dot.

7.3.2 Polarised Emission Spectra

Figure 7.5a plots PL spectra recorded on a dot for emission linearly polarised parallel to the [110] and [1$\bar{1}$0] crystal axes. It can be seen that the XX and X transitions are in fact both linearly polarised doublets and display an equal and opposite splitting, δ_1. As shown in Fig. 7.5b, this derives from an energy splitting of the intermediate exciton state in the biexciton cascade discussed in the previous section. Notice that the sum of the X and XX transition energies for either polarisation in Fig. 7.5a is constant in good agreement with Fig. 7.5b. Typically the exciton splitting ranges between -20 and $+100\,\mu\text{eV}$, and is comparable to the inhomogeneous linewidth of the exciton $\sim(45\pm4)\,\mu\text{eV}$ measured with non-resonant excitation. Despite the substantial inhomogeneous broadening, the splitting may be determined with an accuracy of $\pm0.5\,\mu\text{eV}$ by fitting the measured transition lines with Lorentzian functions [25].

The exciton state is split due to asymmetries of the structural properties of the dot in the layer plane, which lifts its two fold degeneracy. These include the in-built strain in the structure, as well as elongation of the quantum dots

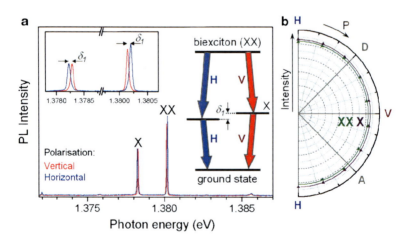

Fig. 7.5. (a) Linearly polarised photoluminescence energy spectra taken from the same quantum dot, highlighting the fine structure splitting δ_1 between the intermediate exciton states. The simple energy level diagram shows the fine structure splitting. (b) The intensity of biexciton and exciton emission as a function of the linear measurement polarisation

along a particular crystallographic axis. Indeed recent micro-structural studies of individual buried quantum dots claim to observe an elongation of their base dimension along the [1$\bar{1}$0] direction [26].

7.3.3 Tuning the Exciton Spin Splitting

As mentioned above the spin splitting of the exciton level has a detrimental effect upon the entanglement of the two photons. Fortunately, however, a number of techniques have been developed for modifying the splitting and tuning its value to be close to zero. These include manipulation of the dot nanostructure, through either controlling the growth conditions [25] or a post-growth anneal step [25, 27, 28], or applying an external perturbation such as an in-plane magnetic [29], electric [30, 31], or strain field [32]. Figure 7.6a plots the exciton spin splitting of a large number of individual quantum dots as a function of their exciton emission energy. It can be seen that despite a large scatter in the points, the splitting tends to decrease with increasing emission energy, tending to zero around 1.4 eV, before turning negative for higher emission energies. Since these energies lie very close to the wetting layer energy of 1.44 eV, this behaviour is interpreted as being due to a reduction in the electron–hole exchange interaction in weakly confined quantum dots as the electron and hole wavefunctions spill into surrounding barrier layers. Indeed expansion of the exciton's wavefunction at high energy has been confirmed by probing the diamagnetic response of individual dots [25]. From Fig. 7.6a it is apparent that by controlling the growth conditions so as to produce dots emitting around 1.4 eV, a sample can be prepared where a large fraction of the dots have close to zero splitting.

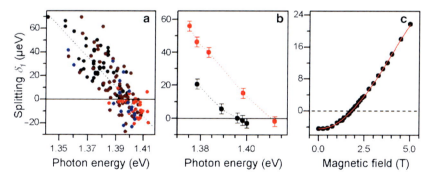

Fig. 7.6. Three techniques to engineer quantum dots with no fine structure splitting δ_1 between intermediate exciton states. (**a**) The splitting of a large number of quantum dots measured as a function of their emission energies. Excitons with $E \simeq 1.4$ eV are found to have very small splittings. (**b**) The splitting of two quantum dots plotted as a function of emission energy after repeatedly annealing the sample, the splitting in both cases is found to pass through zero. (**c**) The splitting as a function of applied in-plane magnetic field, for this specific single dot it is zero at a field of ~ 2 T

It is also possible to tune the splitting of a dot in a particular device to be zero via post-growth annealing [28]. Figure 7.6b plots the splitting of two quantum dots as a function of the exciton emission energy after repeated five minute rapid thermal anneals at 675°C. Annealing has the effect of intermixing the dot and barrier material, resulting in a blue shift of the dot emission. This is accompanied by a reduction in the quantum dot splitting, which may again be explained by a weakening of the electron–hole exchange interaction due to the increased spatial extent of the exciton. The splitting again tends to zero when the emission energy is close to 1.4 eV.

Applying an in-plane magnetic field to the sample allows the splitting to be tuned continuously for a single quantum dot, as shown in Fig. 7.6c. The effect of an in-plane magnetic field is to mix each of the two bright exciton spin states, with one of the two dark states with total angular momentum $+2$ and -2, increasing the separation of the bright and dark excitons. These dark states lie to lower energy than the bright states due to the electron hole exchange interaction. The bright exciton splitting, δ_1, is reduced if the lower energy bright exciton undergoes a stronger interaction with its respective dark state than the higher-energy bright exciton [29]. Applying an in-plane magnetic field is very useful for studying the effect of the splitting on the entanglement in a single sample, as demonstrated in Sect. 7.6.

7.4 Characterising Entanglement

7.4.1 The Correlation Coefficient

The polarisation properties of a photon-pair state can be assessed using the correlation coefficient $E_{\alpha\beta}$, where α and β denote the polarisation detection basis of photons 1 and 2 respectively. Also known as the correlation function, the correlation coefficient is the photon-pair analogue of the degree of polarisation, and forms the basis for Bell's inequality. The correlation coefficient is defined by

$$E_{\alpha\beta} = I(a,b) + I(\bar{a},\bar{b}) - I(a,\bar{b}) - I(\bar{a},b), \quad (7.4)$$

where $I(a,b)$ is the normalised intensity (i.e. probability) of detection of the first and second photon with polarisations a and b, which together with the orthogonal polarisations $\bar{a}$ and $\bar{b}$ form the polarisation bases α and β. Values of $E = 0, 1$ and -1 indicate no correlation, maximum correlation, and maximum anti-correlation respectively.

7.4.2 Determining the Two-Photon Density Matrix

The horizontally (H) and vertically (V) polarised single qubit states are represented using Jones' vector notation as

$$|H\rangle = \begin{pmatrix} 1 \\ 0 \end{pmatrix}, \quad |V\rangle = \begin{pmatrix} 0 \\ 1 \end{pmatrix}. \quad (7.5)$$

The diagonal (D), anti-diagonal (A), right-circular (R) and left-circular (L) polarisation states are defined as follows

$$|D\rangle = (|H\rangle + |V\rangle)/\sqrt{2}, |A\rangle = (|H\rangle - |V\rangle)/\sqrt{2}, \quad (7.6)$$
$$|L\rangle = (|H\rangle + i|V\rangle)/\sqrt{2}, |R\rangle = (|H\rangle - i|V\rangle)/\sqrt{2}.$$

Two-qubit pure states are then expressed using the matrix direct product of the component single qubits

$$|ab\rangle = |a\rangle \otimes |b\rangle. \quad (7.7)$$

The emission of the source can be fully characterised by a two-photon density matrix, $\underline{\underline{\rho}}$, which can represent pure, mixed, classical and entangled states. The generalised form for a two-qubit density matrix is written as

$$\underline{\underline{\rho}} = \begin{pmatrix} \rho_{11} & \rho_{12} & \rho_{13} & \rho_{14} \\ \rho_{12}^* & \rho_{22} & \rho_{23} & \rho_{24} \\ \rho_{13}^* & \rho_{23}^* & \rho_{33} & \rho_{34} \\ \rho_{14}^* & \rho_{24}^* & \rho_{34}^* & \rho_{44} \end{pmatrix}. \quad (7.8)$$

The density matrix is by definition Hermitian with unity trace [33]. Thus the diagonal elements ρ_{ii} are real, and form the relation:

$$\rho_{11} + \rho_{22} + \rho_{33} + \rho_{44} = 1. \quad (7.9)$$

Correlation coefficients $E_{\alpha\beta}$ can be expressed in terms of the elements of the matrix $\underline{\underline{\rho}}$ by determining the terms $I(a,b)$ in (7.4) by projection of the states $|ab\rangle$ onto the matrix $\underline{\underline{\rho}}$. The nine permutations of (r)ectilinear (c)ircular and (d)iagonal polarisation bases α and β form a set of nine additional simple relations. The final six required to fully evaluate $\underline{\underline{\rho}}$ are determined by the degree of polarisation P of the first and second photon in the r, d and c bases. As an example, the rectilinear degree of polarisation of the first photon is given by

$$P_{r1} = I(H,H) + I(H,V) - I(V,H) - I(V,V). \quad (7.10)$$

Note that the terms defining P_{r1} above also determine E_{rr}, and that P_{r1} can equivalently be determined using diagonal or circular bases for photon 2. Thus determination of the degree of polarisations requires no further measurements, and can be averaged over three alternate bases for one of the photons to improve accuracy.

The 16 relationships are readily solved to show that the density matrix can be written as follows:

$$\underline{\underline{\rho}} = \frac{1}{4}\begin{pmatrix} 1 + E_{rr} + P_{r1} + P_{r2} & E_{rd} + P_{d2} & E_{dr} + P_{d1} & E_{dd} - E_{cc} \\ E_{rd} + P_{d2} & 1 - E_{rr} + P_{r1} - P_{r2} & E_{dd} + E_{cc} & -E_{dr} + P_{d1} \\ E_{dr} + P_{d1} & E_{dd} + E_{cc} & 1 - E_{rr} - P_{r1} + P_{r2} & -E_{rd} + P_{d2} \\ E_{dd} - E_{cc} & -E_{dr} + P_{d1} & -E_{rd} + P_{d2} & 1 + E_{rr} - P_{r1} - P_{r2} \end{pmatrix}$$

$$+ \frac{i}{4}\begin{pmatrix} 0 & -E_{rc} - P_{c2} & -E_{cr} - P_{c1} & -E_{cd} - E_{dc} \\ E_{rc} + P_{c2} & 0 & -E_{cd} + E_{dc} & E_{cr} - P_{c1} \\ E_{cr} + P_{c1} & E_{cd} - E_{dc} & 0 & E_{rc} - P_{c2} \\ E_{cd} + E_{dc} & -E_{cr} + P_{c1} & -E_{rc} + P_{c2} & 0 \end{pmatrix} \quad (7.11)$$

It can be seen that the real part of the density matrix depends on five correlations, those with the same bases for the two photons, and also in combinations of rectilinear and diagonal. The imaginary component is independent of all these correlations, and only depends on the remaining four correlations which are combinations that include a single circular basis.

Rectilinear polarisation affects the real diagonal elements, diagonal polarisation affects the remaining real elements, and circular polarisation affects the imaginary elements. It is obvious from inspection that small amounts of polarisation are unlikely to strongly affect the density matrix.

7.4.3 Determining the Entanglement Fidelity

The density matrix is useful for calculating the overlap of the emission with any particular two photon state. Of particular interest is the fidelity of the emission to the expected, maximally-entangled state, Ψ^+, given by (7.2)

$$f^+ = \langle \psi^+ | \underline{\rho} | \psi^+ \rangle. \tag{7.12}$$

It should be noted that the fidelity is occasionally, for example in [34], defined as a probability amplitude rather than density, i.e. by the square root of (7.12). However, in this chapter the more usual convention that the fidelities refer to a probability as per (7.12) is used.

Thus the fidelity f^+ to the maximally entangled state Ψ^+ is

$$f^+ = \langle \psi^+ | \underline{\rho} | \psi^+ \rangle = (\rho_{11} + \rho_{44})/2 + Re\{\rho_{14}\}. \tag{7.13}$$

Substituting from (7.11) into (7.12) reveals

$$f^+ = (E_{\mathrm{rr}} + E_{\mathrm{dd}} - E_{\mathrm{cc}} + 1)/4. \tag{7.14}$$

Thus the fidelity may be determined directly from just the three correlation measurements made with the same bases for the two photons.

No source emitting only classically correlated photon pairs can have a fidelity to Ψ^+ exceeding 0.5. This is readily apparent from the fact that if the source possesses only classical correlations, all of the off-diagonal elements in the density matrix are zero. Thus from (7.12), $f^+ = (\rho_{11} + \rho_{44})/2$, which has an upper value of 0.5 for a perfectly correlated source. The limit for a classical source of $f^+ \leq 0.5$, is a useful test for determining if there are entangled photons in the emission.

7.4.4 Determining the Bell Parameter

The Bell parameter S in the CHSH form [35] is expressed in terms of the correlation function E

$$S = E_{\alpha\beta} - E_{\alpha'\beta} + E_{\alpha\beta'} + E_{\alpha'\beta'} \leq 2. \tag{7.15}$$

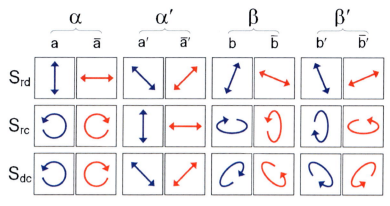

Fig. 7.7. Measurement polarisations (a,b) in the bases (α, β) used to evaluate three Bell parameters in orthogonal planes of the Poincaré sphere

Maximal violation of the inequality is expected for non-orthogonal combinations of bases α, α', β and β'. The optimum basis choice is state dependent, but well known for linear bases and the state $(|HH\rangle+|VV\rangle)/\sqrt{2}$. This case of the bell inequality is denoted as $S_{\rm rd}$, as all measurements are contained within the rectilinear-diagonal plane of the Poincaré sphere. Other equally suitable base choices can be found for other planes in the Poincaré sphere, and are shown in Fig. 7.7. The other choices shown are equivalent to the linear basis measurements in conjunction with a quarter wave plate placed directly in front of the source.

S is readily expressed in terms of the matrix elements via (7.8). For example:

$$S_{\rm rd} = (\rho_{11} - \rho_{22} - \rho_{33} + \rho_{44} + 2Re\{\rho_{23} + \rho_{14}\})/\sqrt{2}. \qquad (7.16)$$

Note that like the fidelity f^+, $S_{\rm rd}$ relies on only diagonal and off diagonal elements of the density matrix, though unlike f^+, the inner elements are as important as the outer elements. Substituting the matrix elements from (7.11) reduces the number of correlations in (7.15) from four to two, with the additional advantage that the same measurement basis is used for both photons. Three different Bell parameters corresponding to measurements in orthogonal planes of the Poincaré sphere are therefore given by

$$\begin{aligned} S_{\rm rc} &= \sqrt{2}(E_{\rm rr} - E_{\rm cc}) \leq 2, \\ S_{\rm dc} &= \sqrt{2}(E_{\rm dd} - E_{\rm cc}) \leq 2, \\ S_{\rm rd} &= \sqrt{2}(E_{\rm rr} + E_{\rm dd}) \leq 2. \end{aligned} \qquad (7.17)$$

The Bell parameters as expressed in (7.17) do not prove non-locality, but this is true also in practice for (7.15) due to loopholes such as finite detector efficiency.

7.4.5 Unpolarised Photon Pair Sources

The most useful sources of entangled photons, and indeed those considered in virtually every theoretical investigation, are unpolarised. In common with most demonstrations of entangled light sources, our source is also found to be unpolarised within experimental error. This is illustrated by Fig. 7.5b which plots the intensity of the X and XX photoluminescence as a function of the linear polarisation detection angle for an entangled light emitting quantum dot. No dependence of the intensity with polarisation angle is observed by eye. The corresponding polarisations are found to be 0 ± 0.005 and 0 ± 0.011 respectively. Emission from all entangled dots described here is also found to be unpolarised from PL with similar precision.

Inspection of (7.11) reveals further symmetry of the density matrix for an unpolarised source by setting all polarisations P to zero. Note that the fidelity and Bell parameters are insensitive to polarisation, and do not include any P terms in their formulation. For the measurements below, the correlation function E is substituted for its unpolarised equivalent $\hat{E}$. Due to the enhanced symmetry of the matrix the number of terms required to determine $\hat{E}$ is reduced from 4 to 2

$$\hat{E}_{\alpha\beta} = I(a,b) - I(a,\bar{b}). \tag{7.18}$$

7.4.6 Photon Correlation Measurements

In order to characterise the entanglement generated by the source, the correlations between the polarisations of the emitted photon pairs are studied. Figure 7.8 depicts the experimental arrangement used to measure polarisation- and time-dependent correlations between the biexciton and exciton photons. Here the emission from the sample is split by a 50:50 beamsplitter and the two arms filtered by monochromators to pass photons only at the XX and X energies. The rate of H- and V-polarised X emission is analysed with a polarising beamsplitter and two single photon detectors, while for the XX transition just the H channel was recorded.

A combination of quarter- and half-wave plates placed in front of both spectrometers allows the correlations to be recorded for any combination of polarisation bases for the two photons. The two measurement bases are denoted, as above, using α and β for the biexciton and exciton photon respectively, consisting of orthogonal polarisations $(a, \bar{a})$ and $(b, \bar{b})$. Measurements are performed for the rectilinear $r = (H, V)$, diagonal $d = (D, A)$ and circular $c = (L, R)$ polarisation bases.

Time correlated single photon counting is used to measure the rate of photon pairs for the selected polarisation combination as a function of the time delay (τ) between the two photons. This measurement yields the second order correlation function, $g^{(2)}_{ab}(\tau)$. The arrangement in Fig. 7.8 allows the second order correlation functions for the two orthogonal polarisations $(b, \bar{b})$

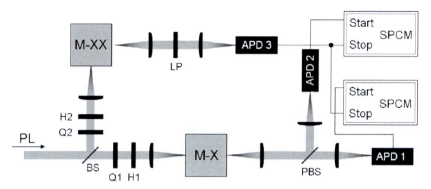

Fig. 7.8. The experimental arrangement used to measure energy- and temporally-resolved emission from single quantum dots along with polarised cross-correlations between different exciton states. To study the biexciton cascade the photoluminescence is divided into two monochromators with a 50:50 beamsplitter (BS), one (M–XX) set to pass emission from the biexciton and the other (M–X) to pass emission from the exciton. Single Photon Counting Modules (SPCMs) record the arrival times of detection events measured by Avalanche Photodiodes (APDs). Polarisation analysing components, Half- (H) and Quarter- (Q) waveplates, a linear polariser (LP) and a Polarising beamsplitters (PBS), allow any combination of polarised correlation measurement to be performed

of the exciton photon, $g^{(2)}_{ab}(\tau)$ and $g^{(2)}_{a\bar{b}}(\tau)$, to be recorded simultaneously, thereby reducing errors due to slow time variations in the experimental set up, such as drift of the laser intensity or sample position.

7.4.7 Determining the Degree of Polarisation Correlation

If the time averaged intensities of the XX and X lines are unpolarised, then the photon pair intensities $I(a,b)$ are proportional to $g^{(2)}_{ab}$. The correlation coefficient $\hat{E}$ in (7.18) describes the degree of polarisation correlation measured in polarisation bases α and β and is given by

$$\hat{E}_{\alpha\beta} = I(a,b) - I(a,\bar{b}) = \frac{g^{(2)}_{ab} - g^{(2)}_{a\bar{b}}}{g^{(2)}_{ab} + g^{(2)}_{a\bar{b}}}, \quad (7.19)$$

which is determined directly by the experimental apparatus in Fig. 7.8.

7.5 Observation of Entanglement

The emission of entangled photon pairs by the biexciton cascade of a single quantum dot was first demonstrated by Stevenson et al. [36, 37]. The breakthrough stemmed from studying a quantum dot with close to zero exciton spin splitting, so that the which-path information that had destroyed the entanglement in previous studies had been erased. The experimental evidence for the generation of entangled photons by quantum dots is reviewed here, as well as the progress in improving the fidelity of these sources.

7.5.1 Correlation in Orthogonal Bases

Figure 7.9 plots the degree of correlation in co-polarised bases measured between the XX and X photons for two different quantum dots, chosen to have exciton spin splittings that are (a) large and (b) close to zero. Measurements are shown for each of the three polarisation bases: rectilinear, diagonal and circular. In the figure the degree of correlation between the photons is shown as a function of the time delay between the photons in units of the laser repetition cycle. It can be seen that apart the pairs generated with zero delay, i.e. emitted in the same cycle, there is no appreciable correlation. This result is anticipated as polarisation correlation between biexciton and exciton photons emitted in different emission cycles is not expected.

Pairs emitted in the same cascade (i.e. zero delay) show strong correlations in the emitted polarisations. For the dot with close to zero exciton spin splitting (Fig. 7.9b), a very striking positive correlation is seen measuring in either of the linear bases, rectilinear or diagonal, with values of $\hat{E}_{rr} = 0.70$ and $\hat{E}_{dd} = 0.61$, respectively. Remarkably the photon pairs appear anti-correlated

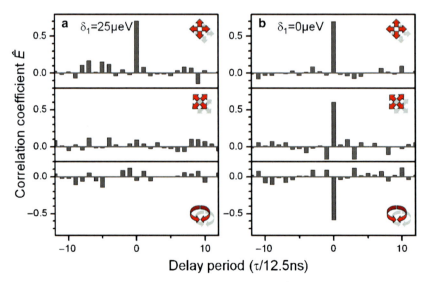

Fig. 7.9. The unpolarised cross-correlation coefficient $\hat{E}$ between the biexciton and exciton photons plotted as a function of the delay between the photons in the rectilinear, diagonal and circular polarisation bases. (**a**) For a quantum dot with a large fine structure splitting δ_1 and (**b**) for a dot with no resolvable splitting

when measuring with circular polarisation basis, yielding $\hat{E}_{cc} = -0.58$. This is exactly the behaviour expected for the entangled state of (7.2). Indeed the strong correlations seen for all three bases in Fig. 7.9b could not be produced by any classical light source or mixture of classical sources and is proof that the source generates entangled photons.

In contrast, a dot with finite splitting shows polarisation correlation for the rectilinear basis only, with no significant correlation for diagonal or circular measurements, see Fig. 7.9a. This confirms that dots with a large exciton spin splitting emit XX and X photons with correlated and well defined polarisations, aligned along either the [110] or [1$\bar{1}$0] crystal axes, as predicted by Fig. 7.5b. These correlations are apparent when measuring in the rectilinear basis, but give random results for diagonal or circular measurements.

The behaviour shown in Fig. 7.9 as been observed for many different quantum dots with close to zero exciton splitting. Figure 7.12, discussed below, shows another example with even stronger correlations. Recently similar behaviour has been reported for quantum dots prepared in another lab. Hafenbrak et al. reported observing strong correlation and anti-correlation in the rectilinear and circular bases, respectively [39].

7.5.2 Independence of Correlation on Linear Polarisation

Figure 7.10 plots the degree of linear co-polarisation correlation as a function of the linear polarisation basis angle. These measurements were recorded

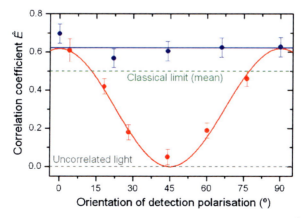

Fig. 7.10. The zero-delay unpolarised cross-correlation coefficient $\hat{E}$ plotted as a function of linear detection basis angle for a quantum dot with a large fine structure splitting (*red*) and a dot with no resolvable splitting (*blue*). The average correlation measured for the dot with no splitting is above the classical limit shown by the *dashed green line*

by rotating the angle of a single half-wave plate, placed directly after the microscope objective. Notice that the degree of correlation is approximately independent of the linear polarisation basis for a dot with close to zero exciton splitting. This is an expected result for photon pairs emitted in the entangled $(|H_{XX}H_X\rangle+|V_{XX}V_X\rangle)/\sqrt{2}$ state, since the linear polarisation measurement of the first photon defines the linear polarisation of the second photon.

For the dot with a finite exciton splitting on the other hand, the degree of correlation varies sinusoidally with the polarisation basis angle, taking a maximum value when the basis angle is parallel to the H-axis and tending to zero when it is aligned with the D-axis. This is again consistent with the emission of classically correlated pairs with polarisations aligned along the H- and V-axis for the dot with finite splitting. The linear polarisation correlation averaged over all bases angles of such a classically correlated source cannot exceed 50%. The fact that the average linear correlation measured for the dot with zero splitting is 62.4% ± 2.4%, ten standard deviations above the 50% limit for classical pairs of photons, can only be explained if the quantum dot emits polarisation entangled photon pairs.

7.5.3 Two-Photon Density Matrix

Figure 7.11 plots the two-photon density matrix measured for the dot with close to zero exciton splitting, along with the associated errors determined from the rate of co-incident counts. The density matrix was derived using (7.11). Unpolarised correlation coefficients $\hat{E}$ were measured as described in

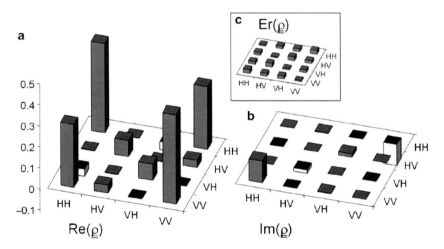

Fig. 7.11. The (a) real and (b) imaginary components of the two-photon density matrix ρ measured for the biexciton emission from a quantum dot with no resolvable fine structure splitting. (c) Counting errors in the measurement, these share a common z-scale with (a) and (b)

Sect. 7.4.7 for the nine combinations of rectilinear, diagonal and circular polarisation basis for each photon.

The strong outer diagonal measurements in the real matrix demonstrate the high probability that the photon pairs have the same linear polarisation. Most significant are the outer off-diagonal elements which are present due to the superposition in the two photon wavefunction and are clear indicators of entanglement. The small imaginary off-diagonal terms indicate a small phase difference between the $|HH\rangle$ and $|VV\rangle$ terms. This is a consequence of a small exciton spin splitting in this dot, as discussed in greater detail below.

The inner diagonal elements represent the probability of measuring pairs with opposite linear polarisations. They have finite strength due to scattering between the exciton spin states during the cascade and uncorrelated emission from layers in the sample other than the dot. These mechanisms degrade the fidelity of the source. All the other elements in the density matrix are zero within experimental error.

The experimental density matrix can be approximately decomposed into two contributions. Firstly ideal entangled pair emission in the state Ψ^+ from the dot, which is characterised by finite elements of height 0.5 in the four corners of the real part of the density matrix. The second contribution, which arises due to the background emission and spin scattering during the cascade, is that of an uncorrelated light source which has finite real elements along the diagonal of the density matrix of height 0.25.

Fitting to the experimental density matrix shows that the ideal emission represents 70% of the pairs generated by the device. The quantum dot emission spectrum measured under identical conditions shows two sharp lines due to the biexciton and exciton transitions superimposed on a broad weak background due to layers other than the dot in the sample. This background contribution is further confirmed by the non-zero value of the exciton self-correlation function at zero delay, shown in Fig. 7.4b. Other layers in the sample contribute to 14% of the collected pairs. It remains that spin scattering of the exciton state during the cascade will effect 16% of the collected pairs. Taking the measured exciton lifetime of 0.84 ns, an estimate of the exciton spin scattering time of approximately 4.4 ns is calculated.

It should be noted that the density matrix in Fig. 7.11, is derived from the experiment without any subtraction of background emission or dark counts. In contrast Akopian et al. [40] artificially subtract a constant value from each element measured for a quantum dot with large exciton splitting, such that the central diagonal elements ρ_{22} and ρ_{33} are close to zero. As discussed above, it is exactly these terms, which limit the fidelity of the source. Cancelling these elements and the subsequent re-normalisation of ρ_{11} and ρ_{44} has the effect of making their density matrix appear entangled. Unfortunately however, such a subtraction is difficult to justify. It cannot be argued that cancelling these elements is equivalent to removing the background emission, since exciton spin scattering also contributes to ρ_{22} and ρ_{33}.

7.5.4 Entanglement Fidelity and Other Numerical Measures of Entanglement

The experimental density matrix may be used to determine the projection of the emission on the expected state Ψ^+. For the data in Fig. 7.11 this yields a fidelity of $f^+ = 0.722 \pm 0.013$. This value exceeds the classical limit of 0.5 by 17 standard deviations and thereby proves with a high degree of certainty that quantum dots emit entangled photon pairs. As discussed above, the fidelity may also be determined directly from the degree of correlation measured in orthogonal bases. Using (7.14) and the data in Fig. 7.9b, yields $f^+ = (0.696 + 0.605 + 0.585 + 1)/4 = 0.722 \pm 0.013$, achieving the same result with minimal effort.

There are several other tests for the presence of quantum correlations in the emission, including the concurrence, the tangle, the Peres criterion and the value of the largest eigenvalue. Table 7.1 tabulates each test applied to the experimental density matrix in Fig. 7.11. Notice that the test for quantum correlations is proven in each case by many standard deviations. The eigenvalue test determines the most probable state of the system, which is found to be the maximally entangled state $(|HH\rangle + e^{0.1\pi i}|VV\rangle)/\sqrt{2}$ with eigenvalue 0.741 ± 0.014. The largest eigenvalue test is similar to the fidelity test, yet makes no assumption about the generated state. Thus for an unpolarised source, an eigenvalue exceeding 0.5 implies the existence of quantum correlations.

7.5.5 Improvement of the Fidelity by Time Gating

Figure 7.12a plots the degree of correlation in orthogonal co-polarisation bases measured by Young et al. on a different sample [38]. For this quantum dot, the degree of polarisation correlation is found to have values of 0.654 ± 0.018, 0.678 ± 0.019 and -0.845 ± 0.029, for rectilinear, diagonal and circular bases,

Table 7.1. A range of entanglement tests performed on the two-photon density matrix. The requirements to prove the state is entangled are quoted as the test limits

Test description	Test limit	Test result
Fidelity with Ψ^+	>0.5	0.722 ± 0.013
Largest eigenvalue	>0.5[a]	0.741 ± 0.014
Concurrence	>0	0.482 ± 0.027
Tangle	>0	0.232 ± 0.026
Average linear correlation	>0.5	0.624 ± 0.024
Peres[b]	<0	-0.241 ± 0.014

[a] For an un-polarised source
[b] The most negative eigenvalue of the partial transpose is used

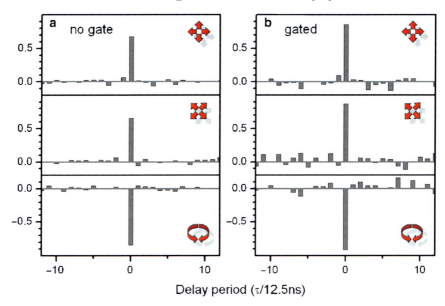

Fig. 7.12. The unpolarised cross-correlation coefficient between the biexciton and exciton photons plotted as a function of the delay between the photons in the rectilinear, diagonal and circular polarisation bases for a quantum dot with no resolvable fine structure splitting. (**a**) With no gating and (**b**) with a temporal gate rejecting photons arriving outside intervals centred on the exciton and biexciton emission

respectively. For this dot the degree of correlation is largest for circular polarisation, while for the dot studied in the previous section, rectilinear polarisation gave the strongest correlations. Clearly the polarisation basis producing the largest degree of correlation varies from dot to dot. The strong circular correlation in the dot may arise from a weak coupling between the intermediate electron–hole states. Using (7.13), the entanglement fidelity can be determined as $f^+ = (0.794 \pm 0.010)$, significantly higher than for the dot studied in the previous section.

The entanglement may be further enhanced by introducing gates, to restrict the times in the emission cycle in which the biexciton and exciton photons are registered. By rejecting the emission from the device just after the excitation laser pulse, a large fraction of the background emission can be rejected as it has a faster radiative decay rate than the dot. Rejecting the emission at long times, on the other hand, will remove pairs which have been subject to entanglement-destroying spin scattering in the exciton state, as well as reduce the contribution of detector dark counts that also degrade the correlations.

Figure 7.12b plots the degree of correlation measured after applying time gates to both the biexciton and exciton emission. The biexciton photons are registered if emitted between 0.6 and 2.6 ns after the peak of the emission intensity, while the exciton photon is collected within a window extending

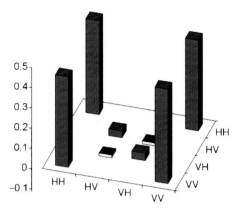

Fig. 7.13. The real component of the partial density matrix constructed from the measurement in Fig. 7.12 as described in the text

−0.4 to 2.6 ns. These time gates are applied by only counting photons which are detected within these windows. However, in principle, the time gate could also be applied by either shuttering the emission or alternatively applying voltages to Stark shift the emission lines away from the collection wavelengths at early and late times in the emission cycle. Such a technique has successfully been applied to improve the two-photon interference of single photons from a quantum dot [41].

Time gating significantly strengthens the polarisation correlations. The degree of correlation in the three polarisation bases in Fig. 7.12b are enhanced to 0.858 ± 0.044, 0.870 ± 0.045 and -0.920 ± 0.073, respectively. This in turn boosts the entanglement fidelity to 0.912 ± 0.024, the highest value recorded to date for a quantum dot source [38].

The three correlation measurements shown in Fig. 7.12b are sufficient to form the partial density matrix in Fig. 7.13. In this density matrix, the elements shown are all fully determined, while the others are expected to have values close to zero. Notice that the central diagonal elements are now further suppressed with respect to that in Fig. 7.11, so that the experimental matrix now more closely resembles the ideal case. This is due to the rejection of a large fraction of the background emission produced at early times in the emission cycle, as well as rejection of late pairs some of which derive from cascades which have been subject to spin scattering in the exciton state.

7.5.6 Violation of Bell's Inequality

Bell's famous inequality [2] was devised to show that information stored locally in the entangled system does not dictate the outcome of measurements. It is a more stringent test for entanglement than the presence of non-classical correlations, requiring a higher fidelity to a maximally entangled state. It is

7 Entangled Photon Generation by Quantum Dots

Table 7.2. Linear polarisation-correlations used to test the CHSH Bell inequality. Basis angles (θ_a, θ_b) correspond to the linear analyser settings used to measure the exciton and biexciton photons respectively. The resulting S_CHSH value >2 indicates a violation of the inequality

Basis angles $(\theta_a, \theta_b$ in °)	Correlation result $|\hat{E}_{ab}|$
0,22.5	0.481 ± 0.050
45,22.5	0.675 ± 0.048
0,67.5	0.611 ± 0.054
45,67.5	0.685 ± 0.041
S_CHSH	$\mathbf{2.45 \pm 0.10}$

an essential threshold for many applications exploiting spatially separated entangled photons, for example for ensuring the security of quantum communications.

The source described in the section above shows a sufficiently high fidelity to violate Bell's inequality [38]. Table 7.2 shows the result of correlation measurements designed to test the four parameter CHSH form of the Bell inequality. This yields a value for the CHSH Bell parameter, $S_\mathrm{CHSH} = 2.45 \pm 0.10$, which violates the inequality $S \leq 2$ by four standard deviations.

As discussed earlier, the Bell parameter may also be estimated from two photon correlation measurements. Using the data in Fig. 7.12b the equivalent Bell parameters are estimated to be, $S_\mathrm{rc} = 2.51 \pm 0.12$, $S_\mathrm{dc} = 2.53 \pm 0.12$, $S_\mathrm{rd} = 2.44 \pm 0.09$. These values agree with one another, and also with CHSH parameter, within experimental error. Clearly they too are in strong violation of the Bell inequality $S \leq 2$.

Though the experimental data presented in this section uses time gating to improve the correlations, this is not necessary to observe violation of Bell's inequality. For this quantum dot, the fidelity is sufficient to violate the inequality with the raw data. Using the ungated data in Fig. 7.12a, $S_\mathrm{rc} = 2.15 \pm 0.05$, $S_\mathrm{dc} = 2.12 \pm 0.05$, $S_\mathrm{rd} = 1.88 \pm 0.04$. Here two of the three estimates clearly violate Bell's inequality.

7.6 Factors Limiting the Entanglement Fidelity

7.6.1 Exciton Spin Splitting

In the previous sections the polarisation correlations for the extreme cases of quantum dots with either large or zero exciton spin splitting have been discussed, demonstrating both classical and quantum correlations in the two cases respectively. In this section the intermediate regime is the focus for which the exciton spin splitting is finite and of order of the homogeneous linewidth.

Not surprisingly the entanglement fidelity is found to be intermediate between the zero and large splitting case. The dependence on the splitting reveals insight on the mechanisms limiting the fidelity. The exciton spin splitting can be varied by applying an in-plane magnetic field to the quantum dot. An in-plane magnetic field mixes the bright and dark exciton states, causing a different energy shift for the two bright spin states and a change in their splitting. In some cases, as shown in Fig. 7.6c, an applied field can invert the sign of the splitting, allowing it to be tuned through zero. For the two dots studied in the section, the splitting is close to zero in the absence of a magnetic field and increases quadratically to around 25 µeV for an in-plane field of 4 T.

Figure 7.14 shows the dependence of the entanglement fidelity to Ψ^+ on the exciton splitting for two different quantum dots, labelled A and B. These dots show maximum fidelities when the splitting is close to zero of $f^+ = 0.75 \pm 0.04$ and $f^+ = 0.74 \pm 0.01$, respectively. Both dots therefore display a fidelity well above the classical limit of 0.5. As expected the fidelity falls as the splitting increases, eventually falling below the classical limit for a splitting of 3–4 µeV. The data in Fig. 7.14 can be well described by Lorentzian fits, shown by the solid lines. These have a full width at half-maximum of 3.3 ± 0.4 and 4.3 ± 0.8 µeV, respectively.

For large exciton spin splittings the quantum dot is expected to emit correlated photon pairs which are aligned either along the H or V-axis, i.e. statistical mixture of $|HH\rangle$ and $|VV\rangle$. Such an idealised source is expected to have a fidelity to Ψ^+ of 0.5. However, in the real source the classical polarisation correlation is degraded by spin scattering in the cascade, as well as the presence of randomly polarised background light. As a consequence of these two mechanisms the experimental fidelity at large splitting is seen to be significantly lower at ∼0.45.

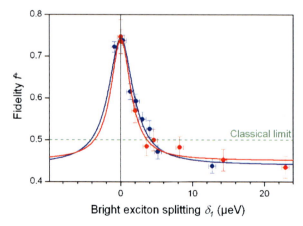

Fig. 7.14. Fidelity of the measured state with Ψ^+ as a function of fine-structure splitting, controlled with an in-plane magnetic field. Data for two different quantum dots is shown (Dot A in *red* and dot B in *blue*). *Solid lines* are Lorentzian fits to the data points

The fidelity of the entangled source, and in particular the fidelity at zero exciton spin splitting, is affected not only by background and spin-scattering, but also potentially by cross-dephasing between the two terms in Ψ^+. This includes any mechanism that changes the phase between $|HH\rangle$ and $|VV\rangle$, but does not flip the polarisation of a photon is considered. However, the model presented in the next section suggests that cross-dephasing may not be significant in these measurements.

7.6.2 Modelling Fidelity Values

Equation (7.3) describes the two-photon state generated by an ideal quantum dot with exciton spin splitting δ_1. Three mechanisms that may degrade the entanglement fidelity are considered: exciton spin scattering, cross-dephasing and uncorrelated background emission from the sample [42]. The fraction of the emission unaffected by these three mechanisms may be written as

$$kg^{(1)}_{H,V} = \frac{k}{1 + \tau_1/\tau_{ss} + \tau_1/\tau_{HV}}, \quad (7.20)$$

where k is the fraction of photon pairs originating exclusively from the dot, $g^{(1)}_{H,V}$ is the first-order cross-coherence, τ_1 is the exciton radiative lifetime, and τ_{ss} and τ_{HV} are the characteristic times for exciton spin scattering and cross-dephasing, respectively. For the terms in the density matrix which are unaffected by cross-dephasing, we also define

$$kg'^{(1)}_{H,V} = \frac{k}{1 + \tau_1/\tau_{ss}}. \quad (7.21)$$

It can be shown that the overall density operator can be written as

$$\rho = \frac{1}{4}\begin{pmatrix} 1 + kg'^{(1)}_{H,V} & 0 & 0 & 2kg^{(1)}_{H,V}z^* \\ 0 & 1 - kg'^{(1)}_{H,V} & 0 & 0 \\ 0 & 0 & 1 - kg'^{(1)}_{H,V} & 0 \\ 2kg^{(1)}_{H,V}z & 0 & 0 & 1 + kg'^{(1)}_{H,V} \end{pmatrix}, \quad (7.22)$$

where $z = (1 + ix)/(1 + x^2)$ and $x = kg^{(1)}_{H,V}\delta_1\tau_1/\hbar$.

Inspection of (7.22) reveals that the off-diagonal corner element, ρ_{14}, which is expected to be strong for a source emitting the entangled state Ψ^+, is real for $\delta_1 = 0$ and develops an imaginary component for increasing splitting. Thus the small imaginary components to ρ_{14} in the experimental density matrix in Fig. 7.11, as well as the data of Hafenbrak et al. [39], may be explained by a small exciton spin splitting in these dots. It also explains the imaginary phase of this term reported by Akopian et al. [40], who studied a quantum dot with large exciton splitting, though its magnitude was artificially enhanced through background substraction.

The fidelity of the source to Ψ^+ is thus:

$$f^+ = \frac{1}{4}\left(1 + kg'^{(1)}_{H,V} + \frac{2kg^{(1)}_{H,V}}{1+x^2}\right). \tag{7.23}$$

The above relation describes a Lorentzian dependence of the fidelity on the splitting, in good agreement with the shape of the experimental curves in Fig. 7.14. As expected the width of the Lorentzian $2\hbar/\tau_1 g^{(1)}_{H,V}$ is determined largely by the radiative lifetime, but is also broadened by the finite rate of spin scattering and cross-dephasing.

According to (7.23), the fidelities at zero and large exciton spin splittings are given by

$$f^+(\delta_1 = 0) = \frac{1}{4}(1 + kg'^{(1)}_{H,V} + 2kg^{(1)}_{H,V}),$$
$$f^+(\delta_1 \to \infty) = \frac{1}{4}(1 + kg'^{(1)}_{H,V}) \tag{7.24}$$

As expected the fidelity at large splittings depends on spin scattering and the fraction of pairs from the dot, while the fidelity at zero splitting depends additionally on cross dephasing. Thus the experimentally determined fidelities can be used to estimate the cross-dephasing rate.

Notice from (7.22) that if the rate of cross-dephasing is zero, such that $g'^{(1)}_{H,V} = g^{(1)}_{H,V}$, the fidelity at zero splitting will be reduced from ideal by three times the reduction at large splitting, i.e.

$$\frac{1 - f^+(\delta_1 = 0)}{0.5 - f^+(\delta_1 \to \infty)} = 3. \tag{7.25}$$

For the experimental data for dots A and B in Fig. 7.14, the measured ratios of (3.9 ± 1.7) and (4.2 ± 1.2) are in agreement, suggesting that the rate of cross-dephasing is low.

The lineshape fits may be used to obtain a more precise estimate for the cross-dephasing time. For dot A in Fig. 7.14 the fidelity in the limit of large exciton splitting is determined to be $f^+(\delta_1 \to \infty) = 0.440 \pm 0.023$. The fraction of pairs originating from the dot may be determined directly from the emission spectrum and second order correlation function to be $k = 0.86 \pm 0.05$. Thus the exciton spin-scattering rate is estimated, using (7.19) and the measured exciton lifetime of 891 ± 11 ps, to be $\tau_{ss} = 6.8$ ns. From the fitted fidelity at zero splitting $f^+(\delta_1 = 0) = 0.743 \pm 0.023$, cross-coherence time is estimated to be $\tau_{HV} = 3.1$ ns, and with uncertainty within the range 1.7–25.2 ns. A similar analysis for dot B also yields $\tau_{HV} > 2.0$ ns. As this is significantly more than the radiative lifetime, it confirms that the rate of cross-dephasing is low.

It is noteworthy that the biphoton cross-dephasing time is much longer than the ordinary exciton dephasing times for the dots. These may be determined from the first-order coherence times measured for the exciton emission from dot A and B of $\tau_2 = 88 \pm 7$ and 110 ± 3 ps, respectively [42]. Exciton dephasing is believed to occur due to spectral jitter of the emission lines, rapid charging and discharging of charge traps within the sample. The fact that the biphoton cross-coherence time is at least an order of magnitude longer than exciton coherence time, demonstrates that the coherence between the HH and VV polarised pairs is not sensitive to dephasing of the exciton state in the dot. This remarkable fact allows high fidelity entangled photons to be prepared by the biexciton cascade.

Thanks to its insensitivity to exciton dephasing, the entanglement is relatively robust to modest increases in the sample temperature. Hafenbrak et al. have reported that the entanglement fidelity falls only slightly from 0.72 at 4 K to 0.68 at 30 K [39]. At higher temperatures the emission is quenched due to the shallow confinement potential of these small quantum dots, designed to allow the use of silicon single photon detectors. An important future challenge will be to generate entangled photons from quantum dots with deeper potential wells and thereby increase the operating temperature.

7.6.3 Excess Charge

The polarisation correlations produced by a quantum dot have been shown to be highly sensitive to the excess charge density in the sample. Figure 7.15a plots the degree of rectilinear polarisation correlation ($\hat{E}_{rr}$) between XX and X recorded as a function of the bias applied to a p-i-n structure designed to allow the excess charge in the vicinity of the dot to be varied. Application of a positive bias injects holes into the quantum dots from the p-type contact, thereby switching the emission spectrum from dominance of negatively charged, to neutral, to positively charged excitons at the highest bias, as shown in Fig. 7.15b.

Notice that for the lowest biases shown, where the emission spectrum is dominated by X$^-$ due to the presence of excess holes in the vicinity of the dot, the degree of correlation is relatively low ($\hat{E}_{rr} \sim 0.13$). However, as the bias is increased and the excess electrons removed, the degree of correlation increases to reach $\hat{E}_{rr} \sim 0.68$. Its value does not appear to decline at the highest bias, for which there are excess holes around the dot. This suggests that the correlation is degraded by exciton spin scattering induced by excess electrons in the vicinity of the dot.

The red line in Fig. 7.15a plots the ratio of the exciton and biexciton decay times as a function of the applied bias. It can be seen that this ratio is also strongly influenced by the charge environment of the dot. At the highest biases, the ratio of the excitons' lifetimes τ_X/τ_{XX}, is expected to be ~ 2 from the fact that the biexciton has two radiative decay channels and which is typical of InAs quantum dots. As the applied bias is reduced, the lifetime ratio increases to $\tau_X/\tau_{XX} = 3.62 \pm 0.02$ due to a significant increase in the exciton decay time. This occurs due to enhanced spin scattering of the exciton state by excess electrons in the vicinity of the dot, allowing the population

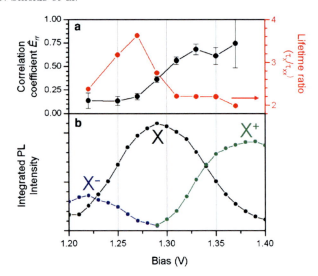

Fig. 7.15. (a) The ratio of the exciton and biexciton state lifetimes and the rectilinear cross-correlation coefficient between the biexciton and exciton photons plotted as a function of the bias applied to a quantum dot in a charge-tuneable structure. (b) Integrated intensities of the three exciton states plotted as a function of the bias applied to the device demonstrating the charge-tuning

of the dark exciton state. Significantly, the lifetime ratio increases over the same bias range as the fall in the degree of correlation, suggesting that excess electron induced scattering is responsible for both.

These results suggest that the strongest correlations are produced if the sample is slightly p-type. This is in fact the usual situation for samples grown by Molecular Beam Epitaxy, due to unintentional background doping with carbon. However, adding excess holes will also increase the fraction of cycles in which an excess hole is captured by the dot and so the neutral biexciton cascade cannot take place. There is therefore an optimum doping density, for which excess electrons have been removed but the emission is still dominated by the neutral exciton recombination.

7.7 Time-Delay Dependent Entanglement

It is widely believed that only quantum dots with close to zero exciton spin splitting emit polarisation entangled photon pairs. However, as pointed out above, this is not correct. Dots with finite exciton spin splitting δ_1 do emit entangled pairs, except in this case the entangled state depends upon the delay in emission time (τ) between the two photons [43]. The splitting introduces a phase term, $e^{i\delta_1\tau/\hbar}$, between the two terms in the rectilinear superposition, as described by (7.3).

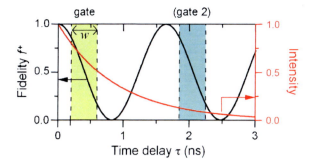

Fig. 7.16. An illustration of the method used to apply a temporal gate with a fixed width w to the emission, the biphoton intensity it intersects with (*red*) and the result of the gate on the fidelity to Ψ^+ (*black*) measured as a function of delay between the start of the gate and the start of the emission window

Figure 7.16 plots the expected sinusoidal variation in the fidelity f^+ with the maximally entangled state Ψ^+ with the delay τ between the photons for an ideal spin split dot. It can be seen that, as expected, for zero delay between the photons, $f^+ = 1$. After a delay of $\tau = \hbar/\delta_1$, the two terms in the rectilinear superposition are in anti-phase and the fidelity to Ψ^+ drops to zero, while the fidelity to the Bell state $\Psi^- = (|H_{XX}H_X\rangle - |V_{XX}V_X\rangle)/\sqrt{2}$ will be unity. Also shown in Fig. 7.16a is the rate of photon pair emission for different delays τ which decays exponentially.

By measuring the time delay for each photon pair, it might be possible to use these time-delay dependent entangled pairs. Time correlated single photon counting allows the time delay between the two photons to be recorded for each pair. The fidelity f^+ with the maximally entangled state Ψ^+ can be determined, for pairs whose temporal separation (τ) falls within a single timing gate $\tau_g < \tau < (\tau_g + w)$. This region is indicated on the predicted exponential decay of the biphoton intensity of Fig. 7.16a by a yellow shaded region.

Figure 7.17a plots the measured fidelity f^+ to Ψ^+ as a function of the gate start time τ_g for a fixed gate width $w = 537\,\text{ps}$. The data was recorded for a dot with an exciton spin splitting of $\delta_1 = 2.5 \pm 0.5\,\mu\text{eV}$. It can be seen that as expected the fidelity varies in an oscillatory fashion with τ_g, due to variation in the phase term in the two-photon wavefunction. The experimental data does not show the largest fidelity at $\tau_g = 0$ due to the finite gate width.

Figure 7.17 compares the observed behaviour to a calculation using the model described in the previous section. For this calculation, the rate of cross-dephasing has been assumed to be negligible ($1/\tau_{HV} = 0$), as suggested by the results in the previous section. The fidelity is computed numerically using a Monte-Carlo approach, to incorporate a Gaussian approximation of the APD jitter observed in experiment. For simplicity, the fraction of uncorrelated light is modelled as time-independent. Despite the limited nature of the model, good agreement is seen between the measured and calculated curves.

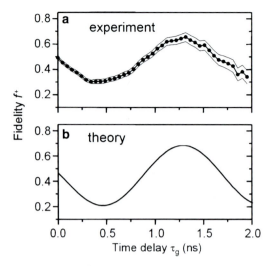

Fig. 7.17. (a) Experimental and (b) calculated fidelity as a function of gate start time with a gate width of 537 ps and a fine structure splitting $\delta_1 = 2.5\,\mu\text{eV}$. Bands denote Poissonian counting errors

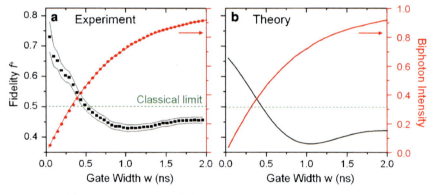

Fig. 7.18. (a) Experimental and (b) calculated fidelity and the fraction of photon pairs retained after temporal post-selection within a gate of width w

Time gating may also be used to greatly enhance the fidelity of pairs produced by spin-split dots to a particular state. Figure 7.18a plots the fidelity f^+ (black points) as a function of the gate width w for a fixed gate start time $\tau_g = 0$, defined as the modal delay between biexciton and exciton photon detection. For a gate width $w = 2\,\text{ns}$, the fidelity f^+ is measured to be 0.46 ± 0.01, which is below the 0.5 maximum achievable fidelity for an unpolarised classical state. However, as the gate width is reduced below $\sim 1\,\text{ns}$, the fidelity begins to increase, up to a maximum of 0.73 ± 0.05 for the smallest gate width of $w = 49\,\text{ps}$, indicating entanglement. This is a consequence of resolving entanglement before the state has significantly evolved over time. Good agreement with the calculated variation is apparent in Fig. 7.18b.

Also shown in Fig. 7.18a, as red points, is the biphoton intensity measured within the gate, normalised to the total biphoton intensity for infinite w. The curve fits excellently to the predicted $1 - e^{(-w/\tau_1)}$, revealing an exciton lifetime τ_1 of $769 \pm 9\,\text{ps}$. It is clear that large increases in fidelity can be achieved without a dramatic effect on the intensity of light collected. Furthermore, the biphoton intensity could be further enhanced by collecting the in-phase pairs emitted after an integer number of 2π phase shifts, as shown schematically by the second shaded region in Fig. 7.16a.

The fidelity is increased for small gate widths because the system post-selects photons in the time-domain that have a similar phase relationship between the two terms in the superposition. In the measurements above, the choice of $\tau_g = 0$ limits the phase acquired in the exciton state close to zero, so collected photons have high fidelity with the symmetric superposition Ψ^+. Similarly enhanced fidelities could be obtained for other values of τ_g with other maximally entangled states with different phase.

Selection in time equivalently reduces which-path information from the polarisation splitting δ_1 in the energy domain. This is because the Fourier transform of a truncated exponential decay results in a broad natural linewidth of the post-selected photons. This is shown in Fig. 7.19 by the Fourier transform of the biphoton decay, truncated after emission time delay τ of 0.39 or 1.0 ns.

In comparison to direct energy-resolved post-selection, resolving in time is more efficient. This is understandable as time-resolved post-selection targets photons at the beginning of the decay cycle, where emission intensity is strongest. In contrast, energy-resolved post-selection targets photons emitted with energies between those of H_X and V_X, where intensity is minimum [40].

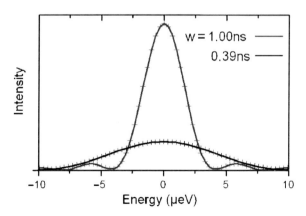

Fig. 7.19. Measured natural linewidth of photon pairs post-selected with a 1 ns (*blue*) and 0.39 ns (*black*) single gate

7.8 Quantum Interferometry

In addition to the more widely investigated quantum information applications, the unusual properties of entangled light have catalysed the emergence of other applications in the field of quantum enhanced metrology. These applications rely upon generation, interference and detection of multi-photon states to enhance the sensitivity and resolution of optical systems, compared to that achievable by using the photons separately. The origin of the enhancement is the multi-photon interference, which can reveal shorter de Broglie wavelength and longer coherence time compared to single photon interference. This approach may be used to multiply the resolution of interferometry and imaging at a fixed wavelength, and has applications in quantum-enhanced lithography [10], low-cell-damage biomedical microscopy [11].

Enhanced frequency interference fringes have been observed using entangled photons generated by parametric down conversion [44,45], or by post-selected measurements using classical light [46,47]. However unlike these approaches, a quantum dot photon pair source has the potential to operate on-demand, without multiple pair generation events, or post selection. Here biphoton interference showing fringes with half the period compared to single photons at the same wavelength is demonstrated [48].

Biphoton interference from a quantum dot device using an interferometer similar to that shown in Fig. 7.20a is measured. The interferometer applies phase control to just one of the two superposed components of the entangled

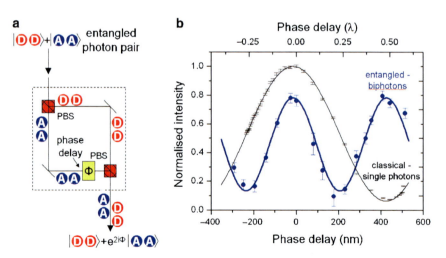

Fig. 7.20. (**a**) A schematic diagram of the effective biphoton interferometer used to measure the biphoton wavelength of the biexciton emission. (**b**) The normalised intensity of classical single photons (*black*) and normalised biphoton intensity of entangled photon pairs (*blue*) for a quantum dot source as a function of the phase delay

two-photon states, before they interfere. This is achieved as follows. Consider an entangled pure input state $(|DD\rangle+|AA\rangle)/\sqrt{2}$. A combination of appropriately configured polarising beamsplitters force D and A polarised photons to take separate paths through the interferometer. As the nature of the entangled state requires both photons to have the same diagonal polarisation, both photons will always take the same random path through the interferometer. Such photons are said to be "path-entangled". Both A polarised photons are delayed by phase ϕ, so that output of the interferometer is $(|DD\rangle + e^{2i\phi}|AA\rangle)/\sqrt{2}$. Interference of the two-photon amplitudes of $|DD\rangle$ and $|AA\rangle$ is achieved by measurement in the rectilinear basis. The biphoton intensity variation with ϕ results in an interferogram, with expected period π, corresponding to the de Broglie wavelength $\lambda/2$, where λ is the average wavelength of the single biexciton and exciton photons. In contrast the single photon input state $|V\rangle = (|A\rangle + |D\rangle)/\sqrt{2}$ has an expected output $(|D\rangle + e^{i\phi}|A\rangle)/\sqrt{2}$. Measuring the intensity in the vertical polarisation would yield an interferogram with period 2π, corresponding to a de Broglie wavelength λ.

The required polarisation dependent phase delay was realised by a liquid crystal with voltage dependent birefringence, and the interferometer was implemented using a collinear geometry, providing exceptional stability compared to a typical independent path Mach–Zehnder design.

Single photon interference fringes from light emitted by a quantum dot, were measured by inserting a linear polariser before the interferometer, to select only vertically polarised photons. Using a single APD, the intensity of the resulting single exciton photon state $|V\rangle$ was measured as a function of the phase delay, and normalised to the maximum. The results are shown in Fig. 7.20b as black points. Clear interference fringes are seen, and the intensity varies as a function of the phase delay in agreement with the fit to expected sinusoidal behaviour, shown by the solid line. The period of the oscillations is determined to be 877 ± 35 nm ($0.99 \pm 0.03\lambda$), approximately equal to the wavelength of the quantum dot emission of ~ 885 nm, as expected.

The linear polariser set before the interferometer was removed, so that entangled photon pairs emitted by the quantum dot could be analysed. The normalised biphoton intensity is equal to $(g_{VV}^{(2)}+g_{HH}^{(2)})/(g_{VV}^{(2)}+g_{VH}^{(2)}+g_{HV}^{(2)}+g_{HH}^{(2)})$, where the denominator and numerator and are proportional to the biphoton generation and detection rates respectively. For an unpolarised source, such as this dot, this is approximated to $g_{VV}^{(2)}/(g_{VV}^{(2)} + g_{VH}^{(2)})$. The second order correlation functions for the $|VV\rangle$ and $|VH\rangle$ two-photon detection bases were measured simultaneously as described above.

The measured normalised biphoton intensity, indicated by blue points in Fig. 7.20b, shows strong interference fringes. The difference in the period of oscillations compared to the classical single photon case is very striking. The fringes fit well to the predicted sinusoidal behaviour shown as a solid line, from which the period of the oscillations is determined to be 442 ± 36 nm

(0.50 ± 0.03λ). The period is equivalent to the de Broglie wavelength of the biphoton, which is in excellent agreement with the two-fold reduction from 885–443 nm expected for an entangled photon pair source.

The shorter de Broglie wavelength for the entangled state compared to the single photon implies that up to a two-fold enhancement of the imaging resolution is possible using biphoton detection. The visibility of the measured biphoton interferogram compared to that for single photon detection demonstrates enhanced phase resolution of the interferometer with a biphoton source [11]. Note that the exciton and biexciton have different wavelengths, but this does not affect the visibility of biphoton interference, as it is the total energy of the two-photon state that defines the de Broglie wavelength.

The interferometer can also be used to control the phase offset of the entangled state. For example, a $\lambda/4$ delay transforms the state $(|HH\rangle + |VV\rangle)/\sqrt{2}$ into $(|HV\rangle + |VH\rangle)/\sqrt{2}$, and the photons are then polarisation anti-correlated in the rectilinear basis. This is an analogous transformation of the entangled state as evolves naturally over time for a dot with non-zero splitting.

7.9 Outlook

Semiconductor quantum dots are perhaps the world's youngest proven technology for entangled light generation. Since their first successful operation in 2006, there has been continuous improvements to the quality of the entangled light generation, and to the understanding of the fundamental concepts that both enable and degrade entanglement. As reviewed here, in the short time since their birth quantum dots can now emit entangled pairs with fidelity >90%, which is suitable for applications.

There are many ways that quantum dot entangled light sources must be improved before their performance will surpass that of their older competitors, particularly parametric down conversion. However the path to improvements seems unusually clear. Though quantum logic circuits can correct for some errors, they must be minimised by further improvements to fidelity. As described above, this is limited chiefly by background light collected from other regions of the sample, which must be reduced. Light collection efficiency can be increased by improving the cavity QED of the device, with recent demonstrations of enhanced spontaneous emission rates implying coupling to the cavity mode of >83% [49,50]. This step is especially important for large scale quantum circuits. Tailoring the wavelength for optical fibre is also desirable, and single quantum dots at this wavelength are now routinely fabricated [51–54]. The operating temperature can be increased to levels compatible with cost-efficient cooling technology by increasing the confinement potential of the quantum dot heterostructure.

Finally, the technology developed for electrically driving a single photon source promises to enable the realisation of an entangled light emitting diode [55]. Should all these improvements come to fruition, the appeal of such a ubiquitous entangled light source is seductive. Quantum information technology can then make a transition from the physics lab, to mainstream technology.

References

1. A. Einstein, B. Podolsky, N. Rosen, Phys. Rev. **47**, 777 (1935)
2. J.S. Bell, Physics **1**, 195 (1964)
3. J.S. Bell, J. Phys. (Paris) **42**, C2-41 (1981)
4. D. Bouwmeester, A.K. Ekert, A. Zeilinger, *The Physics of Quantum Information.* (Springer, Berlin, 2000)
5. A.K. Ekert, Phys. Rev. Lett. **67**, 661 (1991)
6. H.-J. Briegel, W. Dür, J.I. Cirac, P. Zoller, Phys. Rev. Lett. **81**, 5932 (1998)
7. T. Yang, Q. Zhang, T.-Y. Chen, S. Lu, J. Yin, J.-W. Pan, Z.-Y. Wei, J.-R. Tian, J. Zhang, Phys. Rev. Lett. **81**, 110501 (2006)
8. E. Knill, R. Laflamme, G.J. Milburn, Nature **409**, 46 (2001); P. Kok, W.J. Munro, K. Nemoto, T.C. Ralph, J.P. Dowling, Rev. Mod. Phys. **79**, 135 (2007)
9. Rayleigh, Phil. Mag. **8**, 477 (1879)
10. A.N. Boto, P. Kok, D.S. Abrams, S.L. Braunstein, C.P. Williams, J.P. Dowling, Phys. Rev. Lett. **85**, 2733 (2000)
11. V. Giovannetti, S. Lloyd, L. Maccone, Science **306**, 1330 (2004)
12. S. Tanzilli, H. De Riedmatten, H. Tittel, H. Zbinden, P. Baldi, M. De Micheli, D.B. Ostrowsky, N. Gisin, Elec. Lett. **37**, 26 (2001)
13. E. Edamatsu, G. Oohata, R. Shimizu, T. Itoh, Nature **431**, 167 (2004)
14. D. Bouwmeester, J.-W. Pan, K. Mattle, M. Eibl, H. Weinfurter, A. Zeilinger, Nature **390**, 575 (1997)
15. V. Scarani, H. de Riedmatten, I. Marcikic, H. Zbinden, N. Gisin, Eur. Phys. J. D **32**, 129 (2005)
16. A. Aspect, P. Grangier, G. Roger, Phys. Rev. Lett. **47**, 460 (1981)
17. O. Benson, C. Santori, M. Pelton, T. Yamamoto, Phys. Rev. Lett. **84**, 2513 (2000)
18. E. Moreau, I. Robert, L. Manin, V. Thierry-Mieg, J. M. Grard, I. Abram, Phys. Rev. Lett. **87** 183601 (2001)
19. R.M. Stevenson, R.M. Thompson, A.J. Shields, I. Farrer, B.E. Kardynal, D.A. Ritchie, M. Pepper, Phys. Rev. B(R) **66**, 081302 (2002)
20. C. Santori, D. Fattal, M. Pelton, G.S. Solomon, Y. Yamamoto, Phys. Rev. B **66**, 045308 (2002)
21. S.M. Ulrich, S. Strauf, P. Michler, G. Bacher, A. Forchel, Appl. Phy. Lett. **83**, 1848 (2003)
22. V.D. Kulakovskii, G. Bacher, R. Weigand, T. Kmmell, A. Forchel, E. Borovitskaya, K. Leonardi, D. Hommel, Phys. Rev. Lett. **82**, 1780 (1999)
23. O. Stier, M. Grundmann, D. Bimberg, Phys. Rev. B **59**, 005688 (1999)
24. S. Rodt, R. Heitz, A. Schliwa, R.L. Sellin, F. Guffarth, D. Bimberg, Phys. Rev. B **68**, 035331 (2003)
25. R.J. Young, R.M. Stevenson, A.J. Shields, P. Atkinson, K. Cooper, D.A. Ritchie, K.M. Groom, A.I. Tartakovskii, M.S. Skolnick, Phys. Rev. B **72** 113305 (2005)

26. T. Inoue, T. Kita, O. Wada, M. Konno, T. Yaguchi, T. Kamino, Appl. Phys. Lett. **92**, 031902 (2008)
27. W. Langbein, P. Borri, U. Woggon, V. Stavarache, D. Reuter, A.D. Wieck, Phys. Rev. B **69**, 161301 (2004)
28. D.J.P. Ellis, R.M. Stevenson, R.J. Young, A.J. Shields, P. Atkinson, D.A. Ritchie, Appl. Phys. Lett. **90**, 011907 (2007)
29. R.M. Stevenson, R.J. Young, P. See, D. Gevaux, K. Cooper, P. Atkinson, I. Farrer, D.A. Ritchie, A.J. Shields, Phys. Rev. B **73**, 033306 (2006)
30. B.D. Gerardot, S. Seidl, P.A. Dalgarno, R.J. Warburton, D. Granados, J.M. Garcia, K. Kowalik, O. Krebs, K. Karrai, A. Badolato, P.M. Petroff, Appl. Phys. Lett. **90**, 041101 (2007)
31. K. Kowalik, O. Krebs, A. Lematre, S. Laurent, P. Senellart, P. Voisin, J.A. Gaj, Appl. Phys. Lett. **86**, 041907 (2005)
32. S. Seidl, M. Kroner, A. Hgele, K. Karrai, R.J. Warburton, A. Badolato, P.M. Petroff, Appl. Phys. Lett. **88**, 203113 (2006)
33. J. Von Neumann, *Mathematical Foundation of Quantum Mechanics* (Princeton University Press, Princeton, 1955)
34. D. Fattal, K. Inoue, J. Vučković, C. Santori, G.S. Solomon, Y. Yamamoto, Phys. Rev. Lett. **92**, 37903 (2004)
35. J.F. Clauser, M.A. Horne, A. Shimony, R.A. Holt, Phys. Rev. Lett. **23** 880 (1969)
36. R.M. Stevenson, R.J. Young, P. Atkinson, K. Cooper, D.A. Ritchie, A.J. Shields, Nature **439**, 179 (2006)
37. R.J. Young, R.M. Stevenson, P. Atkinson, K. Cooper, D.A. Ritchie, A.J. Shields, New J. Phys. **8**, 29 (2006)
38. R.J. Young, R.M. Stevenson, A.J. Hudson, C.A. Nicoll, D.A. Ritchie, A.J. Shields, Phys. Rev. Lett. **102**, 030406 (2009)
39. R. Hafenbrak, S.M. Ulrich, P. Michler, L. Wang, A. Rastelli, O.G. Schmidt, New J. Phys. **9**, 315 (2007)
40. N. Akopian, N.H. Lindner, E. Poem, Y. Berlatzky, J. Avron, D. Gershoni, B.D. Gerardot, P.M. Petroff, Phys. Rev. Lett. **96**, 130501 (2006)
41. A.J. Bennett, R.B. Patel, A.J. Shields, K. Cooper, P. Atkinson, C.A. Nicoll, D.A. Ritchie, Appl. Phys. Lett. **92**, 193503 (2008)
42. A.J. Hudson, R.M. Stevenson, A.J. Bennett, R.J. Young, C.A. Nicoll, P. Atkinson, K. Cooper, D.A. Ritchie, A.J. Shields, Phys. Rev. Lett. **99**, 266802 (2007)
43. R.M. Stevenson, A.J. Hudson, A.J. Bennett, R.J. Young, C.A. Nicoll, D.A. Ritchie, A.J. Shields, Phys. Rev. Lett. **101**, 170501 (2008)
44. E. Edamatsu, R. Shimizu, T. Itoh, Phys. Rev. Lett. **89**, 213601 (2002)
45. P. Walther, J.-W. Pan, M. Aspelmeyer, R. Ursin, S. Gasparoni, A. Zeilinger, Nature **429**, 158 (2004)
46. G. Khoury, H.S. Eisenberg, E.J.S. Fonesca, D. Bouwmeester, Phys. Rev. Lett. **96**, 203601 (2006)
47. T. Nagata, R. Okamoto, J.L. O'Brian, K. Sasaki, S. Takeuchi, Science **316**, 726 (2007)
48. R.M. Stevenson, A.J. Hudson, R.J. Young, P. Atkinson, K. Cooper, D.A. Ritchie, A.J. Shields, Opt. Express **15**, 6507 (2008)
49. J. Vučković, D. Fattal, C. Santori, G.S. Solomon, Y. Yamamoto, Appl. Phys. Lett. **82**, 3596 (2003)
50. A.J. Bennett, D. Unitt, P. Atkinson, D.A. Ritchie, A. J Shields, Opt. Express **13**, 50 (2005)

51. M.B. Ward, O.Z. Karimov, D.C. Unitt, Z.L. Yuan, P. See, D.G. Gevaux, A.J. Shields, P. Atkinson, D.A. Ritchie, Appl. Phys. Lett. **86**, 201111 (2005)
52. C. Zinoni, B. Alloing, C. Monat, V. Zwiller, L.H. Li, A. Fiore, L. Lunghi, A. Gerardino, H. de Riedmatten, H. Zbinden, N. Gisin, Appl. Phys. Lett. **88**, 131102 (2006)
53. M.B. Ward, T. Farrow, P. See, Z.L. Yuan, O.Z. Karimov, A.J. Bennett, A.J. Shields, P. Atkinson, K. Cooper, D.A. Ritchie, Appl. Phys. Lett. **90**, 063512 (2007)
54. T. Miyazawa, K. Takemoto, Y. Sakuma, S. Hirose, T. Usuki, N. Yokoyama, M. Takatsu, Y. Arakawa, Jpn. J. Appl. Phys. **44**, L620 (2005)
55. Z. Yuan, B.E. Kardynal, R.M. Stevenson, A.J. Shields, C.J. Lobo, K. Cooper, N.S. Beattie, D.A. Ritchie, M. Pepper, Science **295**, 102 (2002)

8

Cavity QED in Quantum Dot–Micropillar Cavity Systems

S. Reitzenstein and A. Forchel

Summary. In this contribution we review our recent work on cavity quantum electrodynamics experiments (cQED) with single quantum dots in high quality micropillar cavities. After a short introduction to the theoretical background of cQED with single two level emitters, important aspects in the growth and patterning of quantum dot–micropillar cavities will be addressed in the second part of this review. In particular, the optimization of both the quantum dot and cavity characteristics will be discussed. Differences between weak and strong coupling are illustrated experimentally in the framework of cQED. Furthermore, the demonstration of the quantum nature of a strongly coupled quantum dot–micropillar system as well as a coherent photonic coupling of quantum dots mediated by the strong light field in high-Q micropillar cavities will be addressed.

8.1 Introduction

Quantum dot–microcavity systems are very attractive for cavity quantum electrodynamics experiments due to the atom like emission properties of the quantum dots (QDs) and the high vacuum field amplitude in the center of high quality, low mode-volume microcavities. In general, cQED describes the interaction of atom like emitters such as semiconductor quantum dots with confined photons and distinguishes between the *weak coupling* regime and the *strong coupling* regime. Depending on the dipole moment of the emitter and the optical quality of the microcavity one can observe either of these regimes. In the weak coupling regime the system experiences a strong dissipation of energy and the emitter's spontaneous emission rate can be tailored due to the Purcell-effect. In contrast, dissipative processes are strongly reduced in the strong coupling regime that is characterized by a coherent exchange of energy between the emitter and the cavity mode. In fact, strong coupling is observed when the emitter–photon coupling rate is larger than any dissipative decay rate in the system. This implies stringent requirements on the technology used to realize high quality quantum dot–micropillar structures for the observation of pronounced cQED effects.

This chapter is organized as follows. Section 8.2 recalls important theoretical aspects of cQED with single emitters in high quality microcavities and addresses the differences between the weak coupling and the strong coupling regime. In Sect. 8.3 recent advances and optimization steps in the fabrication of quantum dot–micropillar cavities will be discussed. The optimization of both, the quantum dots in the active layer as well as the micropillar cavities, aims at the observation of strong coupling between a single QD exciton and the photonic mode of a high quality-factor microcavity. After introducing the experimental setup in Sect. 8.4, examples of weak and strong coupling of QDs in micropillar cavities are presented in Sects. 8.5.1 and 8.5.2, respectively. Here, we will demonstrate and discuss the influence of QD and micropillar design parameters on the coupling strength in different interaction regimes. As we are interested in cQED effects on the single emitter level, it is important to prove that one and only one QD interacts with the cavity mode. In Sect. 8.5.3 we will present photon correlation experiments performed in the strong coupling regime giving clear evidence of the single emitter interaction regime. An interesting aspect of strongly coupled exciton–photon systems is the possibility of coherently coupling quantum dots by a common cavity mode which will be addressed in Sect. 8.5.4 before we conclude this chapter.

8.2 Theoretical Background

Tailoring the spontaneous emission of a discrete quantum emitter is a central topic in cavity quantum electrodynamics [1–3]. First experiments on cQED were conducted in atomic systems using, e.g., dipole traps to precisely localize an atom in the field maximum of the cavity mode [4] (and references therein). The enormous progress in semiconductor technology has facilitated cQED studies also in solid state nanostructures [5–11]. These structures offer the unique capability of permanently positioning the quantum emitter in the center of a microcavity. In this context, self-assembled semiconductor quantum dots have been proven to be very suitable quantum emitters for the observation of cQED. In fact, QDs are highly attractive candidates because they show atom like emission properties and provide high quantum efficiency as well as large dipole moments at comparatively low dephasing rates [12].

In an ideal case, a single quantum emitter is resonantly coupled to a single empty mode of a lossless cavity. Being initially in its excited state the emitter recombines spontaneously after a characteristic time by emitting a photon into the cavity mode. The photon is stored in the cavity mode and can be reabsorbed by the localized quantum emitter which means that the spontaneous emission process becomes reversible and a coherent exchange of energy takes place between the emitter and the cavity mode. The periodic emission and absorption of an photon is described in terms of Rabi oscillations at an angular frequency 2Ω. The Rabi frequency characterizes the strength of the emitter–field coupling and is proportional to the vacuum field amplitude E_{vac}

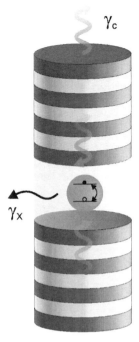

Fig. 8.1. Schematic view of a quantum dot–micropillar cavity system. A single QD acts as a two level emitter which interacts with the confined light field in the center of the micropillar cavity. The system experiences decoherence due to photon losses and dephasing of the emitter characterized by γ_c and γ_x, respectively

at the site of the emitter [6]. E_{vac} depends on the effective mode volume V via $E_{\text{vac}} \propto 1/\sqrt{V}$ which explains the strong interest in low mode volume microcavities such as microdisks, photonic crystal (PC) membrane cavities (see Chap. 9) and micropillar cavities [3]. Here, we will focus on cQED experiments in the quantum dot–micropillar system shown schematically in Fig. 8.1.

In real microcavities, photons are not confined for an infinitive amount of time but leave the cavity after a characteristic time, which is inversely proportional to the quality factor of the latter. Photon losses through leaky modes introduces dissipation in the system and leads to a damping of the Rabi oscillations. Still, it is possible to observe Rabi oscillations if the coupling rate between an atom-like emitter and the photonic cavity mode is larger than the photon loss rate. In this so-called *strong coupling* regime the coupled system is characterized by two split eigenstates separated by the vacuum Rabi splitting (VRS). The term VRS is used because the interaction occurs between a quantum emitter and the *vacuum field* inside an *empty* cavity mode. One refers also to VRS in a classical case when more than a single oscillator is involved, e.g., in quantum wells [13, 14]. However, in this article we will focus on the single emitter case. If irreversible decay dominates and the condition for strong coupling is not fulfilled, the system is described in terms of *weak coupling*. It is characteristic for the weak coupling regime that the spontaneous emission of an emitter can be enhanced or inhibited compared with its vacuum level by tuning the emitter in and out of resonance with cavity quasi modes [5,7,15].

8.2.1 The Strong Coupling Regime

Different theoretical schemes have been applied to describe the interaction of a single QD exciton coupled resonantly to an empty mode of a microcavity [16–19]. For an ideal system in which the decoherence of the emitter and the photon field can be neglected it is possible to describe the nonperturbative dynamics of the emitter–photon coupling by the Jaynes–Cummings Hamiltonian [17]:

$$H = \hbar\omega_0 \hat{\sigma}_3 + \hbar\omega_\mu \left(\hat{a}_\mu^\dagger \hat{a}_\mu + \frac{1}{2} \right) + ig \left(\hat{\sigma}_- \hat{a}_\mu^\dagger - \hat{\sigma}_- \hat{a}_\mu \right). \quad (8.1)$$

Here, $\hat{\sigma}_-$, $\hat{\sigma}_+$, $\hat{\sigma}_3$ represent pseudospin operators for the two-level system with its ground (excited) state $|g>$ ($|e>$). The spectrum of this Hamiltonian consists of a ground state $|g,0>$ and a ladder of doublets $|e,n>$, $|g,n+1>$, $n = 0, 1, \ldots$. At resonance, when $\omega_0 = \omega_\mu$, the coherent coupling gives rise to so-called dressed states which are split by $2g\sqrt{n+1}$ and constitute the famous Jaynes–Cummings ladder [20]. The splitting depends on the number of photons in the cavity mode n and one refers to vacuum Rabi splitting when $n = 0$.

The exciton–photon coupling strength g is defined as the scalar product of the transition matrix element of the dot dipole moment **d** with the local value of the electric field at the dot position in the cavity **E** ($g = |\langle \mathbf{d} \cdot \mathbf{E} \rangle|$). For the case of the emitter (exciton dipole) placed at the maximum of the electric field distribution of a cavity mode, the coupling constant can be expressed in terms of the oscillator strength (OS) $f = 2m\omega_0 d^2/(e^2\hbar)$ as [17]

$$g = \left(\frac{\hbar^2}{4\pi\epsilon_r\epsilon_0} \frac{\pi e^2 f}{m\hat{V}} \right)^{1/2} \quad (8.2)$$

where ϵ_0 and ϵ_r are the vacuum and relative medium permittivity, m is the free electron mass and $\hat{V}$ corresponds to the effective cavity mode volume.

In real cavities decoherence processes occur, effecting either the emitter or the cavity. For instance, photons leak out of the cavity and spontaneous emission is coupled into a continuum of quasi modes. However, if decoherence processes are slow enough to be on the scale of the Rabi period, the coupled system is still in the strong coupling regime and experiences damped Rabi oscillations. Faster decoherence processes lead to overdamping of the coupled system in the weak coupling regime and the emitter relaxes monotonically down to its ground state. In the limit of weak damping the energies of the two eigenmodes at resonance can be expressed by the following formula [16,17,19]:

$$E_{1,2} = E_0 - i(\gamma_c + \gamma_X)/4 \pm \sqrt{g^2 - (\gamma_c - \gamma_X)^2/16}, \quad (8.3)$$

where E_0 denotes the energy of the uncoupled exciton X and photon modes C, while γ_X and γ_c are their full widths at half maximum, respectively.

From (8.3) the vacuum Rabi splitting for the system can be obtained directly as

$$\Delta E = 2\hbar\Omega = 2\sqrt{g^2 - (\gamma_c - \gamma_X)^2/16}. \quad (8.4)$$

According to (8.3), any mode splitting due to strong emitter–photon coupling appears when

$$g^2 - (\gamma_c - \gamma_X)^2/16 > 0. \quad (8.5)$$

Since for high quality (Q) factor microcavities the QD exciton linewidth is in general significantly smaller than the cavity mode linewidth, the condition in (8.5) can be simplified to [9]

$$g > \gamma_c/4. \quad (8.6)$$

This is the threshold condition for strong coupling of a single QD exciton and a microcavity mode. Using (8.2) which tells that $g \propto \sqrt{f/\hat{V}}$ and considering that the cavity Q factor is defined by $Q = E_c/\gamma_c$, one obtains from (8.6) the figure of merit $Q\sqrt{f/\hat{V}}$ for the observation of strong coupling.

For pillar microcavities, the mode volume scales approximately with the square of the cavity diameter d_c, which leads to $Q\sqrt{f}/d_c$ as a quantity for technological optimization. Thus, it is necessary to maximize simultaneously the oscillator strength of the QD exciton transition and the Q-factor to micropillar diameter ratio. The influence of the cavity mode linewidth γ_c and the OS on the coupling behavior of a resonant quantum dot–micropillar cavity system is illustrated in Fig. 8.2 which shows the real part of $E_{1,2} - E_0$ for a micropillar cavity with a diameter $d_c = 1.5\,\mu\text{m}$. Both eigenvalues have the same real part until the threshold condition (8.6) is fulfilled and mode splitting occurs. Due to a different coupling strength of 30, 52, and 67 μeV for $f = 10$, 30, and 50, respectively, the onset of strong coupling shifts toward smaller resonator linewidths, i.e., higher Qs, with decreasing OS which puts more stringent requirements on the pillar quality for QDs with low OS. Typical values for the vacuum Rabi splitting in quantum dot–micropillar cavity systems are in the range of a few tenths of microelectronvolts to about 150 μeV.

8.2.2 The Weak Coupling Regime

In the weak coupling regime, the spontaneous emission process is irreversible and the eigenenergies $E_{1,2}$ in (8.3) have the same real part, so that Rabi oscillations do not occur. Nevertheless, the spontaneous emission process is influenced by the presence of the cavity due to the modified spectral density

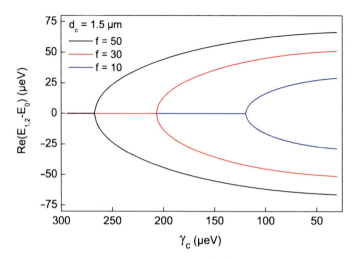

Fig. 8.2. Real part of $E_{1,2} - E_0$ according to (8.3) for different oscillator strengths f as a function of the cavity mode linewidth γ_c of a 1.5 μm micropillar cavity. A splitting of the eigenenergies occurs when the threshold for strong coupling is reached at cavity mode linewidths of 270, 210, and 120 μeV, for $f = 50$, 30, and 10, respectively. Less stringent requirements on the cavity quality ($\gamma_c \propto 1/Q$) are obtained for high oscillator strength emitters due to the correspondingly higher coupling constant

of photonic modes, the amplitude of the vacuum field and the orientation of the emitter's dipole with respect to the polarization of the cavity mode. Depending on the spectral detuning between the emitter and the cavity mode the spontaneous emission rate can be enhanced or inhibited compared to its value in a transparent homogenous medium with refractive index n which is given by

$$\frac{1}{\tau_{\text{free}}} = \frac{|d|^2 \omega^3 n}{3\pi \epsilon_0 \hbar c^3}. \tag{8.7}$$

In the weak coupling limit characterized by $g \ll \gamma_c$ (8.3), the eigenvalue associated with the emitter can be approximated by $E_0 - ig^2/(2\gamma_c)$. This means the emitter decays exponentially at a spontaneous emission rate given by

$$\frac{1}{\tau_{\text{cav}}} = \frac{4g^2 Q}{E_0 \hbar}, \tag{8.8}$$

with $\gamma_c = E_0/Q$. The change of the emitter's spontaneous emission rate due to the presence of localized optical modes in a cavity was initially proposed by Purcell in 1946 [21]. The ratio of the emission rates is therefore called the Purcell factor and can by expressed by

$$F_P = \frac{\tau_{\text{free}}}{\tau_{\text{cav}}} = \frac{3Q(\lambda_c/n)^3}{4\pi^2 V} \times \frac{|\mathbf{d} \cdot \mathbf{f}(\mathbf{r}_e)|^2}{|d|^2}. \tag{8.9}$$

Here, the first term includes the cavity Q-factor as well as the mode volume and therefore corresponds to the technological figure of merit in the weak coupling regime. The second term considers the spatial overlap and the orientation matching between the transition dipole and the electric field of the cavity mode. Taking also a spectral detuning Δ between the cavity mode and the emitter into account leads to the following expression for the Purcell factor [22]:

$$\frac{\tau_{\text{free}}}{\tau_{\text{cav}}} = \frac{3Q(\lambda_c/n)^3}{4\pi^2 V} \times \frac{\gamma_c^2}{4\Delta^2 + \gamma_c^2} \times \frac{|\mathbf{d} \cdot \mathbf{f}(\mathbf{r}_e)|^2}{|\mathbf{d}|^2}. \qquad (8.10)$$

Enhancement and inhibition of spontaneous emission has been observed for various quantum dot–microcavity systems. Initially, the Purcell effect was demonstrated for an ensemble of QDs in micropillar cavities where enhancement as well as inhibition of the spontaneous emission by almost a factor of 10 could be achieved [5,7]. More recent studies, revealed a pronounced Purcell effect also for *single* QDs, e.g., in micropillar and PC membrane cavities [15,23] (see Chap. 9).

8.3 Sample Fabrication

8.3.1 Growth of Planar Quantum Dot–Microcavity Structures

Low mode-volume microcavities with high Q-factors are essential for the observation of pronounced cQED effects with single QD excitons. Several geometries exist for the realization of high-Q microcavities. Promising candidates are photonic crystal membrane structures [24] (see Chap. 9), semiconductor microdisk structures [25] and micropillar cavities [26,27], respectively. While PC membrane cavities have high potential with respect to the integration in photonic circuits, micropillar cavities have advantages with respect to the mode control. In fact, in micropillar cavities large and robust overlap of the central fundamental mode with the QD states can be achieved and the confined mode spectrum can be directly controlled by the post diameter [28,29]. In addition, micropillars facilitate optical excitation and detection perpendicular to the sample surface in optical experiments.

High-Q micropillars with embedded low density self-assembled quantum dots are usually based on planar microcavity structures grown by molecular beam epitaxy (MBE). The planar structures are grown on a GaAs substrate and consist of a bottom distributed Bragg reflector (DBR) followed by a one λ thick GaAs cavity and a top DBR. High reflective DBRs are realized by a stack of alternating quarter-wavelength thick GaAs and AlAs mirror pairs. Here, a combination of GaAs and AlAs is chosen in order to maximize the contrast of the refractive index at the interfaces which guarantees a high reflectivity of the individual mirror pairs [27,30]. The reflectivity of the DBRs as a whole increases with the number of mirror pairs. Therefore, very high-Q planar microcavities consist of more the 30 mirror pairs in each DBR. The

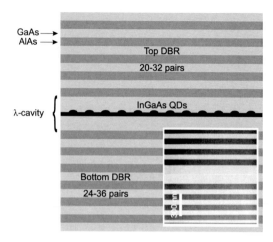

Fig. 8.3. Schematic view showing the layout of a planar microcavity structure. *Inset:* Scanning transmission electron microscope (SEM) image of the center region near the λ-cavity with a layer thickness of about 260 nm

active region consists of a low density layer of InGaAs QDs which is introduced in the middle of the λ-cavity, i.e., at the antinode of the electromagnetic field of the cavity mode. A schematic view of a microcavity structure is shown in Fig. 8.3.

Pronounced light–matter coupling effects are observed for QDs with large oscillator strength in low mode volume, high-Q microcavity systems. Therefore, optimization of the QD growth aims at the realization of large dot structures with a large dipole-moment and large OS, respectively [27]. The QDs are realized in Stranski–Krastanov growth mode [31], where the formation of QDs is initiated by strain relaxation during epitaxial growth of GaInAs on a GaAs substrate as the deposited layer exceeds a critical thickness. As a result, the growth characteristic switches from a two-dimensional to a three-dimensional growth mode. This mechanism allows the formation of high quality, low-dimensional nanostructures as the strain is elastically relaxed without the introduction of crystal defects. In this self-organized process ensemble properties such as the size, area density, and emission wavelength of the QDs can be widely controlled by the growth temperature, growth rate and III/V-ratio. For instance, varying systematically the In content of $In_xGa_{1-x}As$ QDs allows one to address the influence of strain on the shape and the optical properties of the QDs [32] by submonolayer deposition of InAs and GaInAs on GaAs substrate.

Strongly strained $In_{0.6}Ga_{0.4}As$ dots (QD1) form by depositing a nominally 1.4 nm thick InGaAs layer of 7 cycles of 0.1 nm InAs and 0.1 nm $In_{0.20}Ga_{0.80}As$. The strain is reduced for $In_{0.45}Ga_{0.55}As$ dots (QD2) grown by depositing a nominally 2.1 nm thick layer via submonolayer deposition of 11 cycles of 0.07 nm InAs and 0.12 nm $In_{0.125}Ga_{0.875}As$. In addition,

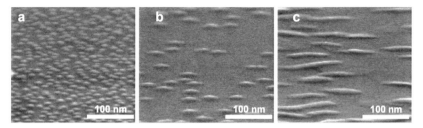

Fig. 8.4. SEM images of uncapped $In_{0.6}Ga_{0.4}As$ (**a**), $In_{0.45}Ga_{0.55}As$ (**b**) and $In_{0.3}Ga_{0.7}As$ (**c**) quantum dot samples. The surface is tilted by 70° to enhance the height contrast

$In_{0.30}Ga_{0.70}As$ dots (QD3) can be realized by depositing a nominally 4.5 nm thick layer via deposition of 30 cycles of 0.03 nm InAs and 0.12 nm $In_{0.12}Ga_{0.88}As$. A growth temperature of 590°C is used for the GaAs buffer, 470°C for sample QD1 and migration enhanced growth at 510°C for the low-strain dot layers QD2 and QD3.

The influence of the strain on the dot nucleation and morphology is illustrated in Fig. 8.4 showing surface SEM images of uncapped samples with different In content. High strain results in diameters between 10 and 15 nm and a rather high dot density of about $1–2 \times 10^{11} cm^{-2}$ for $In_{0.6}Ga_{0.4}As$ QDs. Larger dot diameters of 20–25 nm and lower area density of $1–2 \times 10^{10} cm^{-2}$, crucial for cQED experiments on the single dot level, are realized for sample QD2 by reducing the In content to 45%. The larger diameter of these QDs is partly related to the lower strain due to the lower In content and to a higher growth temperature of 510°C in comparison to 470°C used for QD1. The higher growth temperature facilitates a larger migration length on the surface which reduces the dot density and enlarges the dot diameter [33]. Still larger dot sizes were realized for sample QD3 by decreasing the In content to 30% as can be seen in Fig. 8.4. Due to a further decrease of strain, the island growth is mainly initiated by crystal steps on the surface. As a consequence elongated dot structures with typical lengths of 50–100 nm and widths of about 30 nm form preferentially orientated along the $[0\bar{1}1]$ direction. Due to the larger dimensions, the dot volume is increased by one order of magnitude in comparison with the state of the art circular shaped high In-containing GaInAs or InAs dots with diameters of 15–20 nm. In addition, a further decrease of the dot density down to about $6–9 \times 10^9 cm^{-2}$ is achieved for $In_{0.3}Ga_{0.7}As$ QDs.

8.3.2 Fabrication of High-Q Micropillar Cavities

High Q factor micropillars with small mode volumes and circular cross sections are fabricated by means of electron beam lithography and plasma etching. In a first step, the micropillar cross section is defined into resist (polymethylmethacrylate, PMMA) by high resolution electron beam lithography. After vacuum deposition of the metal (Ni) etch mask and a subsequent lift-off process, reactive ion etching is used to remove the semiconductor material at unexposed regions down to few remaining layers of the lower DBR. This process is performed either with inductively coupled plasma (ICP) or electron cyclotron resonance (ECR) etching. Particular low undercut and very smooth sidewalls are realized by ECR etching. This reduces optical losses due to photon scattering at surface irregularities, and provides strong and uniform optical confinement in the micropillars which is crucial for the realization of cavities with ultrahigh Q-factors. An SEM image of a micropillar with a diameter of 1.8 µm and a height of about 7 µm is shown in Fig. 8.5.

The optical quality of the micropillar cavities is characterized at low temperature by high resolution micro photoluminescence (μPL) spectroscopy (see Sect. 8.4 for details). A μPL spectrum of a 4 µm record high-Q micropillar based on a planar microcavity with 32 (36) mirror pairs in the upper (lower) DBR is presented in Fig. 8.6a. The fundamental, double degenerated HE_{11} mode of the micropillar emits at 1.31832 eV. In addition well resolved higher order modes (EH_{01}, HE_{21}, and HE_{01}) can be identified at about 1.32070 eV. A high resolution spectrum of the fundamental mode is shown in Fig. 8.6b. The resolution limited linewidth of 9.6 µeV corresponds to a Q of $165,000 \pm 8,000$ by taking the spectral resolution (3.5 µeV) of the μPL setup into account [30].

For practical reasons, it is interesting to study the influence of the cavity layout and micropillar geometry on the resonator Q-factor. This allows one to address the influence of intrinsic as well as extrinsic photon losses on the resonator quality and to choose micropillars with the maximum Q/d_c ratio for the observation of pronounced cQED effects (see Sect. 8.2.1). The dependence of the cavity Q on the pillar size in the diameter range between 0.9 and 4 µm is depicted in Fig. 8.7a as a function of d_c for micropillars based on planar microcavities with 20/24 (MC1), 26/30 (MC2) and 32/36 (MC3)

Fig. 8.5. SEM image of a high-Q micropillar cavity with a diameter of 1.8 µm

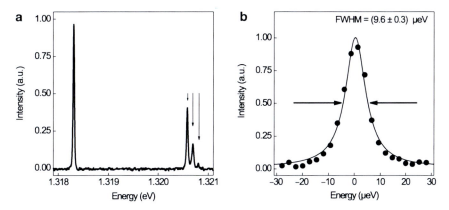

Fig. 8.6. (a) µPL spectrum of a 4 µm micropillar cavity at low temperature (10 K). (b) High resolution spectrum of the fundamental mode at 1.31832 eV. The measured linewidth corresponds to a Q of $165,000 \pm 8,000$ after deconvolution

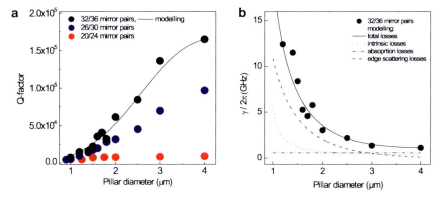

Fig. 8.7. (a) Cavity Q-factor as a function of the pillar diameter for three cavity designs with different number of mirror pairs in the DBRs. (b) Loss rates vs. the pillar diameter for MC3 with 32/36 mirror pairs in the DBRs. The total losses are composed of intrinsic, absorption, and edge scattering losses

mirror pairs in the upper/lower DBR. For a micropillar diameter of 4 µm Qs of 9,300, 110,000, and 165,000 are observed for MC1, MC2, and MC3, respectively. With decreasing d_c, the Qs decrease due to an increasing contribution of intrinsic and extrinsic photon loss channels. Still, the optimized micropillar processing allows one to obtain Q-factors exceeding 50,000 in the 2 µm diameter range (MC3). The ultra high Q-factors of MC3 are related to the very high number of mirror pairs in the DBRs and strong detuning between the cavity modes and the QD emission band. The latter reduces absorption losses in the active layer [30].

Three-dimensional finite-difference time-domain (FDTD) modeling for pillars based on MC3 allows one to assess the relative contributions of intrinsic and extrinsic photon losses as well as the upper limit of Q for the present micropillar structures. In the simulation, an ideal pillar geometry with 12 unetched mirror pairs in the lower DBR as well as layer thicknesses of the cavity $d_{\text{cavity}} = 281$ nm and the DBR mirrors $d_{\text{AlAs}} = 78$ nm and $d_{\text{GaAs}} = 71$ nm were taken into account. In addition, absorption effects as well as edge scattering were neglected to determine the intrinsic Q value of the fundamental cavity mode. The intrinsic Q-factors $Q_{\text{intrinsic}} = 3.2 \times 10^4$, 2.5×10^5, 4.0×10^5, and 5.1×10^5 determined for $d_c = 1, 2, 3,$ and $4\,\mu\text{m}$, respectively, show also a drop of $Q_{\text{intrinsic}}$ with decreasing d_c. This tendency is attributed to additional vertical losses for small pillar diameters. These losses increase with decreasing pillar diameter and are related to a blue shift of the cavity mode away from the center of the photonic band gap in conjunction with a reduction of the latter for small pillar diameters [34]. The calculated values for $Q_{\text{intrinsic}}$ are obtained for ideal structures and can be considered as upper limit values for the given number of mirror pairs. The lower experimental Qs can be explained by additional extrinsic losses related to absorption in the DBRs and in the active layer as well as edge scattering which is most prominent in the small diameter range. For high-Q cavities the losses are additive and the Q-factor can be expressed by [35, 36]

$$1/Q = 1/Q_{\text{intrinsic}} + 1/Q_{\text{edge scattering}} + 1/Q_{\text{absorption}}, \quad (8.11)$$

where $1/Q_{\text{edge scattering}} = \kappa J_0^2(k_t r_c)/r_c$ and $Q_{\text{absorption}} = 4\pi n/\lambda_c \alpha M$ consider losses by edge scattering and a diameter independent material absorption, respectively. κ is a phenomenological proportionality constant and J_0 denotes the Bessel function of the first kind which describes the fundamental radial intensity distribution of a pillar with a radius r_c. λ_c is the resonance wavelength and α corresponds to the material absorption coefficient. The experimental data is well described by (8.11) as can be seen in Fig. 8.7a for MC3. Fitting yields $\kappa = (2.1 \pm 0.5) \times 10^{-5}\,\mu\text{m}^{-1}$. In order to assess the contribution of intrinsic, absorption and edge scattering losses to the total losses for different pillar diameters the loss rates ($\gamma_i \propto 1/Q_i$) are calculated according to (8.11) and plotted as a function of the pillar diameter in Fig. 8.7b. In the small diameter range below $2.5\,\mu\text{m}$ edge scattering losses are dominating while absorption losses become a limiting factor at large diameters, i.e., a further improvement of Q could be achieved mainly by reducing these loss channels.

Aiming at the observation of strong coupling in high-Q micropillar cavities it is important to select micropillars with particular large values of Q/d_c which is the figure of merit from the cavity point of view. In fact, for a given cavity design and QD composition, pillars with the maximum Q/d_c ratio feature the highest VRS. To address this important quantity, Q/d_c is calculated for pillars based on microcavities MC1-3 and plotted in Fig. 8.8 as a function of the pillar diameter. The highest Q/d_c ratio of about $45{,}000\,\mu\text{m}^{-1}$ is obtained for pillars based on MC3. It is interesting to note that for micropillars with

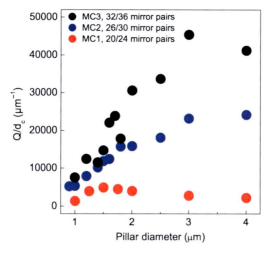

Fig. 8.8. Cavity figure of merit Q/d_c for the observation of strong coupling as a function of the pillar diameter for different cavity designs. For MC1 the optimum pillar diameter is 1.5 µm, while pillars with larger diameters are preferred in case of MC2 and MC3

20/24 mirror pairs in the DBRs a pronounced maximum of $Q/d_c = 4,900$ evolves at a pillar diameter of about 1.5 µm. This means that for this cavity design 1.5 µm micropillars are the best candidates for the observation of strong coupling. In contrast, Q/d_c increases continuously as a function of d_c for pillars based on MC2 and MC3 and the optimum pillar diameter shifts to larger pillar diameters.

8.4 Experimental Setup

cQED effects in the quantum dot–micropillar cavity system are investigated at low temperature by means of micro photoluminescence (µPL) spectroscopy. Individual micropillars are pumped optically using either nonresonant or resonant p-shell excitation of the InGaAs QDs in the active layer. The laser beam is focussed on the sample via a microscope objective to a spot size of approximately 3 µm. PL emission from the micropillar cavities is collected by the same microscope objective, spectrally resolved by a high resolution spectrometer and detected by a nitrogen cooled Si charge coupled device detector (cf. Fig. 8.9). The spectral resolution of up to 3.5 µeV allows one to resolve very sharp spectral features such as the resonator modes of high-Q micropillar cavities.

Time resolved µPL studies are performed using a mode-locked titanium-sapphire laser in conjunction with a multichannel plate (MCP) providing a time resolution better than 100 ps. In addition, photon correlation experiments

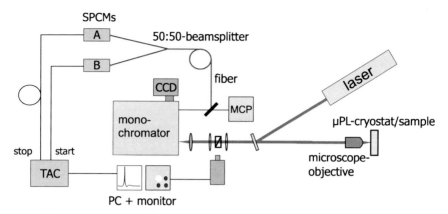

Fig. 8.9. Micro photoluminescence setup (schematic view) used for the investigation of single QD cQED effects in micropillar cavities

are performed with the use of a fiber coupled Hanbury Brown–Twiss setup. For resonance tuning the different temperature dependence of the emission energy of the QD excitons (typically $-0.04\,\mathrm{meV\,K^{-1}}$ at 25 K) and the significantly weaker temperature dependence of the cavity modes (about $-0.005\,\mathrm{meV\,K^{-1}}$ at 25 K) are exploited to precisely tune single QD excitons through resonance with the cavity mode.[1]

8.5 cQED Experiments in Quantum Dot–Micropillar Cavity Systems

cQED effects of single quantum dots in microcavities were first investigated about a decade ago [5, 7]. In the early works enhancement and inhibition of spontaneous emission was observed in the weak coupling regime. The onset of strong coupling could not be reached in these studies mainly due to the lack of low mode volume microcavities with high enough Q-factors [6]. As a result of strong improvement in semiconductor technology it has become feasible to fulfill the condition for strong coupling, i.e., the rate of energy exchange between a resonant emitter and the optical mode of a microcavity exceeds any dissipative decay rate in the system. Recently, strong coupling of single QDs in microcavities has been reported in a number of approaches including micropillar cavities [9], PC membrane cavities [10] (see also Chap. 9), microdisks [11] and microsphere cavities [37]. In this section, we will present recent progress in the observation of strong coupling of single InGaAs QDs in high-Q micropillar cavities.

---

[1] Here, the temperature dependence of the QD exciton is governed by the temperature dependence of the band gap while the much weaker temperature dependence of the refractive index determines the change of the photon mode energy.

8.5.1 Weak Coupling of Single InGaAs Quantum Dots in Micropillar Cavities

The main focus of this review is on strong coupling of InGaAs QDs in micropillar cavities. However, for comparison we will first of all present a case of weak coupling in this system. Weak coupling occurs if dissipation dominates and the rate of exchange of energy between a QD exciton and the cavity mode is smaller than the underlying decay rates (cf. Sect. 8.2.2). In this case the exciton–photon interaction is manifested in an enhancement or inhibition of the exciton recombination rate according to (8.10). Due to the strong decoherence in the system Rabi oscillations are strongly damped and mode splitting does not occur. Instead, the enhancement of the exciton recombination rate is usually accompanied by an increase of the QD emission intensity at resonance with the cavity mode.

Temperature tuning provides a simple and effective way of investigating the exciton–photon coupling character in micropillar cavities. This tuning mechanism is illustrated by the temperature dependent μPL spectra shown in Fig. 8.10a for a 2.5 μm diameter micropillar with a Q of 10,000. In this example an exciton line X of a standard $In_{0.6}Ga_{0.4}As$ self-assembled QD is shifted through resonance with the fundamental cavity mode C by increasing the sample temperature about 15 to 25 K. On resonance an enhancement of the exciton emission intensity occurs that is described in terms of the Purcell effect [21]. Here, the temperature tuning of a single QD line through resonance with the fundamental mode of a 2.5 μm micropillar cavity is presented.

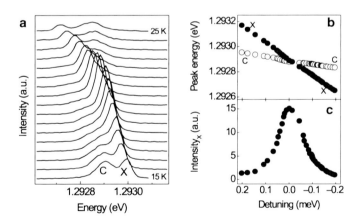

Fig. 8.10. Light–matter interaction in the weak coupling regime. (**a**) Temperature tuning of a single QD exciton X into resonance with the fundamental mode (C) of a 2.5 μm diameter micropillar cavity with a Q of 10,000. (**b, c**) Mode energies and intensity of the QD exciton as a function of detuning between the exciton and photon mode. The weak coupling regime is characterized by a simple crossing behavior of the mode energies and an enhancement of the exciton intensity at resonance

The coupling character is studied in more detail by Lorentzian lineshape fitting of the emission spectra shown in Fig. 8.10a. The variation of the mode energies and the exciton intensity is depicted in Fig. 8.10b,c as a function of detuning between the exciton and cavity mode. As expected for the weak coupling regime, simple crossing of the exciton line and the cavity mode is observed at resonance in Fig. 8.10c. Here, the crossing behavior can be explained in terms of (8.3) when $g^2 < (\gamma_c - \gamma_x)^2/16$ and both eigenenergies have the same real part. The strong enhancement of emission intensity visible in Fig. 8.10c is related to the enhanced recombination rate for the resonant QD exciton due to the Purcell effect. It is important to note that the observed enhancement of emission of about 15 relative to the X intensity at a detuning of 0.2 meV does not directly reflect the Purcell factor of the resonant system. In fact, the variation of the QD occupation probability needs to be taken into account to estimate the experimental Purcell factor from the intensity dependence [38]. From (8.9) we estimate a maximum Purcell factor of 17 for the present quantum dot–micropillar system.

8.5.2 Strong Coupling of Single InGaAs Quantum Dots in Micropillar Cavities

Strong coupling effects are investigated at low temperature by means of high resolution μPL spectroscopy. At first we will present studies performed on structures including large $In_{0.3}Ga_{0.7}As$ QDs in the active layer. These QDs feature a particular large oscillator strength which facilitates the observation of pronounced cQED effects for micropillar cavities with moderate Q-factors. The $In_{0.3}Ga_{0.7}As$ QDs are embedded in micropillars based on 20/24 mirror pairs in the upper/lower DBR (MC1). For these structures a pronounced maximum of $Q/d_c = 4,900\,\mu m^{-1}$, i.e., the figure of merit for the observation of strong coupling, occurs at a pillar diameter of 1.5 μm (cf. Fig. 8.8). In addition to a large Q/d_c-ratio further requirements need to be fulfilled in order to demonstrate strong coupling. Since temperature tuning is used to shift the QD exciton spectrally through resonance with the cavity mode, the tuning range is limited to a few meV in the applicable temperature interval between 4 and about 50 K. Along with the spectral matching, spatial matching with the field in the cavity is required to obtain the maximum coupling strength. This requires the QD to be located spatially in the center of the cavity which introduces some randomness, since it cannot be controlled with the present technology due to the random distribution of the self-assembled QDs in the plane of the active layer. Further technological optimization will probably aim at the controlled positioning of QDs in the field maximum of micropillar cavities in order to ensure a maximum coupling strength [39]. In particular, the growth and integration of prepositioned QDs could help to control the spatial matching with the field of the cavity [40] while a controlled spectral matching might be achieved by long range in situ laser tuning of the QD emission energy [41].

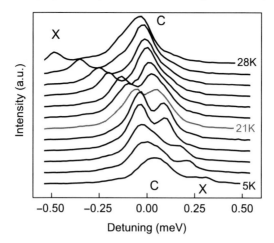

Fig. 8.11. Temperature tuning of a single QD exciton X through resonance with the fundamental mode of a 1.5 μm diameter micropillar cavity with a Q of 7,350. Strong coupling with a VRS of 140 μeV on resonance at about 20 K is identified by the pronounced anticrossing of the excitonic and photonic modes

Temperature tuning of a suitable quantum dot–micropillar structure is demonstrated in Fig. 8.11 showing a series of μPL spectra for temperatures between 5 and 28 K. At low temperature, a single QD exciton (X) is spectrally detuned by about 250 μeV from the fundamental cavity mode (C) of the 1.5 μm micropillar with a Q-factor of 7,350. Increasing the temperature from 5 to 28 K shifts the exciton through resonance with the cavity mode. Interestingly, the two modes do not cross but show a pronounced anticrossing and a splitting of about 140 μeV under resonance condition at about 21 K. Furthermore, the exciton and photon modes exchange their properties, i.e., linewidths and intensities at resonance. This is a clear signature of coherent coupling of two modes as expected in the strong coupling regime [9].

The resonance behavior is analyzed in more detail by plotting the measured peak energies and FWHM values for a 1.5 μm micropillar as function of the temperature in Fig. 8.12. The anticrossing of the excitonic and photonic mode is clearly reflected in the temperature dependence of the peak energies. Here, the strong variation with temperature is related to the excitonic mode while the weaker temperature dependence identifies the cavity mode. The observed mode anticrossing is associated with a change from the exciton like dispersion to a cavity mode like one for the upper branch, whereas the lower branch changes from a cavity mode dispersion to that of an exciton with increasing temperature. At resonance mode mixing occurs due to the coherent interaction of the excitonic and photonic modes which results in a VRS about 140 μeV for this strongly coupled single QD exciton solid state microcavity system [9]. Normal mode mixing at resonance is further reflected in equal mode linewidths on resonance. In fact, it is nicely seen that the larger FWHM of the cavity

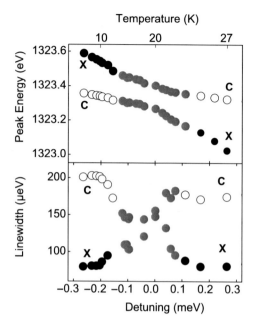

Fig. 8.12. Emission energy and linewidth (FWHM) as a function of the detuning between the exciton (X) and photon mode (C). The strong coupling regime with a VRS of 140 μeV is reflected in the anticrossing behavior of the mode energies and the exchange of linewidths at resonance. The interaction range is indicated by *gray solid circles*

mode (180 μeV for low excitation conditions) and the smaller FWHM of the exciton change significantly with decreasing detuning and become similar in the interaction regime when the system is tuned into resonance.

The mode anticrossing and the exchange of linewidths are clear fingerprints of the single QD strong coupling regime. The coupling strength is expressed in terms of the coupling constant g which can be estimated from the experimental data. Taking the off-resonant cavity linewidth $\gamma_c = 180\,\mu\mathrm{eV}$ and the VRS of 140 μeV into account one obtains a coupling constant of 80 μeV according to (8.4). It is interesting to compare this number with the threshold value $\gamma_c/4 = 45\,\mu\mathrm{eV}$ (8.6) for the observation of strong coupling which clearly confirms that the strong coupling regime has been reached in the present structure. The large coupling strength could be achieved mainly by the elongated $\mathrm{In_{0.3}Ga_{0.7}As}$ QDs in the active layer. These type of QDs were optimized with respect to their OS which can be estimated according to (8.2). For a coupling constant of 80 μeV and an effective mode volume of about $0.3\,\mu\mathrm{m}^3$ we estimate an effective OS of about 50 [9][2], which is about five times higher than the values of about 10 reported for standard self assembled InGaAs QDs [42, 43]. This indicates

---

[2] In this lower limit estimation a possible spatial mismatch of the QD is neglected. Thus, the real oscillator strength might be somewhat larger.

that the larger extension of the $In_{0.3}Ga_{0.7}As$ QDs indeed results in a higher OS of the excitonic transition. Similar oscillator strengths are reported for natural QDs realized by monolayer fluctuations in GaAs/AlGaAs quantum wells [17, 44].

It is clear that QDs with large oscillator strength facilitate the observation of strong coupling. In particular, the threshold value $g = \gamma_c/4$ would not be reached for standard self-assembled InGaAs dots with $f \approx 10$ in micropillars with 20/24 mirror pairs. These structures have cavity mode linewidths of typically 150 µeV at the optimum diameter of 1.5 µm, i.e., no mode splitting is expected to occur according to Fig. 8.2. The lower oscillator strength of smaller InGaAs QDs can be compensated by embedding them in micropillar cavities with significantly higher Q/d_c-ratios since the figure of merit for the observation of strong coupling is given by $Q\sqrt{f}/d_c$ (cf. Sect. 8.2). The Q/d_c-ratio can be significantly increased in micropillars with a larger number of mirror pairs in the DBRs as discussed in Sect. 8.3. In this way it is possible to observe strong coupling also for intermediate and low oscillator strength InGaAs QDs which is demonstrated in Fig. 8.13 for InGaAs QDs with an In content of 45% and 60%, respectively. Figure 8.13a shows an example of strong coupling of a single 45% In QD embedded in an 1.8 µm micropillar with 26/29 mirror pairs in the upper/lower DBR and a cavity Q of 18,000. The temperature dependent series of spectra reveals nicely the anticrossing of the excitonic (X) and the photonic (C) modes associated with a VRS of 65 µeV at the resonance temperature of 26 K. A similar behavior is observed in Fig. 8.13b showing strong coupling of a 60% In QD in a 3 µm micropillar cavity ($Q = 40,000$) based on 29/33 mirror pairs in the DBRs. The high resolution

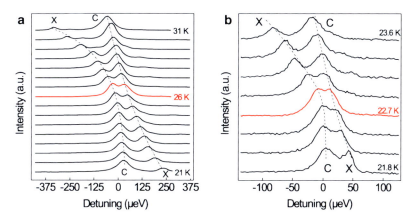

Fig. 8.13. Temperature resonance tuning in the strong coupling regime. (**a**) Strong coupling of a single $In_{0.45}Ga_{0.55}As$ QD exciton (X) and the cavity mode (C) of a 1.8 µm micropillar cavity with a Q of 18,000 (VRS = 65 µeV). (**b**) Strong coupling of a single $In_{0.6}Ga_{0.4}As$ QD exciton (X) and the cavity mode (C) of a 3.0 µm micropillar cavity with a Q of 40,000 (VRS = 23 µeV)

Table 8.1. Comparison of strongly coupled single $In_xGa_{1-x}As$ QD – micropillar cavity systems with different In content x and different numbers of mirror pairs in the upper/lower DBR

Sample #	1	2	3
Number of mirror pairs	20/24	26/30	29/33
In content x (%)	30	45	60
Pillar diameter d_c (µm)	1.5	1.8	3.0
Q-factor	7,350	18,000	40,000
Q/d_c (µm^{-1})	4,900	10,000	13,300
Mode volume (µm^3)	0.3	0.45	1.2
VRS (µeV)	140	65	23
Coupling constant g (µeV)	80	37	12
Oscillator strength (OS) f	≈50	≈20	≈10

spectra allow the detection of a VRS of 23 µeV at resonance (22.7 K). Here, the rather small VRS and the corresponding coupling constant $g = 12$ µeV is related to the large mode volume of the 3 µm micropillar cavity and the small OS of the embedded QDs [30]. Still, the cavity Q-factor of this pillar is high enough so that the light–matter coupling rate exceeds the decoherence rates in the system and the threshold of strong coupling is clearly overcome. A comparison of the presented cases of strongly coupled single $In_xGa_{1-x}As$ QD–micropillar cavity systems is given in Table 8.1. This table summarizes important parameters of the cavities, the experimentally observed VRS as well as the coupling constant g and oscillator strength f for $In_xGa_{1-x}As$ QDs with different In content x.

It is important to note that strong coupling refers in the presented cases to the empty cavity (vacuum) Rabi splitting and the strong coupling occurs due to the interaction of a single QD exciton with the cavity mode. A careful adjustment of the experimental parameters is required to realize this special regime. For example, low excitation powers were applied in order to ensure that multiexciton effects can be ruled out. Still, the QD line could be a superposition of two or more degenerate single exciton transitions related to different dots. However, due to the low spectral density of QD emission lines which is typically about one line per meV, the probability to find two lines emitting within the spectral resolution of 50 µeV is less than 1%. This estimation shows that multi-emitter contributions are very unlikely for the present example of strong coupling. Nevertheless, it is interesting to prove the single emitter regime experimentally by studying the photon statistics of the emitted light which will be addressed in the next section of this review. Also, the average photon number in the cavity mode must be much lower than unity for the investigation of vacuum field Rabi splitting. Since most of the applied laser power (2 µW, 532 nm) is absorbed in the GaAs sections of the upper

Bragg reflector it can be estimated that only a small fraction of about 1 nW reaches the cavity. In the worst case, all the photons reaching the cavity are absorbed there, and the resulting carrier pairs all relax in the dot under consideration. This results in an average photon number of less than 1% in the present cavity providing a photon lifetime of about 5 ps [9]. Therefore, effects due to the population of the cavity with more than one photon at a time are considered to be negligible as required for the observation of vacuum Rabi splitting.

8.5.3 Photon Antibunching in the Strong Coupling Regime

Photon correlation studies reveal important information about the quantum nature of the underlying system. In this section we will apply intensity correlation measurements in order to prove that the vacuum Rabi splitting in a strongly coupled single QD–microcavity system originates from a single quantum emitter interacting with the cavity mode. This regime is very special and represents the most fundamental situation of light–matter interaction in semiconductors. In particular, it needs to be distinguished from the semiclassical case when vacuum Rabi splitting is observed in quantum well–microcavity structures due to the contribution of a large number of quantum well excitons [29, 45].

Nonclassical photon emission from a single quantum emitter leads to a pronounced antibunching behavior in the second order autocorrelation function defined via

$$g^{(2)}(\tau) = \frac{\langle \hat{a}^\dagger(t)\hat{a}^\dagger(t+\tau)\hat{a}(t+\tau)\hat{a}(t)\rangle}{\langle \hat{a}^\dagger(t)\hat{a}(t)\rangle^2}, \qquad (8.12)$$

where $\hat{a}^\dagger$ and $\hat{a}$ are the photon creation and annihilation operators, respectively, at time t (see Chaps. 6 and 11). A pulsed source will have a correlation function consisting of a series of peaks separated by the repetition period T. The area $g^{(2)}(0)$ of the peak around $\tau = 0$, normalized by T, gives an upper bound on the probability that two or more photons are present. Emission from a single quantum emitter is characterized by $g^{(2)}(0) < 0.5$ [46]. In the following we will exploit this relation to prove that the vacuum Rabi splitting of a strongly coupled QD–micropillar cavity system is due to the interaction with a single emitter.

Proving the quantum nature of a strongly coupled QD–micropillar cavity system requires the suppression of the cavity mode illumination by nonresonant QDs. Such background population of the cavity mode is usually present under nonresonant above-band pumping [9–11] and deteriorates the $g^{(2)}(0)$-signal by uncorrelated photon emission [47]. The influence of nonresonant QDs can be strongly suppressed by resonant p-shell pumping of a selected QD interacting with the cavity mode [8, 48] where the p-shell transition is typically located 20–30 nm away from the s-shell (ground-state) transition for the InGaAs QDs with an In content of about 45% used in these studies [49].

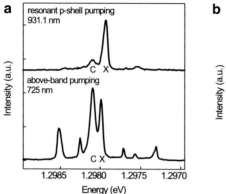
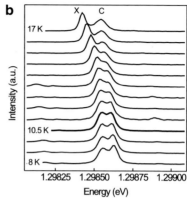

Fig. 8.14. (a) PL spectra of a 1.8 μm micropillar under nonresonant and resonant p-shell excitation at 725 and 931.1 nm, respectively. The micropillar is based on a planar cavity with 26/30 mirror pairs in the upper/lower DBR and InGaAs QDs with an In content of about 45% in the active layer. (b) Temperature dependent PL spectra of a 1.8 μm micropillar under resonant p-shell excitation at 936.25–936.45 nm. Resonance characterized by a VRS of 56 μeV is observed at 10.5 K

This is demonstrated in Fig. 8.14a showing the emission spectrum of a 1.8 μm micropillar under above-band excitation at 725 nm and under resonant p-shell excitation of a selected QD at 931.1 nm. It is clearly seen that resonant pumping leads to a very clean spectrum and reduces background illumination of the cavity mode (C). In fact, the excitonic transition (X) of the selected QD at 1.298 eV (955 nm) dominates the emission spectrum under resonant pumping.

Resonant pumping of a selected QD facilitates the preparation of a system in which single emitter cQED effects can be proven by photon correlation measurements (see Chap. 6). In particular, we are interested in single emitter interaction in a strongly coupled QD–micropillar cavity system. Strong coupling of a resonantly excited QD with the mode of a 1.8 μm micropillar with a Q of 15,200 is demonstrated by temperature tuning as depicted in Fig. 8.14b. The exciton transition (X) of the selected QD shows pronounced photon anticrossing with the cavity mode (C) when the temperature is increased from 8 to 17 K as a fingerprint of the strong coupling regime. Resonance of the two modes is achieved at 10.5 K where a VRS of 56 μeV occurs. This splitting corresponds to a coupling constant of $g = 35$ μeV which is consistent with the value given in Table 8.1 for 45% In QDs and clearly exceeds the threshold value $\gamma_c/4 = 21$ μeV for strong coupling [48].

In principle the VRS observed in Fig. 8.14 could be due to the coherent interaction of n degenerate emitters where the coupling constant scales by $\sqrt{n}$ with the number of emitters [20]. To rule out this possibility and to demonstrate that the system under investigation shows indeed the most fundamental coherent interaction of a *single* quantum emitter with an empty cavity mode, the photon autocorrelation function is measured on resonance. It is necessary

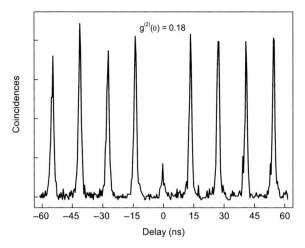

Fig. 8.15. Photon autocorrelation function of a strongly coupled single quantum dot–micropillar cavity system. The strongly reduced number of coincidences at zero delay associated with $g^{(2)}(0) = 0.18$ clearly proves the single emitter regime

to note that in the strong coupling regime the lifetime of the QD exciton is strongly reduced to about 15 ps, which is twice the cavity lifetime [48]. Conventional single photon counting modules with a typical time resolution of several hundred picoseconds cannot resolve photon antibunching at such short timescales under continuous wave excitation [8]. Therefore, pulsed excitation is required to circumvent the problem of limited temporal resolution. Figure 8.15 presents the coincidence counts collected under pulsed resonant excitation from the coupled QD–cavity system at resonance as a function of the delay. The center peak at zero delay is strongly reduced compared to the neighboring peaks which is a clear fingerprint of nonclassical photon emission. The pronounced antibunching of the emitted light with $g^2(0) = 0.18 < 1/2$ unambiguously shows that the coupled QD–cavity system is indeed dominated by the single QD emitter [48]. This is an important result which proves that strong coupling has been achieved due to the interaction between a *single* QD exciton and the photonic mode of a micropillar cavity.

8.5.4 Coherent Photonic Coupling of Quantum Dots

Coherent coupling of quantum dot states is crucial for their application in quantum information processing. For instance a long-range coherent interaction of QD spins mediated by the light field of a high-Q cavity has been proposed as the basis for a quantum computation scheme [50]. This approach has the advantage that QDs can be coupled over a comparatively large distance in the range of several micrometers while a coherent coupling mediated, e.g., by the tunneling of QD excitons, suffers from an interaction length limited to a few nanometers [51,52] (see Chap. 10). The photon mediated interaction of

QDs is also interesting from a fundamental point of view since cQED predicts that the light–matter coupling constant g depends via $g \propto \sqrt{n_E}$ on the number n_E of coherently coupled emitters [20]. In this sense the coherent photonic coupling of two QDs is a natural extension of the single QD coupling effects discussed in the previous sections.

In the following the coherent photonic coupling between two quantum dot excitons and an optical mode of a micropillar cavity will be presented. Such a system can be realized only if the QD exciton transition energies of the involved QDs are sufficiently close, i.e., if they have a spectral separation comparable to the exciton–photon coupling energy. In this case the emitters can be coupled coherently by the large photon field in a micropillar cavity if the conditions for strong coupling are fulfilled. Here, the coherent interaction of more than one emitter is expected to result in a particular large vacuum Rabi splitting on resonance.

As mentioned above it is crucial for the demonstration of coherent photonic coupling of QDs that the respective excitonic transitions exhibit a small but distinct energetic separation δ. This condition is satisfied in the spectrum of a 1.6 μm micropillar with a Q of 9,600 shown in Fig. 8.16a for which two almost degenerate $In_{0.3}Ga_{0.7}As$ excitons (X1 and X2) of different QDs appear close to the cavity mode C ($T = 25\,K$) [53]. For the present case the separation δ between the transition energies of X1 and X2 equals 120 μeV which is indeed in the range of the expected exciton–photon coupling energy of the large 30% In QDs used in these studies. Again, temperature tuning is applied in order to shift the QD excitons through resonance with the common cavity mode. It is interesting to note that no crossing of the lines is observed in the whole temperature range which is a clear sign of the coherent photonic coupling of the two QD excitons. This explains also the change of the highest lying component in the spectrum from an excitonic character at low temperature to an photonic character at high temperatures above resonance. In the same way the lowest lying component changes its character from a photonic mode to an excitonic mode. Thus, unlike the situations discussed in the previous sections, strong coupling involves here simultaneously X1, X2, and C and results in a coherent coupling of the QD excitons mediated by the light field of the cavity mode.

The experimental data presented in Fig. 8.16 was modeled using a coupled oscillator approach [19] in order to obtain information on the underlying coupling mechanisms and the respective coupling constants. The coupled oscillator approach allows one to calculate the emission spectra of the interacting QD–microcavity system for different detunings Δ using the exciton–photon coupling constants g_1 and g_2 associated with X1 and X2, the spectral separation δ of the excitons as well as the mode linewidths as parameters.[3] In this way, quantitative agreement between experimental and calculated spectra can

---

[3] Here, the detuning is defined as the energetic difference $\Delta = E_{X1} - E_C$ of the uncoupled exciton (X1) and cavity modes (C).

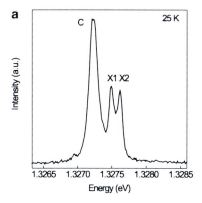
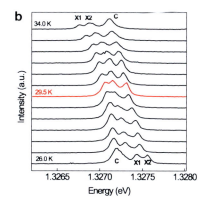

Fig. 8.16. (a) Low temperature (25 K) µPL spectrum of a 1.6 µm micropillar with a cavity Q of 9,600. Two energetically close QD excitons (X1, X2) with a separation $\delta = 120$ can be shifted into resonance with the cavity mode (C) by temperature tuning as shown in panel (b). Here, the coherent photonic coupling of the two excitons is reflected in the anticrossing of all three modes (X1, X2, and C) and leads to a mode splitting of 250 µeV at resonance $T = 29.5$ K

be obtained under variation of g_1 and g_2. Best agreement with the experimental spectra is achieved for coupling constants $g_1 = 66$ µeV and $g_2 = 76$ µeV. This becomes obvious in Fig. 8.17a illustrating the change of emission energies of the three interacting modes as a function of the detuning Δ. The pronounced line anticrossing seen in the experimental data (symbols) is well described by the energy dispersion obtained from the calculated spectra (solid lines) which confirms the interpretation of the experimental data in terms of a coherent photonic coupling of QDs. It is interesting to note that the extracted values of g_1 and g_2 are consistent (cf. (8.2)) with typical oscillator strengths of 40–50 for the large $In_{0.3}Ga_{0.7}As$ QDs used in the present (cf. Table 8.1).

Coherent photonic coupling needs to be distinguished from a sequential strong coupling which can appear when the energetic separation δ between two excitonic transitions is significantly larger than the coupling energy. The energy dispersion of such a system with similar experimental parameters is presented in Fig. 8.17b.[4] In this case, δ is notably larger then the cavity mode linewidth $\gamma_c = 160$ µeV and amounts to 315 µeV. When the excitons are tuned into resonance by increasing the temperature from 22 to 41 K, i.e., by reducing the detuning, the first exciton X1 strongly couples with the cavity mode. A separate anticrossing of X2 and C occurs at $\Delta = -315$ meV. The VRS amounts to about 100 µeV in each case, which is significantly smaller than in the previous case of two QDs. It is clear that these two strong interactions are well separated and represent the behavior of an individual QD exciton

---

[4] Sequential coupling is observed for $In_{0.3}Ga_{0.7}As$ QDs in a 1.6 µm micropillar cavity with a Q of 8,500.

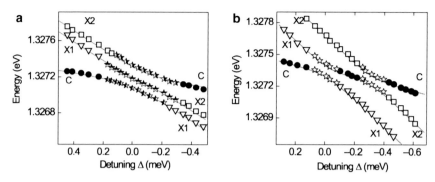

Fig. 8.17. Energy dispersions illustrating the difference between (**a**) coherent photonic coupling and (**b**) sequential strong coupling of QDs. (**a**) Experimental (*symbols*) and calculated (*solid lines*) emission energies of a coherently coupled system. The excitonic (X1, X2, δ = 120 μeV), cavity (C) and mixed modes (*stars*) represent the emission energies of the experimental spectra shown in Fig. 8.16. The smallest separation between the two outermost lines of the triplet is reached at resonance with a splitting of 250 μeV. (**b**) Sequential strong coupling of two QD excitons (X1, X2, δ = 315 μeV) with the cavity mode (C) of a 1.6 μm micropillar cavity. Both, X1 and X2 show a pronounced anticrossing behavior associated with a VRS of about 100 μeV on resonance at Δ = 0 and Δ = −315 meV, respectively. Modeling (*solid lines*) of the experimental data yields coupling strengths of $g_1 = 62\,\mu\text{eV}$ and $g_2 = 56\,\mu\text{eV}$ for X1 and X2, respectively

in strong coupling with an optical mode. The solid lines show again results from calculations which provide coupling coefficients of the two QD excitons of $g_1 = 62\,\mu\text{eV}$ and $g_2 = 56\,\mu\text{eV}$.

For a coherent coupling of two degenerated QD excitons with identical coupling strength g one expects that the VRS amounts to $\sqrt{2} \times 2g$ when the influence of dissipative loss channels is neglected. This means that the VRS would exceed the splitting $2g$ obtained for the single emitter case by a factor of $\sqrt{2}$ [14, 20]. Whereas, the experimental splitting in the coherent coupling case shown in Fig. 8.16 exceeds the values obtained for the vacuum Rabi splitting from individual dots in similar pillars by a factor of about 2. In fact, the observed splitting of 250 μeV is explained in terms of a combination of two contributions: the coherent part due to the cavity–exciton interaction and the sequential part related to the initial splitting δ = 120 μeV of the two QD excitons. In order to address both contributions we first consider the effective Hamiltonian (neglecting dephasing) of three interacting oscillators [53]:

$$H = \sum_{\alpha=C,X1,X2} E_\alpha \hat{a}_C^\dagger \hat{a}_\alpha - (1/2) \sum_{\alpha=X1,X2} g_\alpha \left[\hat{a}_C^\dagger \hat{a}_\alpha + \hat{a}_\alpha^\dagger \hat{a}_C \right] + C_0. \quad (8.13)$$

Here E_C, E_{X1} and E_{X2} denote the energies of the cavity mode and the excitons X1 and X2. The corresponding destruction and creation operators are $\hat{a}_\alpha$ and $\hat{a}_\alpha^\dagger$ and the energy C_0 is defined as $C_0 = (E_C + E_{X1} + E_{X2})/2$. One obtains the following determinant for their eigenvalues:

$$\begin{vmatrix} E_c - E & g_1 & g_2 \\ g_1 & E_{X1} - E & 0 \\ g_2 & 0 & E_{X2} - E \end{vmatrix} = 0. \tag{8.14}$$

In the most simple case of two degenerated emitters with $E_{X1} = E_{X2}$ and $g_1 = g_2$, (8.14) yields the well known result for two coherently coupled oscillators, namely an effective coupling strength $\sqrt{g_1^2 + g_2^2} = \sqrt{2}g$ which exceeds the single dot case by a factor of $\sqrt{2}$ [14]. In the general case when also a sequential contribution to the splitting has to be considered and the total splitting of the three modes is given by

$$\Delta E \approx 2\sqrt{2g^2 + (\delta^2/4) + [(\Delta + \delta/2)^2]/3}, \tag{8.15}$$

where $g_1 = g_2 = g$ is used for simplicity. Equation (8.15) clearly shows that on resonance, i.e., for $\Delta = 0$, both the coherent part g as well as sequential part characterized by the splitting of δ give contributions to the total splitting. The relative influence of both contributions on the total splitting ΔE at resonance is illustrated in Fig. 8.18 in more detail. Here, ΔE is plotted as a function of δ for $g_1 = g_2 = 65\,\mu\text{eV}$. The coherent character of the coupling is

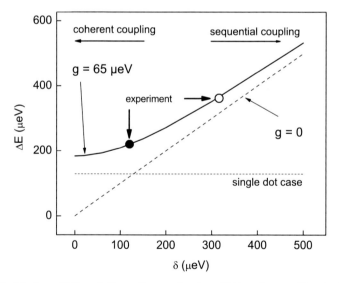

Fig. 8.18. Mode splitting ΔE vs. the excitonic splitting δ according to (8.15) for $g_1 = g_2 = g = 65\,\mu\text{eV}$. The coherent coupling character dominates for small δ while the splitting ΔE approaches the straight line associated with $g = 0$ in the limit of sequential coupling when δ significantly exceeds the coupling energy $2g$. The experimental coupling conditions (cf. Fig. 8.17a,b) are indicated by a *solid circle* ($\delta = 120\,\mu\text{eV}$) and an *open circle* ($\delta = 315\,\mu\text{eV}$), respectively

most pronounced for $\delta = 0$ where ΔE equals to $184\,\mu\text{eV}$. In other words, the splitting on resonance exceeds the value $2g = 130\,\mu\text{eV}$ for the single dot case by a factor of $\sqrt{2}$ as expected. With increasing contribution of the sequential part, i.e., for larger splitting δ, the total splitting increases and approaches the limit of negligible coherent coupling characterized by $g = 0$ (indicated by the dashed line). It is interesting to assess the coupling character present in the experiments depicted in Fig. 8.17a,b with $\delta = 120\,\mu\text{eV}$ and $\delta = 315\,\mu\text{eV}$, respectively, by a comparison with the calculations shown in Fig. 8.18. In case of $\delta = 120\,\mu\text{eV}$ indicated by a solid circle in Fig. 8.18 the coherent part clearly dominates and δ contributes only by approximately 20% to the increase of the splitting in comparison with the single dot case. In contrast, the experimental data associated $\delta = 315\,\mu\text{eV}$ (cf. open circle in Fig. 8.18) shows predominantly sequential character.

The present results show clearly that a dominantly coherent coupling of distinct QDs mediated by the photon mode can be achieved in state of the art micropillar cavities. They imply that a coherent coupling of QD spins for quantum information purposes [50] is feasible in the current semiconductor nanotechnology.

8.6 Conclusion and Outlook

In recent years, tremendous progress has been made in performing cQED experiments on single quantum dots in semiconductor microcavities. In this article, we have reviewed our recent work on the observation of cQED-effects in quantum dot–micropillar cavity systems. Due to the strong advance in the growth and processing of micropillar cavities it has become possible to enter the regime of strong coupling characterized by a coherent interaction between light and matter. In particular, the epitaxial growth of optimized quantum dots with large oscillator strength embedded in highly reflective Bragg reflectors and the subsequent plasma etching of high-Q low mode volume micropillar cavities have been key issues in demonstrating the long sought solid state implementations of strongly coupled cavity mode – two level emitter systems. We have addressed important technological issues in the optimization of quantum dot–micropillar cavity systems which led to the first observation of strong coupling on the single emitter basis. Photon correlation experiments have been presented that unambiguously prove this important single emitter coupling regime. The feasibility of coherent photonic coupling of quantum dots has been demonstrated and could be a versatile basis of quantum information building blocks in the solid state. The present technology also provides a good basis for further studies in the strong coupling regime aiming, e.g., at the observation of quantized field effects such as climbing the Jaynes–Cummings ladder.

Acknowledgments

Many people contributed to this work including C. Böckler, S. Götzinger, T. Heindel, S. Höfling, M. Kamp, L.V. Keldysh, V.D. Kulakovskii, A. Löffler, I.V. Ponomarev, D. Press, T.L. Reinecke, J.-P. Reithmaier, C. Schneider, G. Sek, M. Strauss, Y. Yamamoto. We are grateful for their enthusiasm and skill. This work has been supported by the DARPA QuIST program, the Deutsche Forschungsgemeinschaft via "Research Group Quantum Optics in Semiconductor Nanostructures", the European Commission through the IST Project "QPhoton" and the state of Bavaria.

References

1. Y. Yamamoto, R.E. Slusher, Phys. Today **46**, 66 (1993)
2. P.R. Berman, *Advances in Atomic, Molecular and Optical Physics* (Academic, Boston, 1994), chap. Cavity quantum electrodynamics
3. K.J. Vahala, Nature **424**, 839 (2003)
4. H. Mabuchi, A.C. Doherty, Science **298** (2006)
5. J.M. Gérard, B. Sermage, B. Gayral, B. Legrand, E. Costard, V. Thierry-Mieg, Phys. Rev. Lett. **81**, 1110 (1998)
6. J.M. Gérard, B. Gayral, Physica E **9**, 131 (2001)
7. M. Bayer, T.L. Reinecke, F. Weidner, A. Larionov, A. McDonald, A. Forchel, Phys. Rev. Lett. **86**, 3168 (2001)
8. C. Santori, G.S. M. Pelton, Y. Dale, Y. Yamamoto, Phys. Rev. Lett. **86**, 1502 (2001)
9. J.P. Reithmaier, G. Sek, A. Löffler, C. Hofmann, S. Kuhn, S. Reitzenstein, L.V. Keldysh, V.D. Kulakovskii, T.L. Reinecke, A. Forchel, Nature **432**, 197 (2004)
10. T. Yoshie, A. Scherer, J. Hendrickson, G. Khitrova, H.M. Gibbs, G. Rupper, C. Ell, O.B. Shchekin, D.G. Deppe, Nature **432**, 200 (2004)
11. E. Peter, P. Senellart, D. Martrou, A. Lemaître, J. Hours, J.M. Gérard, J. Bloch, Phys. Rev. Lett. **95**, 067401 (2005)
12. M. Bayer, A. Forchel, Phys. Rev. B **65** (2002)
13. C. Weisbuch, M. Nishioka, A. Ishikawa, Y. Arakawa, Phys. Rev. Lett. **69**(23), 3314 (1992)
14. G. Khitrova, H.M. Gibbs, M. Kira, S.W. Koch, A. Scherer, Nat. Phys. **2**, 81 (2006)
15. G.S. Solomon, M. Pelton, Y. Yamamoto, Phys. Rev. Lett. **86**, 3903 (2001)
16. S. Rudin, T.L. Reinecke, Phys. Rev. B **59**, 10227 (1999)
17. L. Andreani, G. Panzarini, J.M. Gérard, Phys. Rev. B **60**, 13267 (1999)
18. G. Khitrova, H. Gibbs, F. Jahnke, M. Kira, S. Koch, Mod. Phys. 71 **71**, 1591 (1999)
19. L.V. Keldysh, V.D. Kulakovskii, S. Reitzenstein, M.N. Makhonin, A. Forchel, JETP Lett. **84**(9), 494 (2006)
20. S. Haroche, *Cavity Quantum Electrodynamics in Fundamental Systems in Quantum Optics* (Elsevier, New York, 1992), p. 769
21. E.M. Purcell, Phys. Rev. **69**, 681 (1946)

22. B. Gayral, J.M. Gérard, Phys. Rev. Lett. **90** (2003)
23. D. Englund, D. Fattal, E. Waks, G. Solomon, B. Zhang, T. Nakaoka, Y. Arakawa, Y. Yamamoto, J. Vučković, Phys. Rev. Lett. **95** (2005)
24. Y. Akahane, T. Asano, B.S. Song, S. Noda, Nature **425**, 944 (2003)
25. B. Gayral, J.M. Gérard, A. Lemaítre, C. Dupuis, L. Manin, J.L. Pelouard, Appl. Phys. Lett. **75**, 1908 (1999)
26. J.M. Gérard, B. Gayral, J. Lightwave Technol. **17**, 2089 (1999)
27. A. Löffler, J.P. Reithmaier, G. Sek, C. Hofmann, S. Reitzenstein, M. Kamp, A. Forchel, Appl. Phys. Lett. **86**, 111105 (2005)
28. J.P. Reithmaier, H.Z. M. Röhner, F. Schäfer, A. Forchel, Phys. Rev. Lett. **78**, 378 (1997)
29. T. Gutbrod, M. Bayer, A. Forchel, J. Reithmaier, T. Reinecke, S. Rudin, P. Knipp, Phys. Rev. B **57**, 9950 (1998)
30. S. Reitzenstein, C. Hofmann, A. Gorbunov, M. Strauß, S.H. Kwon, C. Schneider, A. Löffler, S. Höfling, M. Kamp, A. Forchel, Appl. Phys. Lett. **90** (2007)
31. L.N. Stranski, L. Krastanov, Akad. Wiss. Lit. Mainz Math. Naturwiss. K1 IIb **146**, 797 (1939)
32. A. Löffler, J.P. Reithmaier, A. Forchel, A. Sauerwald, D. Peskes, T. Kümmell, G. Bacher, J. Crys. Growth **286**, 6 (2005)
33. W. Ma, R. Nötzel, H.P. Schönherr, K.H. Ploog, Appl. Phys. Lett. **79**, 4219 (2001)
34. J. Vučković, M. Pelton, A. Scherer, , Y. Yamamto, Phys. Rev. A **66**(2) (2002)
35. T. Rivera, J.P. Debray, J.M. Gérard, B. Legrand, L. Manin-Ferlazzo, J.L. Oudar, Appl. Phys. Lett. **74**, 911 (1999)
36. D. Englund, J. Vučković, Opt. Express **14** (2006)
37. N.L. Thomas, U. Woggon, O. Schöps, M.V. Artemyev, M. Kazes, U. Banin, Nano Lett. **6**(3) (2006)
38. C. Böckler, S. Reitzenstein, C. Kistner, R. Debusmann, A. Löffler, T. Kida, S. Höfling, A. Forchel, L. Grenouillet, J. Claudon, J.M. Gérard, Appl. Phys. Lett. **92** (2008)
39. A. Badolato, K. Hennessy, M. Atatüre, J. Dreiser, E. Hu, P.M. Petroff, A. Imamoğlu, Science **308**, 1158 (2005)
40. Y. Nakamura, O.G. Schmidt, N.Y. Jin-Phillipp, S. Kiravittaya, C. Muller, K. Eberl, H. Grabeldinger, H. Schweizer, J. Cryst. Growth **339**, 339 (2002)
41. A. Rastelli, A. Ulhaq, S. Kiravittaya, L. Wang, A. Zrenner, O.G. Schmidt, Appl. Phys. Lett. **90** (2007)
42. R.J. Warburton, C.S. Dürr, K. Karrai, J.P. Kotthaus, G. Medeiros-Ribeiro, P.M. Petroff, Phys. Rev. Lett. **79** (1997)
43. J. Johansen, S. Stobbe, I.S. Nikolaev, T. Lund-Hansen, P.T. Kristensen, J.M. Hvam, W.L. Vos, P. Lodahl, Phys. Rev. B **77** (2008)
44. D. Gammon, E.S. Snow, B.V. Shanabrook, D.S. Katzer, D. Park, Science **273**, 87 (1996)
45. R. Houdre, R. Stanley, U. Oesterle, M. Ilgems, C. Weisbuch, Phys. Rev. B **49**, 16761 (1994)
46. P. Michler, A. Kiraz, C. Becher, W.V. Schoenfeld, P.M. Petroff, L. Zhang, E. Hu, A. Imamoğlu, Science **290**, 2282 (2000)
47. P. Michler, A. Imamoğlu, A. Kiraz, C. Becher, M. Mason, P. Carson, G. Strouse, S. Buratto, Phys. Stat. Sol. (B) **1** (2002)
48. D. Press, S. Götzinger, S. Reitzenstein, C. Hofmann, A. Löffler, M. Kamp, A. Forchel, Y. Yamamoto, Phys. Rev. Lett. **98** (2007)

49. J.J. Finley, P.W. Fry, A.D. Ashmore, A. Lemaître, A.I. Tartakovskii, R. Oulton, D.J. Mowbray, M.S. Skolnick, M. Hopkinson, P.D. Buckle, P.A. Maksym, Phys. Rev. B **63**, 161305 (2001)
50. A. Imamoĝlu, D.D. Awschalom, G. Burkard, D.P. DiVincenzo, D. Loss, M. Sherwin, A. Small, Phys. Rev. Lett. **83**, 4204 (1999)
51. M. Bayer, P. Hawrylak, K. Hinzer, S. Fafard, M. Korkusinski, Z.R. Wasilewski, O. Stern, A. Forchel, Science **291**, 451 (2001)
52. H.J. Krenner, M. Sabathil, E.C. Clark, A. Kress, D. Schuh, M. Bichler, G. Abstreiter, J.J. Finley, Phys. Rev. Lett **94**, 057402 (2005)
53. S. Reitzenstein, A. Löffler, C. Hofmann, A. Kubanek, M. Kamp, J.P. Reithmaier, A. Forchel, V.D. Kulakovskii, L.V. Keldysh, I.V. Ponomarev, T.L. Reinecke, Opt. Lett. **31**, 1738 (2006)

9

Physics and Applications of Quantum Dots in Photonic Crystals

Dirk Englund, Andrei Faraon, Ilya Fushman, Bryan Ellis, and Jelena Vučković

Summary. Photonic crystals provide an engineerable electromagnetic environment for controlling the interaction between light and matter. In this chapter, we discuss quantum dot-embedded photonic crystal devices for classical and quantum information processing. For classical applications, we discuss high-speed, low-power lasing dynamics, and carrier-induced switching in photonic crystal cavities. For quantum information applications, we describe photonic crystals for controlling the spontaneous emission rate of embedded quantum dots in the weak-coupling regime. Using a set of tools to fine-tune the interaction between quantum dots and cavities, we also demonstrate strong coupling between single dots and photonic crystal cavities. We probe these systems by photoluminescence (PL) and by a technique for coherent optical dipole access in a cavity (CODAC). With these steps, and a toolkit of independent control of quantum dots and photonic crystal components, we are seeing the beginning of coherent control of quantum systems on photonic crystal chips that promises to contribute to secure long-range communication and quantum computing.

9.1 Introduction and Overview

Recent years have witnessed dramatic practical and theoretical advancements toward creating the basic components of quantum information processing (QIP) devices. Cavity quantum electrodynamics (QED)-based approaches are particularly promising [1, 2]. Quantum networks consisting of distributed quantum nodes would combine the ease of storing and manipulating quantum information in atoms [2], ions [3, 4], or quantum dots (QDs) [5], with the advantages of transferring information via photons, using coherent interfaces [6–8]. So far, the demonstrations of basic building blocks of such networks have relied on atomic systems [9–12]. A solid-state implementation of these pioneering approaches would open new opportunities for scaling the network into practical and useful QIP systems. Among the proposed solid-state implementations, systems comprised of QDs in photonic crystals are some of the most promising largely because the smallo volumes of photonic crystal cavities enable strong QD-cavity field interaction [13–15]. They are also suited for monolithic on-chip integration [16]. Derivates of these same systems with a higher QD density are also appealing for classical information processing (e.g., in optical interconnects), including high-speed lasers [17] and modulators [18] operating at low power.

In this chapter, we discuss our recent work on photonic crystals with embedded QDs for applications in quantum and classical information processing. We begin with a brief discussion of the fabrication of QD-coupled photonic crystals in Sect. 9.2, noting some of the major limitations such as lithography fabrication errors [19] and QD inhomogeneous broadening, but leaving QD growth and photonic crystal designs to other texts [20–23]. To address fabrication limitations, we have developed a set of postprocessing tools to selectively tune QDs [24] and cavities [25]; these are described in the Sect. 9.2.2.

With this groundwork, we advance to experimental studies of cavity QED of single QDs in photonic crystals. In Sect. 9.3, we describe single QDs weakly coupled to small-volume photonic crystal cavities. We show that depending on the degree of coupling to a cavity with moderate quality factor, the spontaneous emission (SE) rate of photonic crystal-embedded QDs can be either decreased by up to factor 5 or increased up to factor 8 in the experiment [13]. In cavities with a high quality factor, we show the ability to initiate vacuum Rabi oscillations between a single QD and the cavity field. Then we describe our first steps in integrating multifunctional devices into QIP circuits on photonic crystal chips in Sect. 9.3.1. We demonstrate generation and transfer of single nearly indistinguishable photons from one cavity into a waveguide, which can be either directly out-coupled via an on-chip grating, or transferred to a second cavity which could represent another node in a quantum network [16].

Such a quantum network requires a way to coherently access individual QDs; in Sect. 9.3.2 we describe a technique that allows us to coherently probe and manipulate QDs coupled to photonic crystal cavities [26], a technique which we call coherent optical dipole access in a cavity (CODAC). Combined with CODAC, the cavity/QD system provides a giant optical nonlinearity that enables nonclassical generation of light and may yield two-qubit quantum gates for use in a quantum computer or repeater.

The giant optical nonlinearity is also promising as an all-optical switch for classical information processing. Indeed, many of the quantum devices described here are structurally similar to "classical" devices, and we will describe applications for ultrafast carrier-induced switching (Sect. 9.4.1) and lasing (Sect. 9.4.2).

9.2 Quantum Dot–Photonic Crystal System

In this section, we discuss QDs coupled to photonic crystal cavities in the strong and weak coupling regimes, including a set of tools to mitigate shortcomings in fabrication.

9.2.1 Quantum Dot–Photonic Crystal Circuits: Design and Fabrication

We consider photonic crystal membranes in GaAs, containing a central active layer of self-assembled InAs QDs. The wafers were grown by our collaborators N. Stoltz of the P. Petroff group at UCSB and B. Zhang of the Y. Yamamoto group at Stanford, using Stransky–Krastanow-mode molecular beam epitaxy [23, 27]. A detailed discussion of self-assembled QD growth is given in Chap. 1. In our wafers, the active GaAs layer is typically $d \approx 150$–160 nm thick. We write photonic crystal patterns by electron beam lithography into a polymer resist layer on the sample (Fig. 9.1a). After this mask is transferred into the sample by plasma dry etching, a wet etching step removes a sacrificial $Al_xGa_{(1-x)}As$ layer, resulting in a suspended PC membrane. Some of the samples we will discuss also contain a Distributed Bragg Reflector (DBR) of alternating $GaAs/Al_xGa_{(1-x)}As$ layers. The DBR is beneath the membrane and improves free-space outcoupling efficiency [28] (see Fig. 9.1b,c). An example of a fabricated three-hole (L3) photonic crystal

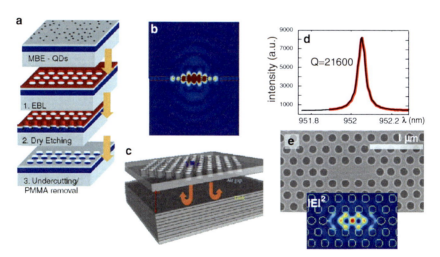

Fig. 9.1. Photonic crystal fabrication and design. (**a**) Fabrication. (**b**) The free-standing membrane radiates up and down, but efficiency can be improved with bottom DBR mirror (**c**). (**d**) Three-hole defect cavity with $Q \sim 21,600$. (**e**) Scanning electron microscope (SEM) image of a three-hole defect cavity and the simulation of the electric field pattern

cavity is presented in Fig. 9.1e,f. It is a slightly modified version of the cavity in [29] with a predicted Q value of 145×10^3. The spectrum from the same type of cavity, shown in Fig. 9.1d, is obtained by illuminating a broad distribution of embedded InAs QDs and indicates a cavity quality (Q) factor of 21.6×10^3, below the design value. Our and other groups have observed that cavities are so far limited by material absorption and fabrication imperfections to tens of thousands in active GaAs structures in the near-infrared [16]. These Q values are, however, more than sufficient to reach the strong coupling regime between cavities and QDs [22].

We choose a QD distribution centered near 940 nm. This distribution is reasonably far from the InAs wetting layer, yet still within the detector sensitivity of our streak camera and silicon avalanche photo diodes. The QD emission is inhomogeneously distributed about the center wavelength with linewidths $\Delta\lambda_{\rm QD,inh} \sim 20\text{--}30$ nm, depending on the sample.

9.2.2 Local Quantum Dot and Cavity Tuning on a Photonic Crystal Chip

The radiative behavior of QDs greatly depends on their spatial and spectral alignment to modes in the photonic crystal [13]. Therefore the inhomogeneous broadening and random positioning creates a major difficulty for single QD/photonic crystal devices. Let us consider the PL observed from a diffraction limited spot $A_{\rm f}$ on the sample which contains the cavity. For QDs with a random spatial distribution, the probability that a QD in $A_{\rm f}$ is also within the cavity area $A_{\rm mode} \approx V_{\rm mode}/d$ is $P_{\rm s} \approx A_{\rm mode}/A_{\rm f}$. For an L3 cavity with $V_{\rm mode} \approx 0.74(\lambda/n)^3$, slab thickness $d \approx 0.6\lambda/n$, and a focal spot near the diffraction limit with an objective lens NA of $\sim$0.6, this spatial coupling probability is $P_{\rm s} \sim 1/30$. We bring this probability near unity by using QD densities of 10–100 QDs in the focal area. However, spatial alignment is useful for better spatial coupling between the QD and the cavity, and may be crucial for building integrated networks consisting of a multitude of QD-coupled cavities. Spatial alignment may be achieved by prepatterning QDs, as discussed in Chap. 2, or by identifying QDs and pattering cavities around them. The latter technique was achieved by positioning photonic crystal cavities on dots identified by scanning electron or atomic force microscopy [15, 30].

Because of the spectral inhomogeneous broadening of the quantum dots, the spectral alignment between quantum dots and resonators brings another difficulty in realizing integrated devices. With an inhomogeneous linewidth $\Delta\lambda_{\rm QD,inh} \sim 20$ nm and $Q \sim 10^4$ (linewidth ~ 0.1 nm), the spectral coupling probability to a single-exciton transition is only about 1/200. The odds become even worse when one tries to interact two distant QDs of homogeneous linewidth $\Delta\lambda_{\rm QD} \sim 0.01$ nm in a quantum network. In fact, the ratio $\Delta\lambda_{\rm QD}/\Delta\lambda_{\rm QD,inh} \sim 10^{-4}$ is similar in other solid state emitters and points to a universal problem faced by solid-state approaches. This problem can be partially addressed by reducing the inhomogeneous linewidth by postgrowth thermal treatments of QDs [31–33], as discussed in Chap. 2. However, matching dots to a common transition wavelength requires a way of tuning individual dots. A few techniques can be used to modify the wavelength of InAs QDs: Stark shift [34], Zeeman shift [35], temperature tuning [36] and strain tuning [37]. In the next section, we describe a local laser heating technique which enables independent control of QDs on chip, employing structures with high-Q cavities whose temperature is controlled by laser beams [24]. Our in situ technique allows extremely precise spectral tuning of InAs QDs by up to 1.8 nm and of cavities up to 0.4 nm (four cavity linewidths). Given one cavity-coupled QD, the technique improves the probability of matching a second dot from $\Delta\lambda_{QD}/\Delta\lambda_{QD,inh} \sim 10^{-4}$ to $\sim 1\,\text{nm}/\Delta\lambda_{QD,inh} \sim 1/20$. It therefore forms an essential step toward creating on-chip QIP devices.

Local Laser Heating for Tuning Quantum Dots and Cavities

The local laser heating (LLH) technique allows independent thermal control of spatially separated regions on photonic crystal membranes. The regions (12 μm long, 4 μm wide, 150 nm thick) are isolated from the rest of the sample by air trenches, as shown in Fig. 9.2a. Six narrow bridges provide structural support.

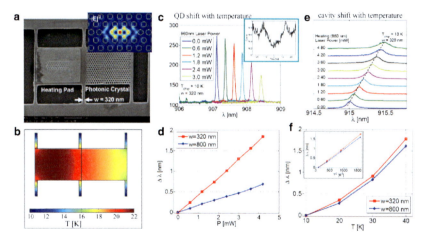

Fig. 9.2. (a) Scanning electron micrograph of the fabricated structure showing the PC cavity, the heating pad and the connection bridges of width w. The temperature of the structure was controlled with a laser beam (960 or 980 nm) focused on the heating pad. Assuming uniform heating of the pad results in the temperature profile in (b), simulated by a finite element model. (c) The single-exciton PL from a QD positioned in the photonic crystal cavity red-shifts with increasing power of the heating laser, as shown here for a structure with $w = 320$ nm. (d) Exciton-tuning for cavities with $w = 320, 800$ nm. (e) The photonic crystal cavity, seen here dominating the PL under strong above-band excitation, red-shifts with temperature at roughly one-third the rate as the QD in the same structure with $w = 320$ nm. (f) QD temperature tuning by changing the temperature of the entire chip by heating the cryostat. The *inset* shows that the detuning is linear in T^2

The thermal conductivity of narrow (≈100 nm), cold (4–10 K) GaAs bridges is reduced by up to four orders of magnitude with respect to the bulk GaAs [38], thus improving the thermal insulation. We tested two devices with connection bridges of the same length (2 μm) but different widths: $w = 320$ nm and $w = 800$ nm. A focused laser beam heats a pad next to the photonic crystal cavity. The pad is deposited with (20 nm Cr / 15 nm Au) layers for improved absorption, and the beam's wavelength ($\lambda_h = 960$ nm) is tuned above the QD absorption wavelength to minimize background PL in single QD measurements. As described in [24], we expect the temperature of the membrane to increase by a few tens of Kelvin for milliwatts of pump power, as shown in Fig. 9.2b.

This and later experiments were performed in a continuous flow liquid helium cryostat maintained between 5 and 50 K, and imaged with a confocal microscope setup. A Ti:sapph laser at 855 nm excites QD PL. The QD emission wavelengths of single lines redshift up to $\Delta\lambda_{QD} = 1.8$ nm with heating power as shown in Fig. 9.2c. At high power, we observe QD linewidth broadening from 0.04 to 0.08 nm; this is also observed when the full sample is heated and is associated with phonon broadening. As the thermal conductance of the bridges is proportional to their width, the temperature change of the structure is inversely proportional to w. This is observed in the comparison of the shifts for $w = 320$ and 800 nm in Fig. 9.2d.

To show the compatibility of the LLH technique with single photon measurements and QIP, we verify single photon emission under pulsed excitation (13-ns repetition). This is measured by the autocorrelation function $g^{(2)}(0)$ which gives the probability of generating more than one photon in a given pulse, normalized by the probability for an equally bright Poisson-distributed source (see also the discussion in Chap. 6). For short time scales, this function is measured from the coincidence rate between the two detectors of a Hanbury-Brown and Twiss interferometer (HBT in Fig. 9.10). A start–stop scheme measures time delay $t' = t_1 - t_2$ between detection events. The inset in Fig. 9.2c shows antibunching when the dot is detuned by 0.8 nm.

The actual temperature dependence of the QD shift was determined by changing the temperature of the whole chip. The results are plotted in Fig. 9.2f and indicate that a shift of 1.8 nm corresponds to a temperature of 40 K, so the LLH also heated the structure to the same temperature. The QD shift shows a quadratic dependence on temperature (Fig. 9.2f inset), which is expected since the band gaps of GaAs/InAs shift quadratically with temperature in this interval [39]. The observed linear dependence of the QD shift on the heating power (Fig. 9.2d) implies a linear relation between the power of the heating laser and T^2. For a fixed thermal conductivity and fixed absorption coefficient of the material, one expects a linear dependence between the temperature of the structure and the heating power. However, the thermal conductivity of micro and nanoscale structures is temperature dependent and may explain the T^2 dependence observed in our experiments. We plan a more detailed study of the device's thermal properties in future experiments.

Due to the temperature dependence of the refractive index, the cavity also shifts as seen in Fig. 9.2e, though at three times lower rate. Over a shift of $\Delta \lambda = 0.48$ nm, the Q drops from 7,600 to 4,900. Our LLH technique is fully reversible, though similar permanent techniques are also possible [40].

Postfabrication Tuning by Photorefractive Top Layers

While the LLH technique allows some fine-tuning of photonic crystal cavity resonances, a second tuning method is required to tune QDs and cavities individually. Several other techniques have been demonstrated to tune PC cavities, including chemical digital etching [41], gas deposition [42, 43], thin-film deposition [44], and evanescent tuning with a nanowire [45]. However, to tune different cavities and QDs on a chip into resonance, we again require a technique that again acts locally and preferably in situ.

For that purpose, we developed a technique to locally change the refractive index in planar optical devices by photodarkening of a thin chalcogenide glass top layer [25]. This local index tuning (LIT) technique permanently changes the resonance of GaAs-based photonic crystal cavities operating near 940 nm by up to 3 nm.

Chalcogenide glasses change their optical properties when illuminated with light above their band gap. This effect has been used to tune optical devices such as quantum cascade lasers [46]. The tuning of PCs directly fabricated in chalcogenide glasses was demonstrated in [47], but most applications rely on PCs fabricated in group IV and III–V semiconductors.

Here we use As_2S_3 films with thickness between 30 and 100 nm. These are deposited onto the fabricated photonic crystals with L3-type cavities using thermal evaporation. The glass films contain disconnected molecular cage-like structures [48] that form in the vapor phase and are then frozen into the deposited film. Optical excitation near the band edge polymerizes these structures in an extended glass network. This rebonding process is accompanied by an increase in the refractive index (from 2.31 to 2.43 at $\lambda = 1,550$ nm) and a small decrease of the material volume [49]. The change is permanent unless the glass is annealed above the glass transition temperature [46].

We will describe low-temperate tests with QDs in the deposited structure shown in Fig. 9.3a, though we note that the LIT technique also works at room temperature. The As_2S_3 layer is exposed near its 527-nm bandgap using a 543 nm HeNe laser focused to $\sim 1\,\mu m^2$ (Fig. 9.3b). The thickness of the As_2S_3 influences both the quality factor of the cavity and the maximum tuning range. We tested thicknesses of 30, 50, 60, and 100 nm, and found 60 nm to offer the best compromise between optical quality and tunability.

During the HeNe laser exposure, the cavity resonance shifted by up to 3 nm as shown in Fig. 9.3c. At $1\,\mu W$ focused on a spot size of $\sim 1\,\mu m^2$, this took ~ 20 min. The saturation time decreases with increased power. In this trial, the photoexposure decreased the Q factor from 7,000 to 4,650, probably through loss in the glass induced by nanofractures.

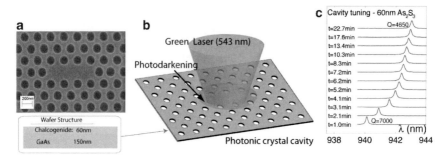

Fig. 9.3. (a) Photonic crystal structure (thickness 150 nm), shown before deposition of 60-nm thick As_2S_3. (b) The cavity is tuned by photodarkening the chalcogenide layer with a green laser (continuous-wave, wavelength 543 nm). (c) Spectra showing the shift of the cavity resonance due to photodarkening of the 60-nm thick chalcogenide layer

For cavity QED experiments, we also studied the effect of LIT on QDs. In Fig. 9.4 we show the tuning of a cavity with 50 nm of As_2S_3 that contains several weakly coupled QD lines. Early in the exposure, these lines shift in the characteristic fashion shown in Fig. 9.4: QDs first rapidly red-shift, then relax to a constant frequency. As coupled QDs tune across the cavity spectrum, they become bright due to the enhanced emission rate and outcoupling efficiency [13]. The cavity itself steadily red-shifts, even after the QDs have stopped tuning. We attribute the initial QD tuning behavior to strain in the sample. First strain builds up as the sample is cooled to ∼10 K because the thermal expansion coefficient of the As_2S_3 layer is larger than that of GaAs. The strain increases when the chalcogenide layer shrinks during exposure, reaching a maximum after about 36 s in Fig. 9.4. Then the strain appears to relax as the QDs settle to a fixed emission wavelength. We attribute the relaxation to nanofracturing of the chalcogenide glass as the compressive pressure exceeds the material's tensile strength. Once the strain is relaxed, only the cavity continues to red-shift. In applications where QDs must be tuned independent of the cavity, it might be necessary to first relax the strain or induce fracturing of the chalcogenide layer in some other way. However, we note that this ability to induce local strain may actually be useful for other applications, such as creating locally strained quantum wells [50], QDs, or impurities [51].

The photosensitive exposure is confined to the focus area of ∼1 μm^2. This locality allows for independent tuning of interconnected optical components on photonic crystal chips. While the LIT technique complements the laser heating technique mentioned above for our single QD experiments, we believe it has general utility for integrated nanophotonic circuits for classical and quantum information processing, including applications such as filtering, multiplexing, optical storage, fine-tuning of modulators and lasers, and local tuning of distinct PC cavities on integrated photonic chips for quantum optics research.

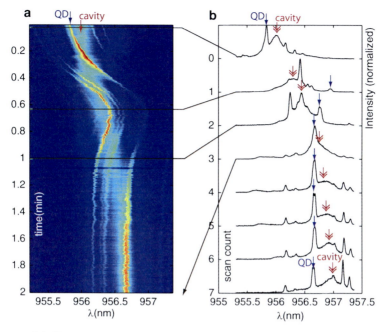

Fig. 9.4. (a) Spectra showing cavity and QD shifting, as a function of exposure time. The QD lines first shift rapidly, presumably through changing material strain induced by the chalcogenide layer. Soon after, the QD lines become stationary, while the cavity continues to red-shift. This data set was taken on a sample with 50 nm of As_2S_3. (b) Individual scans of QD/cavity tuning show that after strain relaxation, the cavity can be shifted independently of the QDs. Scans 4–7 were taken for $t > 2$ min when the pump power was temporarily increased to speed up the chalcogenide exposure

9.2.3 Cavity QED with Quantum Dots in Photonic Crystals

Before we advance to experiments on strong and weak coupling in the QD/PC system, we will briefly review some relevant elements of cavity QED. Consider a QD at position **r** in a resonant cavity mode with electric field energy density $\frac{1}{2}\varepsilon|E(\mathbf{r})|^2$. The energy density is maximized at $\mathbf{r}_M$ inside the high-dielectric ε_M. The coupling strength between this dot and the electromagnetic field is then given by $g(\mathbf{r}) = g_0 \psi(\mathbf{r}) \cos(\xi)$, where $g_0 = \frac{\mu}{\hbar}\sqrt{\frac{\hbar\omega}{2\varepsilon_M V_{\mathrm{mode}}}}$, $\psi(\mathbf{r}) = \frac{E(\mathbf{r})}{|E(\mathbf{r}_M)|}$, and $\cos(\xi) = \frac{\boldsymbol{\mu}\cdot\hat{\mathbf{e}}}{\mu}$ [20]. Here V_{mode} is the cavity mode volume, defined as $V_{\mathrm{mode}} = \frac{\int_V \varepsilon(\mathbf{r})|E(\mathbf{r})|^2 d^3\mathbf{r}}{\varepsilon_M |E(\mathbf{r}_M)|^2}$. The electric field orientation at the location **r** is denoted as $\hat{\mathbf{e}}$, $\boldsymbol{\mu}$ is the dipole moment of the dot, and $\mu = |\boldsymbol{\mu}|$. Therefore, the coupling $|g(\mathbf{r})|$ reaches its maximum value of g_0 when the QD is located at the point $\mathbf{r}_M$ where the electric field energy density is maximum, and when its dipole moment is aligned with the electric field.

The losses of the system can be described in terms of the *cavity field decay rate* $\kappa = \omega/2Q$ and the *dipole decay rate* γ. The rate γ includes QD dipole decay to modes other than the cavity mode and to nonradiative decay routes. Two regimes of exciton/cavity field coupling are distinguished: the *strong-coupling regime*, for $|g(\mathbf{r})| > \kappa, \gamma$, and the *weak-coupling regime*, for $|g(\mathbf{r})| < \kappa$ or γ [52].

In the strong-coupling regime, the QD-field coupling rate g exceeds the photon loss rates κ, γ. In this case, the QD and cavity can no longer be treated as distinct entities and polariton modes of the dot and the cavity field are formed. These polaritons are visible in the emission spectrum of the cavity-dot system, which splits about the cavity resonance by a frequency $2g$, referred to as the vacuum Rabi splitting. We will assume that the QD dipole decay rates into modes other than the resonant cavity mode and into nonradiative routes are not modified in the presence of a cavity, and estimate γ from the homogeneous linewidth of the QD without a cavity.

On the other hand, in the weak-coupling case, irreversible decay dominates. The QD/cavity system decoheres before the cavity photon is reabsorbed into the QD. The total SE rate of a QD at position $\mathbf{r}$, spectrally detuned from the cavity resonance wavelength by $\lambda - \lambda_{\text{cav}}$, equals the sum of rates into cavity modes and all other modes in the PC, $\Gamma = \Gamma_{\text{cav}} + \Gamma_{\text{PC}}$.

In the Wigner–Weisskopf approximation, the SE rate is directly proportional to the local density of states (LDOS) [53]. The cavity density of states follows a Lorentzian [53], so the SE rate enhancement into the cavity, normalized by the natural emission rate Γ_0, is

$$F_{\text{cav}} = \Gamma_{\text{cav}}/\Gamma_0 = F_{\text{cav},0} \left(\frac{\mathbf{E}(\mathbf{r}) \cdot \boldsymbol{\mu}}{|\mathbf{E}_{\max}||\mu|}\right)^2 \frac{1}{1 + 4Q^2 \left(\frac{\lambda}{\lambda_{\text{cav}}} - 1\right)^2}. \tag{9.1}$$

The factors $\left(\frac{\mathbf{E}(\mathbf{r}) \cdot \boldsymbol{\mu}}{|\mathbf{E}_{\max}||\mu|}\right)^2$ and $1/(1 + 4Q^2(\frac{\lambda}{\lambda_{\text{cav}}} - 1)^2)$ describe the spatial and spectral mismatch of the emitter dipole $\boldsymbol{\mu}$ to the cavity field $\mathbf{E}$. For perfect spatial and spectral alignment, the rate enhancement into the cavity is given by

$$F_{\text{cav},0} \equiv \frac{3}{4\pi^2} \frac{\lambda_{\text{cav}}^3}{n^3} \frac{Q}{V_{\text{mode}}}. \tag{9.2}$$

The SE rate of a QD uncoupled to the cavity, Γ_{PC}, is given by the LDOS of non-cavity PC modes. It is diminished in the PC bandgap by a factor $F_{\text{PC}} \equiv \Gamma_{\text{PC}}/\Gamma_0$. In a planar triangular-lattice PC, we approximate in Sect. 9.2.3 by simulation and measurement that the average $\langle F_{\text{PC}} \rangle \approx 0.2$. The total SE rate modification is then

$$F = \frac{\Gamma}{\Gamma_0} = F_{\text{cav}} + F_{\text{PC}}. \tag{9.3}$$

This equation describes the SE rate modification measured in Sect. 9.2.3.

9 Physics and Applications of Quantum Dots in Photonic Crystals

The strong and weak coupling regimes in the solid state are not only of academic interest; they can also be used in the construction of efficient sources of single photons with high indistinguishability and numerous QIP devices, as described later in this chapter.

Weak Coupling Regime and Purcell Effect

Over the past decade, photonic resonators with increased LDOS have been exploited to enhance emission rates for improving numerous quantum optical devices (e.g., [54,55]). Single photon sources, in particular, stand to see large improvements [56]. While more attention has been given to *increasing* the emission rate, the reverse is also possible in an environment with a decreased LDOS.

We will show now that a PC with a modified single-defect cavity can significantly increase or decrease the SE rate of embedded self-assembled InAs QDs. The cavity is shown in Fig. 9.5 and described in [19,22]. Cavity resonances of fabricated structures were measured by PL spectroscopy at 5 K. Although we have achieved Q values for this single-defect design of up to 8,100, such high-Q cavities have the disadvantage that spectral coupling is unlikely. To analyze a larger data set, we focused in this weak-coupling study on low-Q cavities. For the moment we consider a structure with $Q \approx 200$, seen in the strongly pumped cavity spectrum in the inset of Fig. 9.6a. This x-dipole PC mode (Fig. 9.5a) is linearly polarized.

The low-intensity spectrum reveals a single QD exciton line A, matched with the cavity in wavelength and polarization (Fig. 9.6a). The emission following pulsed excitation gives a lifetime of $\tau_A = 650$ ps, measured with a streak camera (Fig. 9.6b). Compared to the excitonic lifetime in bulk semiconductor, which has a distribution of $\tau_0 = 1.7 \pm 0.3$ ns, this indicates a rate enhancement by $F_A = 2.6 \pm 0.5$. In striking contrast is the lifetime of QD excitons *not* coupled to a cavity. The cross-polarized emission, A90°, has a lifetime of 2.9 ns, far longer than in bulk. As can be seen in Fig. 9.6a, this line

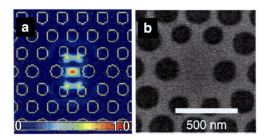

Fig. 9.5. FDTD-assisted design of the photonic crystal cavity. The periodicity is $a = 0.27\lambda_{\mathrm{cav}}$, hole radius $r = 0.3a$, and thickness $d = 0.65a$. (a) Electric field intensity of x-dipole resonance. (b) SEM image of fabricated structure

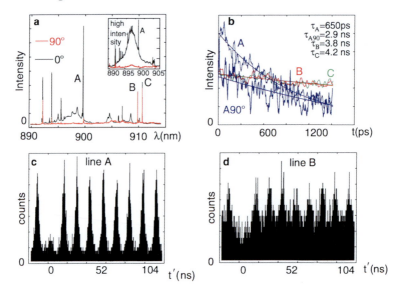

Fig. 9.6. SE rate modification. (**a**) The high pump-power spectra (*inset*) reveals a polarized resonance with $Q \approx 200$. The low-power ($\sim 5\,\mathrm{W\,\mu m^{-2}}$) spectra show coupled single exciton line A that matches, and lines B and C that don't match the cavity polarization and frequency. (**b**) Lifetime is shortened for the coupled QD line A and extended for the decoupled ones, including the near-degenerate orthogonal emission line A90°. (**c, d**) Autocorrelation histograms for lines A and B

is much weaker than A due to low out-coupling and SE rate. Lines B and C, which are spectrally far detuned, show lifetimes extended even further to 3.8 and 4.2 ns, respectively.

We verified the single-emitter nature of the emission lines by second-order coherence measurements, as described in Sect. 9.2.2. The coincidence histogram for line A is shown in Fig. 9.6c and indicates antibunching of $g_A^{(2)}(0) \approx 0.14$ at zero delay time. Since $g_A^{(2)}(0) < \frac{1}{2}$, the fluorescence is indeed from a single emitter. Lines B (Fig. 9.6d) and C are also antibunched with $g^{(2)}(0) < \frac{1}{2}$.

This pattern of short-lived coupled, and long-lived decoupled excitonic lines was observed in many other structures, as detailed in [13]. They are summarized in Fig. 9.7a. To explain these increased lifetimes, we again turn to Eq. (9.3). The first term for emission into the cavity mode vanishes as the QD exciton line is far detuned from λ_{cav}. The second term F_{PC} is reduced below unity due to the diminished LDOS in the PC for emission inside the PC bandgap, leading to longer lifetime. This SE rate modification is illustrated in Fig. 9.7b. Equation (9.2) indicates that for this single-defect cavity design with Q of several thousands, F can reach several hundred, reaching the strong coupling regime. We will discuss this regime in Sect. 9.2.3.

We modeled the lifetime modification by FDTD analysis of a classical dipole in the PC with the x-dipole cavity. The quantum electrodynamic and

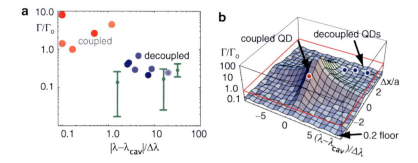

Fig. 9.7. Summary of SE rate measurements. (**a**) Experimental (*circles*) and calculated (*bars*) data of Γ/Γ_0 of single QD exciton lines vs. spectral detuning (normalized by the cavity linewidth $\Delta\lambda$). Coupling was verified by spectral alignment and polarization matching. (**b**) Illustration of the predicted SE rate modification in the PC as function of normalized spatial and spectral misalignment from the cavity (a is the lattice periodicity). This plot assumes $Q = 1,000$ and polarization matching between the emitter dipole and cavity field

classical treatments of SE emission yield proportional results, so that the true SE rate is related to the classical dipole radiation power by $\Gamma_{SE}^{PC}/\Gamma_{SE}^{bulk} = P_{classical}^{PC}/P_{classical}^{bulk}$ for bulk GaAs and the PC [57]. We simulated ensembles and single emitters to obtain the average SE rate modification and its variance. These results, plotted in Fig. 9.7a in green, agree well with our experimental observations and confirm earlier theoretical predictions of SE lifetime suppression inside the PC bandgap of a similar structure [58].

The observations of up to eight-fold enhancement and five-fold suppression of SE rate make PCs very interesting as materials for controlling properties of embedded emitters. For example, the QD-cavity coupled system promises to increase out-coupling efficiency and photon-indistinguishability of single photon sources. The quenching of SE may find applications in QD-based photonic devices including QIP (e.g., quantum memory in cavity-detuned states). It also shows that reported QD lifetime reduction due to surface proximity effects [59] should not limit the performance of PC–QD single photon sources. The simultaneous enhancement of coupled and suppression of uncoupled SE rates results in very high cavity coupling efficiency of QD emission into the cavity mode, with $\beta = \frac{F_{cav}}{F} \approx 1$, even for moderate cavity Purcell enhancement.

Strong Coupling Regime

Several groups including ours have in recent years observed vacuum Rabi splitting in the PL spectrum of a QD tuned through a PC cavity [14, 15, 26]. We present here PL measurements on a QD that is tuned through a strongly coupled photonic crystal cavity using the LLH technique of Sect. 9.2.2

(temperature tuning of the whole sample shows the same result). These results are analogous to the measurements of a single QD exciton strongly coupled to a micropost cavity [60], as discussed in Chap. 8. Later in Sect. 9.3.2 we will also present a method for directly and coherently probing the strong coupling regime. The structure consists of an L3 cavity as that shown in Fig. 9.2a. It has $Q = 1 \times 10^4$, determined by PL with the QD tuned off-resonance.

The QD/PC cavity system is pumped with a continuous-wave laser beam at 780 nm, above the GaAs bandgap. For low excitation powers, the QD PL in Fig. 9.8c increases linearly with pump power, indicating a single exciton line. As the QD is temperature-tuned through the cavity, clear anticrossing between the QD and the cavity lines is observed: the QD splits the cavity spectrum into two polariton peaks (with frequencies $\omega_\pm$) when it becomes resonant with the cavity. The frequencies of the two polaritons are derived from the eigen energies of the Jaynes–Cummings Hamiltonian [53], and are given by

$$\omega_\pm = \frac{\omega_c + \omega_d}{2} - i\frac{\kappa + \gamma}{2} \pm \sqrt{g^2 + \frac{1}{4}(\delta - i(\kappa - \gamma))^2}, \qquad (9.4)$$

where ω_c denotes the cavity frequency, ω_d the QD frequency, $\delta = \omega_d - \omega_c$ the QD/cavity detuning, cavity field decay rate $\kappa/2\pi = 16$ GHz (linewidth 0.1 nm), Rabi frequency $g/2\pi = 8$ GHz (from Rabi splitting of $2g$ corresponding to 0.05 nm), and the dipole decay rate without the cavity $\gamma/2\pi \approx 0.1$ GHz. As $g \approx \kappa/2$, the cavity/QD system operates at the onset of strong coupling [52].

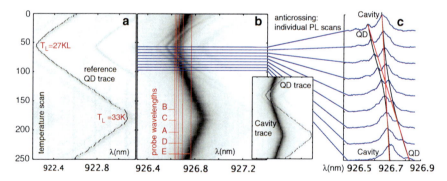

Fig. 9.8. PL of a single QD tuned through strong coupling to a PC cavity by local laser heating. (**a**) A reference QD is used to trace the frequency of the strongly coupled QD, as the two dots show very similar temperature tuning behavior. (**b**) The PL shows the strongly coupled QD PL shows the strongly coupled QD tuned in and out of resonance with the PC cavity. In the reflectivity measurements, the above-band pump is switched off and the cavity/QD system probed at different detunings of the reflected laser beam from the point of anticrossing (lines A–E). *Inset:* QD and cavity traces. (**c**) Individual PL cross-sections show anticrossing between QD and cavity, with measured Rabi splitting of 0.05 nm (corresponding to $2g$, where the coupling strength $g/2\pi = 8$ GHz). The *red lines* shows the expected wavelengths of the uncoupled QD and cavity

To accurately interpret the PL data, we require the wavelengths of the cavity and strongly coupled QD. Since direct tracking of the dot is difficult in the anticrossing region, we instead track a nearby QD that precisely follows, at a fixed offset, the coupled dot's trajectory (see Fig. 9.8a). The calculated trajectory of the strongly coupled QD and the cavity are marked in the inset of Fig. 9.8b.

9.3 Quantum Information Applications

Recent years have witnessed dramatic practical and theoretical advancements toward creating the basic components of QIP devices. One essential element is a source of single indistinguishable photons, which is required in quantum teleportation [61], linear-optics quantum computation [62], and several schemes for quantum cryptography [63]. Sources have been demonstrated from a variety of systems [64] including semiconductor QDs [23] whose efficiency and indistinguishability can be dramatically improved by placing them inside a microcavity [54]. A second major component is a quantum channel for efficiently transferring information between spatially separated nodes of a quantum network [1]. This network would combine the ease of storing and manipulating quantum information in QDs [5], atoms or ions [3, 4], with the advantages of transferring information between nodes via photons, using coherent interfaces [6–8].

9.3.1 On-chip Integrated Single Photon Source

Single photons are generated from the InAs QD on demand by pulsed excitation and spectral filtering of exciton-complex lines [20]. The efficiency of such a system is poor, since the majority of emitted photons are lost in the semiconductor substrate. As described in Sect. 9.2.3, photon extraction efficiency and emission rate are improved by using cavity QED effects [20].

An on-chip photonic network consisting of spatially separated cavities and QDs would greatly improve gate and transfer efficiencies in a quantum network [1], as sketched in Fig. 9.9a. We will now describe a basic building block of such a quantum network involving the generation and transfer of single photons on a photonic crystal (PC) chip. A cavity-coupled QD single photon source is connected through a $25\,\mu$m channel to an otherwise identical target cavity so that different cavities may be interrogated and manipulated independently (Fig. 9.9b). This system provides a source of single photons with a high degree of indistinguishability (mean wavepacket overlap of $\sim 67\%$), 12-fold SE rate enhancement, SE coupling factor $\beta \sim 0.98$ into the cavity mode, and high-efficiency coupling into a waveguide. These photons are transferred into the target cavity with a target/source field intensity ratio of 0.12 ± 0.01 (up to 0.49 observed in structures without coupled QDs), showing the system's potential as a fundamental component of a scalable quantum network for building on-chip QIP devices.

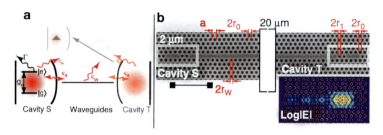

Fig. 9.9. Coupled cavities system. (**a**) Basic network consisting of two cavities and one cavity-coupled QD. The QD is coupled to a cavity (rate g_0) and decays with SE rate Γ. The cavity, in turn, is coupled to a waveguide and leaky modes at field coupling rates $\kappa_\parallel$ and $\kappa_\perp$, respectively. The waveguide field decay rate is κ_W. (**b**) Fabricated structure: source (S) and target (T) cavities of identical design are connected via the 25 μm waveguide. *Inset:* Electric field pattern

The structure consists of two L3 cavities [29], butt-coupled and connected via a 25-μm long closed portion of a waveguide (Fig. 9.9b). The detailed design is described in [16]. The cavity and waveguide field decay rates can be expressed as a sum of vertical, in-plane, and material loss, respectively: $\kappa = \kappa_\perp + \kappa_{\parallel,wg} + \gamma$. Removed from the waveguide, the "bare" outer cavities radiate predominantly in the vertical direction at rate $\kappa_\perp$, as in-plane losses can be suppressed with enough PC confinement layers. Introducing an open waveguide coupled to the cavity creates additional loss $\kappa_{\parallel,wg}$.

In a waveguide of finite extent, the continuum of modes in the waveguide band breaks up into discrete resonances. For photon transfer, one of these must be coupled to the outer cavities. Assuming the two cavities are spectrally matched and detuned from the waveguide resonance by Δ, and that material losses γ are negligible [19], the field amplitudes in the source and target cavities (S,T) and waveguide (W) evolve according to

$$\begin{aligned}
\dot{c}_s(t) &= -i\kappa_\parallel c_w(t) - \kappa_\perp c_s(t) + p(t), \\
\dot{c}_t(t) &= -i\kappa_\parallel c_w(t) - \kappa_\perp c_t(t), \\
\dot{c}_w(t) &= -i\kappa_\parallel c_s(t) - i\kappa_\parallel c_t(t) - (\kappa_W + i\Delta)c_w(t).
\end{aligned} \quad (9.5)$$

Here we assume equal coupling rates for the outer cavities, based on their near-identical SEM images and Q values, which fall within a linewidth of each other in most structures. The constant κ_W denotes the waveguide loss rate (other than loss into the end-cavities), and $p(t)$ represents a dipole driving the source cavity. It will suffice to analyze this system in steady-state since excitation of the modes, on the order of the exciton lifetime $\tau \sim 100\,\mathrm{ps}$, occurs slowly compared to the relaxation time of the photonic network, of order $\tau = \omega/Q \sim 1\,\mathrm{ps}$ for the cavity and waveguide resonances involved. Then the amplitude ratio between the S and T fields is easily solved as $c_s/c_t = 1 + \kappa_\perp(\kappa_W + i\Delta)/\kappa_\parallel^2$.

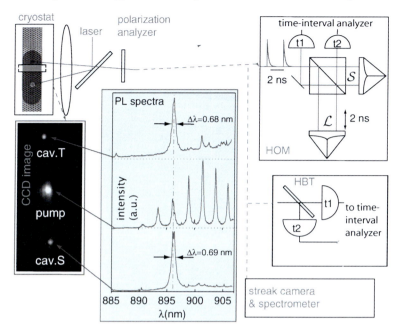

Fig. 9.10. Experimental setup. The sample is addressed with a confocal microscope with a steerable pump beam (spot diameter ∼1 μm) and movable probe aperture (selection region diameter 2.9 μm). PL is directed to the 0.75 m spectrometer (with LN-cooled Si detector), 0.75 m streak camera, or Hanbury-Brown–Twiss or Hong–Ou–Mandel setups. The inset shows the overlapping resonances of the terminated waveguide (extended cavity) and the source/target cavities

We now focus on a particular system that showed high coupling between the two cavities, while also exhibiting large QD coupling in the source cavity. Measurements were done with the sample at 5 K and probed with a confocal microscope setup as shown in Fig. 9.10. A movable aperture in the microscope image plane, together with a steerable pump beam, allow independent adjustment of the pump and observation regions. The structures show close spectral match (see inset of Fig. 9.10). Comparing the emission intensities from S and T cavities gives the transfer efficiency $|c_t/c_s|^2 = 0.12 \pm 0.01$. Other instances of a slightly modified cavity/waveguide design, with one-hole waveguide–cavity separation, yielded photon transfer ratios as large as $|c_t/c_s|^2 = 0.49$ [16]. Optimizing the geometry can improve transfer efficiency further [65]. The cavities/waveguide system strongly isolates transmission to the cavity linewidth, as seen from the transmission spectrum denoted "ST" in Fig. 9.11a for above-band excitation in S and observation from T.

In Fig. 9.11b, we show a single-exciton transition coupled to cavity S at 897 nm. The transition is driven quasiresonantly through a higher-order excited QD state with a 878 nm pump from a Ti:sapph laser (the quasiresonant

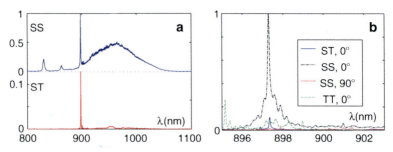

Fig. 9.11. Cavity–cavity coupling via a waveguide. (**a**) Broad emission in cavity S (plot "SS") is filtered into the target cavity (plot "ST"). (**b**) When the QD exciton at 897.3 nm in cavity S is pumped (at 878 nm, 460 µW, 1µm focal spot), the emission is observed from S ("SS") and T ("ST"). The cross-polarized spectrum from S shows nearly complete quenching of QD emission ("SS, 90°"). The line at 897.3 nm is only observed when S is pumped

excitation regime was discussed in Chap. 6). We measure a lifetime of 116 ps (Fig. 9.12b) and estimate Purcell enhancement $F = 12$ and SE coupling factor into the cavity mode $\beta = F/(F + F_{PC}) \sim 0.98$ [13].

We characterized the exciton emission by measurements of the second-order coherence and indistinguishability of consecutive photons. When the QD in cavity S is pumped quasiresonantly, then photons observed from S shows clear antibunching (Fig. 9.12a), with $g^2(0) = 0.35 \pm 0.01$; this value is relatively high because of a large QD density of $\sim 200\,\mu m^{-2}$.

Because of the shortened lifetime of the cavity-coupled QD exciton, the coherence time of emitted photons becomes dominated by radiative effects which can improve the photon indistinguishability [54, 56, 66]. We measured the indistinguishability using the Hong–Ou–Mandel (HOM) type setup sketched in Fig. 9.10. This measurement was discussed in greater detail in Chap. 6. Briefly, the QD is excited twice every 13 ns, with a 2.3 ns separation. The emitted photons are collected from the source cavity and directed through a Michelson interferometer with a 2.3 ns time difference. The two outputs are collected with single photon counters to obtain the photon correlation histogram shown in the inset of Fig. 9.12b. The five peaks around delay $\tau = 0$ correspond to the different possible coincidences on the beamsplitter of the leading and trailing photons after passing through the long or short arms $\mathcal{L}$ or $\mathcal{S}$ of the interferometer. If the two photons collide and are identical, then the bosonic symmetry of the state predicts that they must exit in the same port. This photon bunching manifests itself as *anti*-bunching in a correlation measurement on the two ports. This signature of photon indistinguishability is apparent in Fig. 9.12b in the reduced peaks near zero time delay. Following the analysis of Santori et al. [54], the data (inset Fig. 9.12b) indicate a mean wavefunction overlap of $I = 0.67 \pm 0.18$, where we adjusted for the imperfect visibility (88%) of our setup and subtracted dark counts in the calculation. Even with higher SE rate enhancement, we expect that $I \lesssim 0.80$ for resonantly

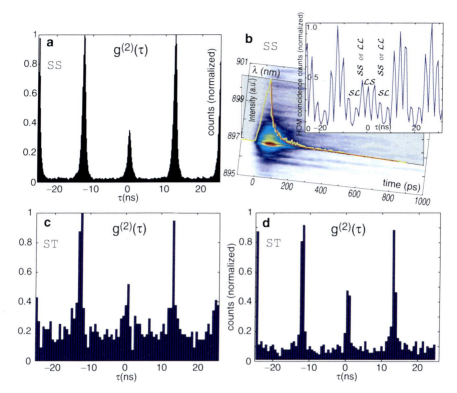

Fig. 9.12. Single photon source characterization. (**a**) Autocorrelation data when cavity S pumped and collected. (**b**) Streak camera data indicate exciton lifetime $\tau = 116$ ps. The rise-time is measured at 23 ps with a lower-density grating with higher time response (data not shown). *Inset:* Two-photon interference experiment (Fig. 9.10). Colliding indistinguishable photons interfere, resulting in a decreased area of peak $\mathcal{LS}$. The area does not vanish largely because of nonzero $g^{(2)}(0)$ of the source. (**c**) Autocorrelation data when cavity S pumped and T is collected (with grating filter). (**d**) Cavity S pumped and T collected directly (no grating filter)

excited QDs [28] because of the finite relaxation time, measured here at 23 ps by the streak camera. This limit to the two-photon indistinguishability was also discussed in Chap. 6 [see in particular (6.16)].

The single photon transfer is described by (9.5), where cavity S is now pumped by the QD single exciton. Letting $g(t), e(t)$ represent the amplitudes of states $|g, c_s = 1, c_w = 0, c_t = 0\rangle, |e, 0, 0, 0\rangle$ corresponding to the QD in the ground and excited states with one or no photons in the source cavity, we have

$$p(t) = -ig_0 e(t), \qquad (9.6)$$
$$\dot{e}(t) = -\frac{\Gamma}{2} e(t) - ig_0 g(t).$$

318   D. Englund et al.

Here, g_0 is again the QD-field coupling strength and Γ the QD SE rate. In the present situation, where the structure's coupling rates $\kappa_\perp, \kappa_\parallel \sim 1/1\,\mathrm{ps}$ greatly exceed the exciton decay rate $\sim 1/116\,\mathrm{ps}$, the steady-state results apply, and the signal from cavity T mirrors the SE of the single exciton coupled to cavity S. Experimentally, we verified photon transfer from S to T by spectral measurements as in Fig. 9.11b: the exciton line is observed from T only if S is pumped. It is not visible if the waveguide or cavity T itself are pumped, indicating that this line originates from the QD coupled to cavity S and that a fraction of the emission is transferred to T. This emission has the same polarization and temperature-tuned wavelength dependence as emission from S. Photon autocorrelation measurements on the signal from T indicate the antibunching characteristic of a single emitter when S is pumped (Fig. 9.12c). The signal-to-noise ratio is rather low because autocorrelation count rates are $(|c_t/c_s|^2)^2 \sim 0.014$ times lower than for collection from S. Nevertheless, the observed antibunching does appear higher, in large part because the background emission from cavity S is additionally filtered in the transfer to T, as shown in Fig. 9.11b. Indeed, this filtering through the waveguide/cavity system suffices to bypass the spectrometer in the HBT setup (a 10-nm bandpass filter was used to eliminate room lights). The count rate is about three times higher while antibunching, $g^2(0) = 0.50 \pm 0.11$, is still clearly evident (Fig. 9.12d). The largest contribution to $g^2(0)$ comes from imperfectly filtered PL near the QD distribution peak seen in Fig. 9.11a. This transmission appears to occur through the dielectric band and could be eliminated. The on-chip filtering will be essential in future QIP applications and should also find uses in optical communications as a set of cascaded drop filters.

9.3.2 Coherent Optical Dipole Access in a Cavity

Several proposals for scalable quantum information networks and quantum computation rely on direct probing of the cavity-QD coupling by means of resonant light scattering from strongly or weakly coupled QDs [1, 2, 5, 67–69]. Such experiments were performed in atomic systems [10–12] and superconducting circuit QED systems [70]. In 2007, we showed that this interaction can also be probed in solid-state systems and that the QD strongly modifies the cavity transmission and reflection spectra, as predicted [68]. A similar result was simultaneously reported by Srinivasan and Painter in a QD strongly coupled to a microdisk resonator [71]. When the QD is coupled to the cavity, a weak laser beam that is resonant with its transition is prohibited from coupling to the cavity. We observe this effect at different detunings between the dot and cavity. As the average probe photon number approaches unity inside the cavity, we observe a giant optical nonlinearity as the QD saturates. This technique, which we call CODAC, represents a major step toward quantum devices based on coherent light scattering and large optical nonlinearities from QDs in photonic crystal cavities.

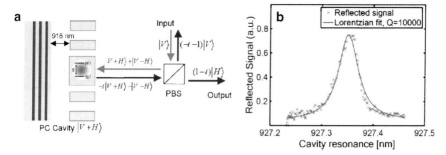

Fig. 9.13. Coherent optical probing. (a) The QD/cavity system is probed with a vertical ($|V\rangle$-polarized) probe laser directed onto the linearly polarized cavity oriented at 45° ($|V + H\rangle$). QD and cavity are tuned by LLH with the 980-nm laser. Upon interaction with the cavity, the $|H + V\rangle$ component of the probe beam is reflected with a frequency-dependent coefficient $-t(\omega)$. The $|V - H\rangle$ component reflects directly with a π phase shift. The polarizing beam splitter (PBS) passes $|H\rangle$, giving a signal that is proportional to $|1 - t|^2$ on the detector (see (9.7)). (b) Measured reflectivity of the empty cavity

In the experiment, a narrow-bandwidth laser beam is scanned through the resonance of the L3 photonic crystal cavity described in Sect. 9.2.3 ($\lambda_{\text{cav}} \sim 926$ nm, $Q = 1.0 \times 10^4$) which is strongly coupled to a QD with polariton splitting of 0.05 nm. The principle of the measurement is illustrated in Fig. 9.13a. Since only a fraction of the probe signal couples into the cavity, the signal reflected by the cavity is monitored in cross-polarization to reduce background [72]. This is analogous to observing the transmission through a polarizing cavity inserted between two crossed polarizers. A GaAs/AlAs DBR underneath the PC membrane effectively creates a single-sided cavity system and enhances collection efficiency of the probe beam. The horizontal $|H\rangle$ component of the scattered probe beam then carries the cavity reflectivity R as given by Eq. (9.7).

The reflectivity is measured by scanning the narrow-linewidth probe laser beam through the cavity resonance, as shown in Fig. 9.13a–b. In this way, we greatly exceed the 0.03 nm resolution of the spectrometer in order to sample the narrow spectral features of the system (i.e., 0.05 nm Rabi splitting). To avoid difficulties related to laser spectral and power stability, we keep the laser wavelength fixed and instead scan the cavity and QD using the LLH technique [24] described in Sect. 9.2.2. The reflectivity signal from a cavity without coupled QDs is shown in Fig. 9.13b. The half-wave plate in Fig. 9.13a orients the probe to 45° to the cavity to maximize the visibility in the reflected signal.

The reflectivity of the QD/cavity system is probed at five different spectral detunings $\Delta\lambda = \lambda - \lambda_{\text{SC}}$ of the probe laser (a narrow linewidth diode laser) from the intersection of QD and cavity, as shown in Fig. 9.8b. The incident power is in the weak excitation limit corresponding to fewer than one photon

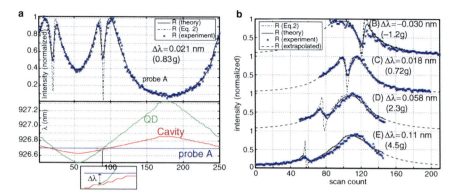

Fig. 9.14. QD-controlled cavity reflectivity at different probe wavelengths A–E, as indicated in Fig. 9.8b. (**a**) Reflectivity spectrum of the probe laser as function of QD and cavity detunings. The probe laser is detuned by $\Delta\lambda = 0.021$ nm (corresponding to $\Delta\lambda = 0.83g$) from the anticrossing point between QD and cavity (see *inset*). Ideal theoretical plots are calculated from (9.7) and convolved to model experimental noise (*solid lines*). (**b**) Probe laser at various detunings $\Delta\lambda$ from the anticrossing point samples different QD-cavity detunings

inside a cavity per cavity lifetime (3 nW measured before the objective lens), as required for probing the Vacuum Rabi splitting. For each reflectivity scan, a corresponding PL scan tracks the QD and cavity wavelengths. Figure 9.14 plots the reflectivity signal as a function of temperature scan. Here the temperature is tuned to sweep the QD and cavity back and forth through the probe laser. This data forms a central result of the measurement: as the single QD sweeps across the cavity, it coherently scatters the probe laser to interfere with the cavity signal. Instead of observing a Lorentzian-shaped cavity spectrum as in Fig. 9.13b, a drop in the reflected signal is observed at the QD wavelength with a linewidth of g^2/κ, as expected [68].

The reflected signal from the cavity is derived following [68] and [2]. The spectrum of the reflected probe signal after the polarizing beam splitter is then given by

$$R = \eta \left| \frac{\kappa}{i(\omega_c - \omega) + \kappa + \frac{g^2}{i(\omega_d - \omega) + \gamma}} \right|^2, \quad (9.7)$$

where η accounts for the efficiency of coupling to the cavity. We fit this relation to the observed spectrum, using the above-mentioned cavity/QD parameters, together with the tracked QD and cavity wavelengths shown in the bottom of Fig. 9.14a. The experimental data show smoother features than the plot of Eq. (9.7) based on tracked QD and cavity lines (dashed line). We attribute this difference to spectral fluctuations in the QD and cavity which are below the resolution limit of the spectrometer, but are greater than the linewidth of the probe beam. These fluctuations arise from instabilities in the power of the heating laser of $\sim 0.7\%$. When these are taken into account in our fit by convolving the theoretical prediction with a Gaussian (FWHM = 0.005 nm), the theoretical model matches the data (black fits). The fits yield values for coupling strength g and cavity Q that agree with PL measurements in above-band pumping given in Sect. 9.2.3. The reflectivity data for the other probe wavelengths, shown in Fig. 9.14b, capture the QD at various detunings from the cavity/QD intersection ranging from $-1.2g(-0.03\,\text{nm})$ to $4.5g(0.11\,\text{nm})$. The reflected probe drops toward zero precisely where the QD crosses its wavelength, and the depth and shape of the drop changes with cavity detuning, as predicted [68]. The spectrum is not explained by an alternate model of an absorbing QD [73]. These measurements also point to one of the advantages of the solid-state cavity QED system: it is possible to capture the spatially fixed QD in various states of detuning, while atomic systems are complicated by moving emitters.

9.3.3 Giant Optical Nonlinearity

In Fig. 9.15, we explore the nonlinear behavior of another strongly coupled QD/PC cavity system as a function of power P_{in} of the probe laser beam. This system shows the same coupling strength as the first, with $g/2\pi = 8\,\text{GHz}$ and $Q = 1 \times 10^4$, and is probed here when the QD is detuned by $\Delta\lambda = -0.012\,\text{nm}$ (corresponding to $-g/2$) from the anticrossing point. P_{in} is increased from the low-excitation limit at 5 nW before the objective (corresponding to average cavity photon number $\langle n_{\text{cav}} \rangle \approx 0.003$ in a cavity without QD) to the high-excitation regime with $P_{\text{in}} \approx 12\,\mu\text{W}$ (corresponding $\langle n_{\text{cav}} \rangle \approx 7.3$). Here, $\langle n_{\text{cav}} \rangle$ is estimated as $\eta P_{\text{in}}/2\kappa\hbar\omega_c$, where $\eta \approx 1.8\%$ is the coupling efficiency into the cavity at this wavelength. Figure 9.15a shows the QD-induced reflectivity dip vanishing as P_{in} is increased roughly by three orders of magnitude. We modeled the saturation behavior by a steady-state solution of the quantum master equation following [74] using independently measured system parameters. Figure 9.15a also plots the calculated normalized reflected intensity as a function of the cavity and QD tuning with temperature (solid line). We see very good agreement when the solution is convolved with the Gaussian filter accounting for spectral fluctuations arising from heating noise, as explained above. The full data set is summarized in Fig. 9.15b, where we plot the reflectivity R at the QD detuning $\Delta\lambda = -0.012\,\text{nm}$, normalized by the reflectivity value R_0 for an empty cavity at the same wavelength as the probe laser (i.e., for $g \to 0$). Our results agree with the theoretical model (solid curve) and previous measurements in atomic systems [75]. Due to the spectral fluctuations, the reflectivity does not approach zero at low power, as it would in the ideal system (dashed curve). Saturation begins at $\sim 1\,\mu\text{W}$ of incident power (measured before the objective), corresponding to $\langle n_{\text{cav}} \rangle \approx 1/2$. Taking into account the coupling efficiency η, this implies a saturation power inside the cavity of only $\sim 20\,\text{nW}$, in agreement with previous predictions for giant optical nonlinearity in a microcavity [76]. We have verified in separate measurements that in the weak coupling regime, saturation of the dip occurs when the probe flux reaches one photon per modified QD lifetime. We furthermore

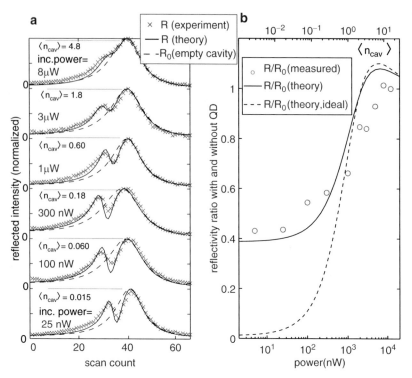

Fig. 9.15. Nonlinear response of a QD–PC cavity QED system. (**a**) The QD-induced feature in cavity reflectivity saturates with increasing probe power (measured before the objective lens). The data are fit by a numerical solution to the full master equation, again convolved to account for thermal fluctuations (*solid curves*). Also plotted is the expected reflectivity R_0 when the QD is removed (*dashed curve*). (**b**) Reflectivity at $\Delta\lambda = -0.012$, normalized by empty-cavity reflectivity at the same wavelength, as a function of probe laser power. Saturation begins near $1\,\mu\mathrm{W}$ of input power, corresponding to intracavity photon number $\langle n_{\mathrm{cav}} \rangle \approx 1/2$. The *dashed curve* shows the reflectivity ratio if no thermal fluctuations were present

verified that the QD-induced reflectivity dip vanishes controllably when excitons are (incoherently) generated by excitation with an above-GaAs-bandgap laser beam.

The coherent QD access enabled by the CODAC technique [26] is essential for QIP in solid-state systems, as it opens the door to high-fidelity controlled phase gates [2], single photon detection [12], coherent transfer of the QD state to photon state [1], and quantum repeaters employing nondestructive Bell measurements with the addition of a third long lived QD level [68]. The observed giant optical nonlinearity has promising applications for generation of nonclassical squeezed states of light [10, 77], nondestructive photon number state measurements [78], and optical signal processing.

9.4 Optoelectronics Applications

The devices presented so far have close derivates for classical applications such as short-range communication in optical interconnects or all-optical signal processing. These devices would operate with a higher QD density than the devices mentioned above.

9.4.1 Ultrafast Quantum Dot–Photonic Crystal Cavity Optical Switch

The photonic crystal cavity is not only suited for emitting light, but carriers induced inside the cavity can also switch a light beam coupled through the cavity. The induced carriers temporarily change the material's refractive index, which results in a shift of the cavity resonance. Such nonlinear optical switching in photonic networks is a promising approach for ultrafast low-power optical data processing and storage. In addition, it might find applications in optical data processing which will be essential for optics-based QIP systems. We observed direct nonlinear optical tuning of photonic crystal PC cavities containing QDs, with a switching time of $\sim$50 ps [18]. Switching via free-carrier generation is limited by the lifetime of free carriers and depends strongly on the material system and geometry of the device. In our case, the large surface area and small mode volume of the PC reduce the lifetime of free carriers in GaAs.

The experimental data obtained from an L3 cavity are shown in Fig. 9.16. We used moderate energy (120 fJ) pulses to shift the cavity by one-half linewidth. Stronger excitation results in higher shifts as indicated by an extremely asymmetric spectrum shown in the inset of Fig. 19.16(d), where 1.4 pJ were used. However, prolonged excitation at this power leads to a reduction in Q over time.

9.4.2 Quantum Dot–Photonic Crystal Laser Dynamics

The enhanced QD SE rate and coupling efficiency β decrease the turn-on time and lasing threshold in PC cavity lasers [79]. Above threshold, higher pump powers lead to faster decay times due to increased stimulated emission rates. Small mode volume PC cavities can be used to achieve large photon densities and speed up this process. Compared to other types of lasers such as VCSELs, PC lasers have lower driving power, increased relaxation oscillation frequency, and potentially faster electrical modulation speed due to lower device capacitance and resistance.

Compared to quantum wells, QD gain media permit lower thresholds due to a reduced active area (though this also lowers saturation power), lower nonradiative surface recombination, and greater temperature stability. The

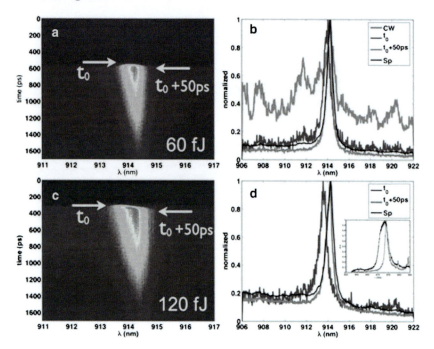

Fig. 9.16. Experimental result of free-carrier cavity tuning. (**a, b**) Wavelength vs. time plots of the cavity, while the cavity is pumped. (**c, d**) Normalized spectra of the cavity at different time points from the data (**a, b**). In (**a**), the cavity is always illuminated by a light source and pulsed with a 3 ps Ti:sapph-generated pulse. Panel (**b**) shows the normalized cavity spectrum at the peak of the free carrier distribution $t = 0$ and 50 ps later. We verify that the cavity tunes at the arrival at the pulse by combining the pulsed excitation with a weak cw above-band pump. The emission due to the cw source is always present, and this very weak emission is reproduced in panel (**b**) as the broad background with a peak at the cold cavity resonance in (**b**)

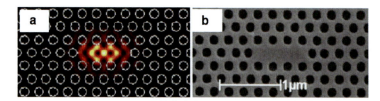

Fig. 9.17. Three hole defect cavity for a QD–PC laser. (**a**) |E|-field pattern. (**b**) Fabricated structure in GaAs with a central InGaAs QD layer

speed of QD lasers is set by the smaller of carrier relaxation rate $1/\tau_{E,f}$ and relaxation oscillation rate ω_R. In ultrasmall, high-Q photonic crystal cavities, the lasing response is sped up through the Purcell effect, while large β and thus higher efficiency and lower threshold are achieved. We investigated these aspects in the 135 nm thick GaAs PC membrane shown in Fig. 9.17, containing a high density (600 μm^{-2}) layer of InGaAs QDs.

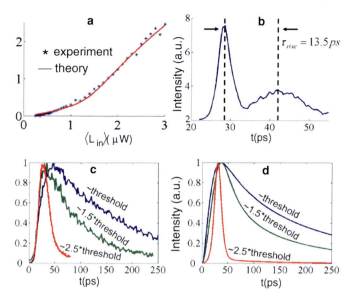

Fig. 9.18. GaAs PC laser with InAs QD gain. (**a**) Measured lasing curve and fit by a rate equations model. (**b**) The measured turn-on delay between pump (first peak) and laser response (second peak) is limited by carrier relaxation. (**c**) Measured large-signal modulation speed increases with pump power. (**d**) Corresponding fit by rate equations

We measure a gradual onset of lasing near 1 µW, as shown in Fig. 9.18a. From fits to the lasing curve, we estimate a SE coupling factor $\beta \sim 0.2$. Streak camera measurements of the QD PL rise time in the unpatterned sample indicate that the carrier relaxation time $\tau_{E,f} \sim 10$ ps for a wide range of pump powers. We also find that resonant pumping of higher-order confined states of the QDs (such as p-level states) does not appreciably lower $\tau_{E,f}$. Because the carrier capture time is longer than the cavity photon lifetime, it ultimately determines the maximum modulation bandwidth. This is what we observe in Fig. 9.18b which shows a delay of 13.5 ps (at five times threshold) and does not drop below 12 ps for higher powers. Simulations support this observation as rise time is limited by the carrier capture time. In our cavity-QED-enhanced structure, the relaxation-time limit is rapidly reached in the high-β case.

Once lasing is reached, stimulated emission causes fast carrier recombination. We measured a minimum decay time of 8.5 ps at pump powers around five times threshold (Fig. 9.18c). For higher pump powers the laser response appears largely unchanged, presumably due to carrier saturation. We again model the system with a rate equations model, employing a linear gain model and parameters given in [17]. Figure 9.18d shows the simulated laser response at various pump powers, demonstrating good agreement between theory and experiment.

Though the present work predicts that large-signal modulation in these present PC lasers employing conventional self-assembled In(Ga)As QDs is limited to ∼30 GHz due to relaxation dynamics, advances in QD growth could lead to higher performance. Recent advances in tunnel injection quantum dots show relaxation times of 1.7 ps at room temperature, and were recently demonstrated in ridge waveguide lasers with 25 GHz small-signal modulation bandwidth [80]. This bandwidth may be significantly improved using a PC laser cavity. In addition, reduction in the inhomogeneous linewidth broadening and reduction in hot-carrier effects and associated gain compression [81], will improve PC QD laser efficiency and speed.

Practical PC lasers will require electrical pumping. Many groups are currently pursuing electrical pumping; one electrically driven device has already been demonstrated recently [82]. For high-speed electrical modulation, the challenge is to keep RC time constants small, where C and R are the capacitance and resistance of the laser. It appears that very fast electrical pumping of nanocavity lasers is possible, as a recent experiment achieved time constants below 10 ps using micron-scale contacts with sub-femtofarad capacitance [83]. If fast electrical pumping can be achieved, then PC crystal lasers may well fill a growing need for integrated, ultrafast optical communication.

9.5 Conclusions

In conclusion, photonic crystals grant tremendous control over embedded emitters and guided photons. This control makes them promising for a new generation of low-power, ultrafast communication devices. For the same reasons, QDs coupled to photonic crystals represent a promising architecture for creating quantum networks. This approach has enabled sources of on-demand, single, nearly indistinguishable photons. We also showed that the PC system is well suited for integrating several photonic functions into photonic crystal networks, which promise to greatly improve coupling efficiencies of quantum networks. Recently we achieved coherent probing and a giant optical nonlinearity with QDs strongly coupled to PC cavities. With these steps, and a toolkit of independent control of QDs and photonics on the chip, we are hopeful to be entering a new era of coherent control of on-chip quantum systems that may contribute to quantum computing and secure long-range communication.

References

1. J.I. Cirac, P. Zoller, H.J. Kimble, H. Mabuchi, Phys. Rev. Lett. **78**(16), 3221 (1997)
2. L.M. Duan, H.J. Kimble, Phys. Rev. Lett. **92**(12), 127902 (2004). doi:10.1103/PhysRevLett.92.127902

3. C. Monroe, D.M. Meekhof, B.E. King, W.M. Itano, D.J. Wineland, Phys. Rev. Lett. **75**(25), 4714 (1995)
4. J. Chiaverini, D. Leibfried, T. Schaetz, M.D. Barrett, R.B. Blakestad, J. Britton, W.M. Itano, J.D. Jost, E. Knill, C. Langer, R. Ozeri, D.J. Wineland, Nature **432**, 602 (2005)
5. A. Imamoglu, D.D. Awschalom, G. Burkard, D.P. DiVincenzo, D. Loss, m. Sherwin, A. Small, Phys. Rev. Lett. **83**(20), 4204 (1999)
6. D.A. Fattal, Single photons for quantum information processing. Ph.D. thesis, Stanford University (2005)
7. W. Yao, R.B. Liu, L.J. Sham, Phys. Rev. Lett. **95**, 030504 (2005)
8. B.B. Blinov, D.L. Moehring, L.M. Duan, C. Monroe, Nature **428**, 153 (2004)
9. Q. Turchette, C. Hood, W. Lange, H. Mabuchi, H.J. Kimble, Phys. Rev. Lett. **75**, 4710 (1995)
10. K.M. Birnbaum, A. Boca, R. Miller, A.D. Boozer, T.E. Northup, H.J. Kimble, Nature **436**, 87 (2005)
11. A. Rauschenbeutel, G. Nogues, S. Osnaghi, P. Bertet, M. Brune, J.M. Raimond, S. Haroche, Phys. Rev. Lett. **83**, 5166 (1999)
12. G. Nogues, A. Rauschenbeutel, S. Osnaghi, M. Brune, J.M. Raimond, S. Haroche, Nature **400**, 239 (1999)
13. D. Englund, D. Fattal, E. Waks, G. Solomon, B. Zhang, T. Nakaoka, Y. Arakawa, Y. Yamamoto, J. Vučković, Phys. Rev. Lett. **95**, 013904 (2005)
14. T. Yoshie, A. Scherer, J. Hendrickson, G. Khitrova, H.M. Gibbs, G. Rupper, C. Ell, O.B. Shchekin, D.G. Deppe, Nature **432**, 200 (2004)
15. K. Hennessy, A. Badolato, M. Winger, D. Gerace, M. Atature, S. Gulde, S. Falt, E.L. Hu, A. Imamoglu, Nature **445**, 896 (2007)
16. D. Englund, A. Faraon, B. Zhang, Y. Yamamoto, J. Vučković, Opt. Express **15**, 5550 (2007)
17. B. Ellis, I. Fushman, D. Englund, B. Zhang, Y. Yamamoto, J. Vučković, Appl. Phys. Lett. **90**, 151102 (2007)
18. I. Fushman, E. Waks, D. Englund, N. Stoltz, P. Petroff, J. Vučković, Appl. Phys. Lett. **90**(9), 091118 (2007)
19. D. Englund, J. Vučković, Opt. Express **14**(8), 3472 (2006)
20. J. Vučković, C. Santori, D. Fattal, M. Pelton, G. Solomon, Y. Yamamoto, in *Optical Microcavities*, edited by K. Vahala (World Scientific, Singapore, 2004)
21. D. Englund, I. Fushman, J. Vučković, Opt. Express **12**(16), 5961 (2005)
22. J. Vučković, Y. Yamamoto, Appl. Phys. Lett. **82**(15), 2374 (2003)
23. P. Michler (ed.), *Single Quantum Dots: Fundamentals, Applications, and New Concepts*. Topics in Applied Physics (Springer, Berlin, 2003)
24. A. Faraon, D. Englund, I. Fushman, N. Stoltz, P. Petroff, J. Vučković, Appl. Phys. Lett. **90**, 213110 (2007)
25. A. Faraon, D. Englund, D. Bulla, B. Luther-Davies, B.J. Eggleton, N. Stoltz, P. Petroff, J. Vučković, Appl. Phys. Lett. **92**, 043123 (2008)
26. D. Englund, A. Faraon, I. Fushman, N. Stoltz, P. Petroff, J. Vučković, Nature **450**(6), 857 (2007)
27. P.M. Petroff, S.P. DenBaars, Superlattices Microstruct. **15**(1), 15 (1994)
28. J. Vučković, D. Englund, D. Fattal, E. Waks, Y. Yamamoto, Physica E **32**, 466 (2006)
29. Y. Akahane, T. Asano, B.S. Song, S. Noda, Nature **425**, 944 (2003)
30. A. Badolato, K. Hennessy, M. Atatüre, J. Dreiser, E. Hu, P. Petroff, A. Imamoglu, Science **308**(5725), 1158 (2005)

31. S. Malik, C. Roberts, R. Murray, M. Pate, Appl. Phys. Lett. **71**(14), 1987 (1997)
32. R. Santoprete, P. Kratzer, M. Scheffler, R.B. Capaz, B. Koiller, J. Appl. Phys. **102**(2), 023711 (2007)
33. A. Rastelli, S. Ulrich, E.M. Pavelescu, T. Leinonen, M. Pessa, P. Michler, O. Schmidt, Superlattices Microstruct. **36**, 181 (2004)
34. A. Högele, S. Seidl, M. Kroner, K. Karrai, R.J. Warburton, B.D. Gerardot, P.M. Petroff, Phys. Rev. Lett. **93**(21), 17401 (2004)
35. D. Haft, C. Schulhauser, A. Govorov, R. Warburton, K. Karrai, J. Garcia, W. Schoedfeld, P. Petroff, Physica E **13**, 165 (2002)
36. A. Kiraz, P. Michler, C. Becher, B. Gayral, A. Imamoglu, L. Zhang, E. Hu, W. Schoenfeld, P. Petroff, Appl. Phys. Lett. **78**(25), 3932 (2001)
37. S. Seidl, M.K. amd Alexander Högele, K. Karrai, R.J. Warburton, A. Badolato, P.M. Petroff, Appl. Phys. Lett. **88**, 203113 (2006)
38. W. Fon, K. Schwab, J. Worlock, M. Roukes, Phys. Rev. B **66**, 045302 (2002)
39. C. Thurmond, J. Electrochem. Soc. **122**, 1133 (1975)
40. A. Rastelli, A. Ulhaq, S. Kiravittaya, L. Wang, A. Zrenner, O. Schmidt, Appl. Phys. Lett. **90**, 073120 (2007)
41. K. Hennessy, A. Badolato, A. Tamboli, P. Petroff, E. Hu, M. Atature, J. Dreiser, A. Imamoglu, Appl. Phys. Lett. **87**, 021108 (2005)
42. S. Mosor, J. Hendrickson, B.C. Richards, J. Sweet, G. Khitrova, H.M. Gibbs, T. Yoshie, A. Scherer, O.B. Shchekin, D.G. Deppe, Appl. Phys. Lett. **87**(14), 141105 (2005)
43. K. Srinivasan, O. Painter, Appl. Phys. Lett. **90**(3), 031114 (2007)
44. S. Strauf, M.T. Rakher, I. Carmeli, K. Hennessy, C. Meier, A. Badolato, M.J.A. DeDood, P.M. Petroff, E.L. Hu, E.G. Gwinn, D. Bouwmeester, Appl. Phys. Lett. **88**, 043116 (2006)
45. C. Grillet, C. Monat, C.L. Smith, B.J. Eggleton, D.J. Moss, S. Frédérick, D. Dalacu, P.J. Poole, J. Lapointe, G. Aers, R.L. Williams, Opt. Express **15**(3), 1267 (2007)
46. S. Song, S.S. Howard, Z. Liu, A.O. Dirisu, C.F. Gmachl, C.B. Arnold, Appl. Phys. Lett. **89**, 041115 (2006)
47. M.W. Lee, C. Grillet, C.L. Smith, D.J. Moss, B.J. Eggleton, D. Freeman, B. Luther-Davies, S. Madden, A. Rode, Y. Ruan, Y. hee Lee, Opt. Express **15**(3), 1277 (2007)
48. A. Schulte, C. Rivero, K. Richardson, K. Turcotte, V. Hamel, A. Villeneuve, T. Galstain, R. Vallee, Opt. Commun. **1**(198), 125 (2001)
49. S. Kugler, J. Hegedus, K. Kohary., J Mater Sci: Mater Electron **18**, S163 (2007)
50. J.C. Sturm, H. Manoharan, L.C. Lenchyshyn, M.L.W. Thewalt, N.L. Rowell, J.P. Noël, D.C. Houghton, Phys. Rev. Lett. **66**(10), 1362 (1991)
51. C. Santori, P. Tamarat, P. Neumann, J. Wrachtrup, D. Fattal, R.G. Beausoleil, J. Rabeau, P. Olivero, A.D. Greentree, S. Prawer, F. Jelezko, P. Hemmer, Phys. Rev. Lett. **97**(24), 247401 (2006)
52. H.J. Kimble, in *Cavity Quantum Electrodynamics*, edited by P. Berman (Academic, San Diego, 1994)
53. M.O. Scully, M.S. Zubairy, *Quantum Optics* (Cambridge University Press, Cambridge, 1997)
54. C. Santori, D. Fattal, J. Vučković, G.S. Solomon, Y. Yamamoto, Nature **419**(6907), 594 (2002)
55. J. McKeever, A. Boca, A.D. Boozer, J.R. Buck, H.J. Kimble, Nature **425**, 268 (2003)

56. A. Kiraz, M. Atatüre, I. Imamoglu, Phys. Rev. A **69**, p.032305 (2004)
57. Y. Xu, J. Vučković, R.K. Lee, O.J. Painter, A. Scherer, A. Yariv, J. Opt. Soc. Am. B **16**(3), 465 (1999)
58. R.K. Lee, Y. Xu, A. Yariv, J. Opt. Soc. Am. B **17**(8), 1438 (2000)
59. C.F. Wang, A. Badolato, I. Wilson-Rae, P.M. Petroff, E. Huc, J. Urayama, A. Imamoglu, Appl. Phys. Lett. **85**(16), 3423 (2004)
60. J.P. Reithmaier, G. Sek, A. Loffler, C. Hofmann, S. Kuhn, S. Reitzenstein, L.V. Keldysh, V.D. Kulakovskii, T.L. Reinecke, A. Forchel, Nature **432**, 197 (2004)
61. D. Bouwmeester, J. Pan, K. Mattle, M. Eibl, H. Weinfurter, A. Zeilinger, Nature **390**, 575 (1997)
62. E. Knill, R. Laflamme, G.J. Milburn, Nature **409**, 4652 (2001)
63. M.A. Nielsen, I.L. Chuang, *Quantum Computation and Quantum Information* (Cambridge University Press, Cambridge, 2000)
64. P. Grangier, B. Sanders, J. Vučković, New J. Phys. **6** (2004)
65. A. Faraon, E. Waks, D. Englund, I. Fushman, J. Vučković, Appl. Phys. Lett. **90**, 073102 (2007)
66. S. Laurent, S. Varoutsis, L. Le Gratiet, A. Lemaître, I. Sagnes, F. Raineri, A. Levenson, I. Robert-Philip, I. Abram, Appl. Phys. Lett. **87**, 3107 (2005)
67. L. Childress, J.M. Taylor, A.S. Sorensen, M.D. Lukin, Phys. Rev. A **72**, 052330 (2005)
68. E. Waks, J. Vučković, Phys. Rev. Lett. **96**, 153601 (2006)
69. T.D. Ladd, P. van Loock, K. Nemoto, W.J. Munro, Y.Yamamoto, New J. Phys. **8**, 184 (2006)
70. D.I. Schuster, A.A. Houck, J.A. Schreier, A. Wallraff, J.M. Gambetta, A. Blais, L. Frunzio, J. Majer, B. Johnson, M.H. Devoret, S.M. Girvin, R.J. Schoelkopf, Nature **445**, 515 (2007)
71. K. Srinivasan, O. Painter, Nature **450**, 862 (2007)
72. H. Altug, J. Vučković, Opt. Lett. **30**(9), 982 (2005)
73. B.D. Gerardot, S. Seidl, P.A. Dalgarno, M. Kroner, K. Karrai, A. Badolato, P.M. Petroff, R.J. Warburton, Appl. Phys. Lett. **90**, 221106 (2007)
74. S.M. Tan, J. Opt. B: Quantum Semiclass. Opt. **1**, 424 (1999)
75. C.J. Hood, M.S. Chapman, T.W. Lynn, H.J. Kimble, Phys. Rev. Lett. **80**(19), 4157 (1998)
76. A. Auffeves-Garnier, C. Simon, J. Gerard, J.P. Poizat, Phys. Rev. A **75**, 053823 (2007)
77. J.E. Reiner, W.P. Smith, L.A. Orozco, H.J. Carmichael, P.R. Rice, J. Opt. Soc. Am. B **18**, 1911 (2001)
78. N. Imoto, H. Haus, Y. Yamamoto, Phys. Rev. A **32**, 2287 (1985)
79. H. Altug, D. Englund, J. Vučković, Nat. Phys. **2**, 484 (2006)
80. S. Fathpour, Z. Mi, P. Bhattacharya, J. Phys. D Appl. Phys. **38**(13), 2103 (2005)
81. D.R. Matthews, H.D. Summers, P.M. Smowton, M. Hopkinson, Appl. Phys. Lett. **81**(26), 4904 (2002)
82. H. Park, S. Kim, S. Kwon, Y. Ju, J. Yang, J. Baek, S. Kim, Y. Lee, Science **305**, p. 1444 (2004)
83. R. Schmidt, U. Scholz, M. Vitzethum, R. Fix, C. Metzner, P. Kailuweit, D. Reuter, A. Wieck, M.C. Hübner, S. Stufler, A. Zrenner, S. Malzer, G.H. Döhler, Appl. Phys. Lett. **88**(12), 121115 (2006)

10

Optical Spectroscopy of Spins in Coupled Quantum Dots

Matthew F. Doty, Michael Scheibner, Allan S. Bracker, and Daniel Gammon

Summary. Vertically stacked pairs of quantum dots can be coupled together by the tunneling of either electrons or holes. Such pairs of coupled dots provide a unique opportunity to study the interactions between spins. We review experimentally measured spectra of coupled quantum dots and explain the interactions that give rise to the spin fine structure. The formation of molecular states through tunnel coupling also leads to unique properties for spins confined in these molecular states. We review and explain resonant changes in the single-spin g factor for holes in the molecular states of coupled quantum dots. The orbital character of molecular states is determined by spin–orbit interactions.

10.1 Introduction

The coupling of single quantum dots (QDs) has been an exciting area of research for many years. [1,2] The interest in this area has been motivated both by curiosity about the fundamental physical mechanisms of coupling and by possible applications in complex new devices. Because of lattice mismatch and strain, sequentially grown layers of self-assembled InAs quantum dots align vertically [3–5]. Using this technique, pairs of dots can be grown with a separating barrier of controllable height. These coupled quantum dots, also called quantum dot molecules (QDMs), have been studied by a variety of techniques [6]. The first optical studies of single QDMs found spectroscopic evidence of coupling between the dots through magneto-PL experiments [7]. This coupling is mediated by electron or hole tunneling between the dots and appears in PL spectra as anticrossings, which are described in Sect. 10.3.2. More detailed studies became possible when the QDMs were embedded into a diode structure that enabled tuning of the relative energy levels of the two dots by an applied electric field [8, 9]. In 2006 Stinaff et al. identified the spectroscopic signatures of coupling in charged QDMs [10]. Since then, substantial progress has been made in understanding the properties of QDMs [11–13], including the origins of spin fine structure [14, 15] and the properties of spins delocalized over the entire QDM [16, 17]. In parallel, there has been substantial progress in calculating the optical spectra of QDMs [18–28].

In this chapter we review the formation of coupled quantum dots, and discuss methods for controllably selecting tunnel coupling of electrons or holes. We review the study of individual QDMs by photoluminescence spectroscopy and provide theoretical models that explain the spin fine structure observed in the spectra. We then present techniques and results of magnetophotoluminescence spectroscopy, which can be used to split spin states that are degenerate at zero magnetic field. The magneto-optical spectroscopy reveals the formation of molecular orbital states with electric-field-dependent g factors. The contribution of the spin–orbit interactions, which can be used to control spin-dependent tunneling, are presented. We conclude with a discussion of prospects and directions for future research.

10.2 Growth and Characterization of Quantum Dot Molecules

InAs QDs are formed during molecular beam epitaxial deposition of InAs on GaAs, as described in detail in Chap. 2. The lattice mismatch between InAs and GaAs leads to the formation of randomly distributed InAs clusters, a process called Stranski–Krastanov growth. These InAs clusters can be either pyramidal or lens shaped and can have a large inhomogeneous energy distribution. Some control over the distribution can be obtained by a "cap and flush" technique [29–34]: the InAs clusters are partially covered with a defined thickness of GaAs, followed by raising the temperature of the entire wafer, which removes the exposed InAs. The result is a pancake-shaped dot of defined thickness. This truncation in the z direction provides the strongest confinement and controls the center emission wavelength of the dot ensemble. To make QDMs, a barrier layer (usually GaAs) of defined thickness is grown on top of the truncated dot. The strain induced by the buried InAs creates preferential nucleation sites for a second layer of QDs [3–5], which can again be truncated to a specific height. Consequently, it is possible to grow two dots that are stacked on top of one another in the growth direction with individual control over both the truncation height of each dot and the barrier thickness.

The truncation of the vertical height of the QDs determines the energies of the lowest confined states. Even if the dots are grown under nominally identical conditions, strain and asymmetry generally lead to nondegenerate dots with slightly different energy levels. A schematic depiction of a QDM where the dots have slightly different sizes is shown in Fig. 10.1a and can be compared to a cross-sectional scanning tunneling microscope image of such a QDM in Fig. 10.1b. The difference in confined energy levels is shown by the schematic band diagram of Fig. 10.1c. The energy levels can be tuned into resonance in situ by embedding the QDMs in a Schottky diode structure and applying an electric field in the growth direction. When energy levels

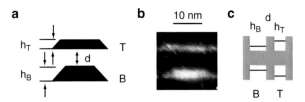

Fig. 10.1. (a) Schematic depiction of a QDM with nondegenerate dots. h_B (h_T) is the height of the bottom (top) dot, d the thickness of the barrier. (b) Cross-sectional STM image of such a QDM. (c) Schematic band diagram showing the relative energy levels of nondegenerate dots. Figure adapted from [35]

are resonant, charge carriers can tunnel between the two dots. Because the dots have nondegenerate energy levels, electron and hole energy levels are not simultaneously on resonance. The particular charge carrier that tunnels is determined by the relative sizes of the two dots and the sign of the applied electric field [35].

QDMs can be studied with photoluminescence spectroscopy techniques developed for single QDs. Like QDs, ensembles of QDMs have a very large distribution of parameters. Consequently, resolving spin fine structure requires isolating single QDMs. One way of isolating individual QDMs is to pattern an aluminum shadow mask on the sample surface. A shadow mask has narrow (~1 μm diameter) apertures, so with an appropriate dot density on average only one dot lies within a single aperture. The sample is illuminated with a continuous wave laser that excites an electron–hole pair. This illumination is not resonant with the energy levels of the QDM; it can be tuned anywhere from above the GaAs bandgap to somewhat below the InAs wetting layer. The excited electron–hole pair relaxes into the low-energy states of the QDM before recombining to emit a photon. The emitted PL is collected, dispersed with a monochromator, and measured, for example with a charge coupled device (CCD) camera.

To generate spectral maps, PL spectra are sequentially accumulated at various values of the applied electric field and then assembled to plot PL intensities as a function of both applied electric field (x-axis) and energy (y-axis). Spectral maps generally contain PL from multiple optically generated charge states. Optical charging occurs when an electron (hole) tunnels out of the QDM before the injected electron–hole pair recombines. The hole (electron) left behind remains in the dot until either an electron (hole) tunnels back into the dot or the excess hole (electron) tunnels out. Using optical charging, we typically observe positively charged excitons in n-type diodes and negatively charged excitons in p-type diodes. This turns out to be useful because we tend to see the entire spectral patterns for these charged features. We have also observed electrically charged excitons (for example negatively charged excitons in n-type diodes), but we typically only observe fragments of the spectral patterns before the charge state changes.

10.3 Photoluminescence Signatures of QDMs

10.3.1 Coulomb Doublets

In Fig. 10.2 we compare a spectral map for a single QD with one for a QDM. Figure 10.2a shows several discrete PL lines from a single QD; each line arises from a unique charge configuration. Figure 10.2b shows similar discrete lines from a QDM, but now each line appears split into a doublet, as shown in Fig. 10.2c. This doublet structure illustrates an important feature of QDMs: charges can be distributed over both QDs in the molecule. If all charges are located in the same dot (for example the bottom), the Coulomb interactions induce energy shifts that mirror those of single QDs. If one of the charges (for example a hole) is located in the top dot, the overall Coulomb energy shift is determined primarily by the electrons and holes within the bottom dot. The presence of the additional hole in the top dot introduces a small energy shift (about 0.5 meV in a QDM with 6 nm barrier). The magnitude of this Coulomb shift decreases with increasing barrier thickness.

We can describe the location of charges using a simple notation. We will consider only the lowest confined energy level for electrons and holes in each dot. There are two possible spatial locations for electrons and two possible spatial locations for holes. We will label the spatial distribution of charges as $\binom{a,b}{c,d}$ where a(b) is the number of electrons in the bottom (top) dot and c(d) is the number of holes in the bottom (top) dot. Using this notation, we can label each element of the doublet structure in Fig. 10.2c. We see that the

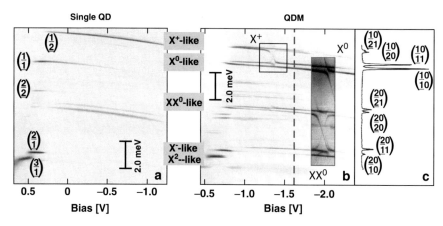

Fig. 10.2. Spectral map of (**a**) a single QD and (**b**) a QDM. (**c**) Line cut showing a PL spectra of a QDM at a single value of the applied bias. The single QD is 2.5 nm high, while the QDM is composed of two nominally symmetric QDs that are 2.5 nm high and separated by a 6 nm GaAs barrier. Both the single QD and the QDM are embedded in n-type diodes

doublets always consist of two elements ($\begin{pmatrix} n_e,0 \\ n_h,0 \end{pmatrix}$ and $\begin{pmatrix} n_e,0 \\ n_h,1 \end{pmatrix}$) that differ only by the presence or absence of an additional hole in the top dot.

The presence of charge carriers in both dots also leads to two classes of excitonic recombination: direct and indirect. When the electron and hole that recombine to emit a photon are located in the same dot, for example $\begin{pmatrix} 1,0 \\ 1,0 \end{pmatrix}$, the resulting PL line is called a "direct" exciton. Direct excitons have a weak energy dependence on the applied electric field that results from the single-dot Stark shift. Direct excitons appear in spectral maps as nearly horizontal lines. The observed PL doublets are all direct excitons. In contrast, "indirect" excitons consist of electrons and holes located in different dots (for example $\begin{pmatrix} 1,0 \\ 0,1 \end{pmatrix}$). Indirect excitons have a strong energy dependence on applied electric field (F) because of the spatial separation between the dots. Indirect excitons therefore appear in spectral maps as diagonal lines. The slope of the line is determined by the thickness of the barrier (d) separating the two dots:

$$\frac{\Delta E}{\Delta F} = ed, \tag{10.1}$$

where e is the charge of the electron. In samples with d between 2 and 20 nm ed ranges from about 0.3–2 meV kV^{-1} cm, as shown in Fig. 10.3. To differentiate between direct and indirect excitons, we underline the numbers that indicate the spatial location of the recombining electron and hole.

10.3.2 Anticrossings

PL doublets are the signature of Coulomb interactions in QDMs when the energy levels of the electrons and holes in the two dots are out of resonance. When the energy levels are tuned into resonance, electrons or holes can tunnel between the two dots. Coherent tunneling couples the states in

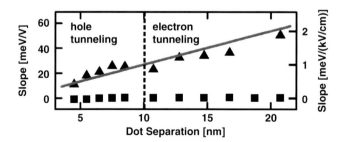

Fig. 10.3. Slope ($\Delta E/\Delta F$) of PL lines as a function of barrier thickness. *Squares* indicate the slopes of direct excitons, while *triangles* indicate the slopes of indirect excitons. The *solid line* is a predicted energy dependence, p, of indirect excitons calculated from the dot geometry: $p = ed$

each dot and results in the formation of molecular orbitals delocalized over both dots and the intervening barrier. These molecular states are constructed from the symmetric and antisymmetric combination of the basis states, i.e., $\frac{1}{\sqrt{2}}\left(\binom{1,0}{1,0} + \binom{1,0}{0,1}\right)$ and $\frac{1}{\sqrt{2}}\left(\binom{1,0}{1,0} - \binom{1,0}{0,1}\right)$. The formation of these molecular orbitals results in anticrossings between the direct and indirect excitons as a function of applied electric field. The anticrossings can be seen in photoluminescence spectral maps [8], for example the boxed regions of Fig. 10.2b.

Anticrossings in raw spectral maps are difficult to interpret for two reasons. First, because optical charging is a random process, a spectral map contains many different overlapping charge configurations. The particular energy of each charge configuration is determined by Coulomb interactions. Specific charge configurations can be identified from their relative energies and spin fine structure [15]. Second, the intensities of photoluminescence lines can vary by several orders of magnitude as a function of applied electric field.

The simplest anticrossing arises for the case of the neutral exciton. In Fig. 10.4 we show the anticrossing of a neutral exciton where the hole tunnels through a 2 nm GaAs barrier. The direct and indirect PL lines are indicated by insets on the left-hand side of Fig. 10.4, which show the spatial distributions of the electron and hole. The insets on the right illustrate the formation of bonding and antibonding molecular states at the anticrossing.

We can model the anticrossing of the neutral exciton in a QDM using a simple matrix Hamiltonian. We consider the case in which the electron remains trapped in the bottom dot and the hole can be located in either the top or

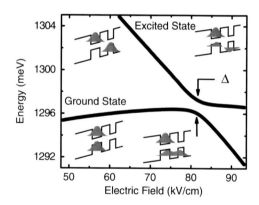

Fig. 10.4. Energies of observed photoluminescence lines for the neutral exciton show an anticrossing as a function of applied electric field due to holes tunneling through a 2 nm barrier. The QDs are both nominally 2.5 nm high and embedded in an n-type diode. The anticrossing energy gap has magnitude Δ. *Insets on the left* show the spatial distributions of electrons and holes for a direct (lower) and indirect (upper) exciton. *Insets on the right* illustrate the formation of delocalized molecular states with bonding and antibonding character. Figure taken from [17]

the bottom dot. There are two basis states: $\begin{pmatrix} 1,0 \\ 1,0 \end{pmatrix}$ and $\begin{pmatrix} 1,0 \\ 0,1 \end{pmatrix}$. In this basis, the Hamiltonian is

$$\widehat{H}^{X^0} = \begin{pmatrix} E_0 & -t_{X^0} \\ -t_{X^0} & E_0 - edF \end{pmatrix} \tag{10.2}$$

with E_0 the exciton energy and t_{X^0} the tunneling matrix element. t_{X^0} is determined by the overlap of the hole wavefunctions in the two dots and consequently depends on barrier thickness. For relatively small barriers, the tunneling matrix element can also be influenced by Coulomb interactions with any electrons, which distort the wavefunction amplitudes. Consequently in samples with a thin barrier t_{X^0} differs from t_h, the tunneling of a single hole [10]. $-edF$ captures the change in energy with applied electric field (F) for the hole in the top dot separated from the bottom dot by an effective barrier thickness d.

Equation (10.2) considers the states of QDMs using atomic-like basis states coupled together via tunneling. This method is quite similar to the Heitler–London method that can be used to analyze the states of simple diatomic molecules like H_2 [36]. An alternative way to consider such coupled states is the *molecular orbital* method [37]. The molecular orbital approach starts by considering the single-particle eigenstates of the molecule: the bonding and antibonding states. The state of a charged molecule is then determined by filling these molecular orbitals with particles beginning with the lowest energy state. In the hydrogen molecule, the approach that determines the proper ground state depends on the separation between the nuclei, which determines the degree of tunnel coupling between the two nuclear potentials. In natural molecules, of course, the nuclear spacing cannot be varied. In QDMs, however, the degree of tunnel coupling is varied by the applied electric field. For this reason, the states of QDMs evolve continuously between atomic-like and molecular orbitals as a function of applied electric field.

In (10.2) the basis states are simply the two spatial locations of the particle: a hole in the bottom dot or a hole in the top dot. The electron always remains in the bottom dot. Away from resonance (large magnitude of F), the tunnel coupling is minimal and the eigenvalues of (10.2) are essentially just the energies of the two basis states. This is the atomic-like or Heitler–London limit. On resonance ($F = 0$), the eigenvalues of (10.2) are $E_0 - t_{X^0}$ and $E_0 + t_{X^0}$. The eigenstates that correspond to these values are, respectively, the symmetric (bonding) and antisymmetric (antibonding) combination of the two basis states. The relaxation between molecular orbital states can be studied by time-resolved PL [11].

In analogy with natural molecules, one would expect the bonding orbital to have lowest energy. This is the case when t_{X^0} is positive, as it is for natural molecules. However, in QDMs spin–orbit interactions can reverse the sign of t_{X^0} and create an antibonding ground state [17]. We will discuss this effect in detail in Sect. 10.6. For the moment we shall assume that t_{X^0} is positive and the bonding orbital has lowest energy. This case is illustrated in Fig. 10.4.

10.3.3 Excited State Anticrossings

Although our matrix model includes only the lowest energy level in each dot, several higher energy levels are confined in the dots. We find that the PL signatures of anticrossings and spin interactions for the ground state are well understood by this model without the inclusion of higher energy states. The higher energy levels do appear in the spectra as additional anticrossings. Because it is energetically unfavorable for a particle to remain in an excited energy level, we rarely observe PL emitted from a configuration that has a particle in an excited state. However, when an excited state of one dot is in resonance with the ground state of the second dot, tunnel coupling again induces a mixing of the states, which appears as an anticrossing energy gap. In Fig. 10.5 we show a series of anticrossing energy gaps that arise as the ground state hole level of the bottom dot crosses a series of excited hole levels in the top dot. This technique allows for the in situ mapping of excited electron or hole levels in each QD [13].

10.3.4 Controllable Coupling of Electrons or Holes

Thus far we have discussed only the case in which holes tunnel between the two dots. Hole coupling occurs when the top dot of a nominally identical pair of dots is slighter thinner in the growth direction than the bottom dot. Using the cap and flush technique, it is possible to intentionally grow QDMs with different truncation heights for each dot [35]. This intentional difference is larger than the random differences in energy levels for the two dots and can be used to controllably choose whether electrons or holes tunnel. An example

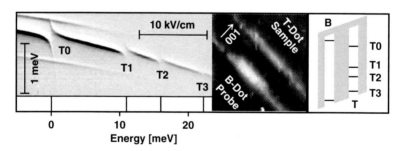

Fig. 10.5. Series of excited state anticrossings for the neutral exciton. The anticrossings arise as the ground state of the hole in the bottom dot is tuned into resonance with ground and excited hole states in the top dot. The QDM is composed of two nominally symmetric QDs that are 2.5 nm high and separated by a 6 nm GaAs barrier, as shown by the cross-sectional STM image in the *middle panel*. The QDM is embedded in an n-type diode. The energy scale on the x-axis in the *left panel* is calculated from the applied electric field by using the measured dipole moment ($\Delta E/\Delta F$) for this QDM. The *right panel* shows how this measured spacing determines the energy levels in the QDM valence band. Figure adapted from [13]

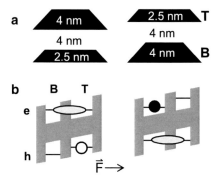

Fig. 10.6. (a) Schematic depiction of two QDMs with dots truncated to different heights. (b) When the taller dot is on top, a positive electric field induces electron tunneling. If the smaller dot is on top, a positive electric field induces hole tunneling. Figure adapted from [35]

is shown in Fig. 10.6. In Fig. 10.6a we show two QDMs; each has one dot truncated to 2.5 nm and one dot truncated to 4 nm. The dots are separated by 4 nm, but the order of the dots is reversed in the two examples. In Fig. 10.6b we show schematic band diagrams of each QDM. The weaker confinement in the taller dot creates electron (hole) energy levels that lie closer to the bottom (top) of the conduction (valence) band. As a consequence, when the taller dot is on top, a positive applied electric field brings the electron levels into resonance. Similarly, when the taller dot is on the bottom a positive applied electric field brings the hole levels into resonance.

In Fig. 10.7a,b we show the calculated relative energy levels of the X^0 states as a function of applied electric field for the two configurations of QDMs shown in Fig. 10.6. There are four anticrossings in each panel. Two anticrossings are for a tunneling electron with the hole confined in either the top or bottom dot. The other two anticrossings are for a tunneling hole with the electron in either the top or bottom dot. Electrons have a lighter effective mass than heavy holes, and therefore have a higher tunneling rate. Consequently, for the same barrier thickness, tunnel coupling of electrons creates a substantially larger anticrossing energy gap. In Fig. 10.7a the taller dot is on top. At negative electric fields hole levels come into resonance and the anticrossing energy gaps are small. At positive electric fields the electron levels come into resonance and the anticrossing energy gaps are larger. In Fig. 10.7b the taller dot is on the bottom and the situation is reversed.

In real samples, the doped substrate makes it impossible to obtain PL spectra at both positive and negative electric fields. For an n-doped substrate, negative applied electric fields flood the QDM with electrons. The situation is similar for positive electric fields applied to p-doped substrates. The choice of substrate therefore restricts a given sample to half of the applied electric fields displayed in Fig. 10.7a,b. Additionally, the asymmetric energy levels and

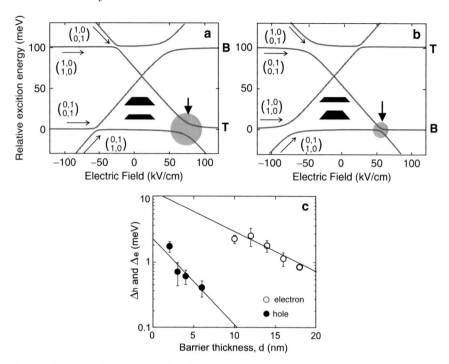

Fig. 10.7. Calculated energy levels for the two QDMs with 4 nm barrier depicted in Fig. 10.6a: (**a**) 4 nm dot above 2.5 nm dot and (**b**) 2.5 nm dot above 4 nm dot. (**c**) Measured anticrossing energies for electrons and holes as a function of barrier thickness. Figure adapted from [35]

applied electric field create a substantial energy difference for the two possible spatial locations of the nontunneling particle. So, for example, in a QDM on an n-doped substrate with the taller dot on the bottom the electron is trapped in the bottom dot while the hole can be in either the bottom or top dot. This restricts real samples to the lower half of the calculated energy levels in Fig. 10.7a,b. On n-doped substrates, only the two anticrossings indicated by the shaded circles in Fig. 10.7a,b are accessible.

Figure 10.7a,b demonstrate that the charge carrier that tunnels can be selected by choosing the relative heights of the two dots and the substrate doping. In Fig. 10.7c we show the measured tunneling matrix elements for electrons and holes in samples with a variety of barrier thicknesses. The tunneling matrix elements are extracted from PL spectra of the singly charged exciton, which will be described further below. The tunneling matrix elements for electrons are nearly an order of magnitude larger than those for holes due to the smaller electron effective mass. The solid line fit to the tunneling of holes in Fig. 10.7c neglects spin–orbit interactions that can have a significant impact on the tunnel coupling. We will discuss the spin–orbit contributions to tunnel coupling in Sect. 10.6.

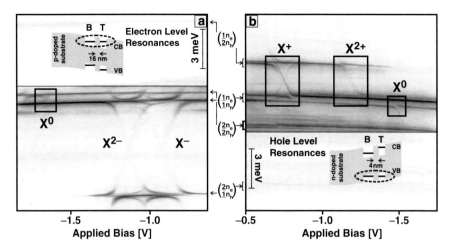

Fig. 10.8. Spectral maps comparing the anticrossings of electrons (**a**) and holes (**b**). In (**a**), the bottom dot is 3.5 nm while the top dot is 2.5 nm. In (**b**), the bottom dot is 4 nm while the top dot is 2.5 nm. Figure adapted from [38]

Using asymmetric molecules to controllably select electron and hole tunneling allows us to compare the observed anticrossing features in the two cases. In Fig. 10.8 we show measured spectral maps for tunneling of electrons (a) and holes (b). In each case, there are families of direct PL lines that have a weak energy dependence on electric field. These families are labeled as, for example, $\left(\begin{smallmatrix}1,n_e\\1,n_h\end{smallmatrix}\right)$. Each family has the same total number of charges in the bottom dot, with differing numbers of electrons or holes in the top dot. The energy of each family is determined by the Coulomb interactions within the bottom dot. The additional charges in the top dot introduce small Coulomb energy shifts between the members of each family, as discussed above in reference to Fig. 10.2.

In each panel of Fig. 10.8 we have labeled spectral features that arise from the neutral (X^0), singly charged (X^+ or X^-) and doubly charged (X^{2+} or X^{2-}) excitons. For the neutral exciton, we observe a simple anticrossing as described above. For the singly and doubly charged excitons, we observe x-shaped patterns.

10.3.5 x Patterns

The x patterns shown in Fig. 10.8 span the energy between two different families of direct transitions. For example, the X^+ x pattern spans the energy between the $\left(\begin{smallmatrix}1,0\\2,0\end{smallmatrix}\right)$ and $\left(\begin{smallmatrix}1,0\\1,1\end{smallmatrix}\right)$ families. These two states are different spatial distributions with the same total charge. Because recombination of an electron–hole pair results in an emitted photon with energy equal to the difference between the optically excited (initial) and optical ground (final) states,

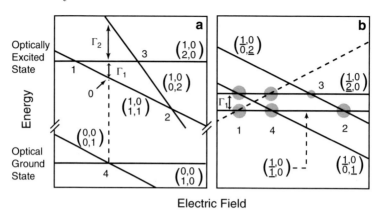

Fig. 10.9. Schematic optical ground and excited state energy levels (**a**) and transition spectrum (**b**) for the positively charged exciton without tunnel coupling

tunnel coupling in both the initial and final states combine to produce the x-shaped pattern. The fine structure of the x patterns conveys information on spin and charge interactions in both the initial and final states. In this section we review the formation of these x patterns. For the sake of clarity and continuity, we will continue to present only the case of tunneling holes where the electron is trapped in the bottom dot. We will start by considering only charges and then add spin interactions in the next section.

In Fig. 10.9a we show schematic energy levels for the optical ground and optically excited states of the singly positively charged exciton, X^+. For schematic purposes, we have temporarily neglected the effects of tunnel coupling. The optical ground state has only a single hole, which must be located in either the bottom or top dot. We reference all energies to the energy of a single hole in the bottom dot, so this state ($\binom{0,0}{1,0}$) has no dependence of energy on applied electric field. If the hole is in the top dot ($\binom{0,0}{0,1}$) the energy of the state depends linearly on the applied electric field. The optically excited states contain one additional electron–hole pair and therefore have an overall energy offset given by the electron–hole Coulomb interaction. The dependence of energy on applied electric field for these states is similar to the optical ground states. $\binom{1,0}{2,0}$ has no energy dependence on applied electric field because all charges are in the bottom dot. The energy of $\binom{1,0}{1,1}$ has a linear dependence on applied electric field because one hole is in the top dot. The energy of $\binom{1,0}{0,2}$ has a linear dependence on electric field with twice the slope because two holes are in the top dot.

In the absence of Coulomb interactions, all three optically excited states in Fig. 10.9a would be degenerate at the point labeled 0. When the electron and both holes are all in the bottom dot, the Coulomb interactions between the electron and the second hole raise the energy of the $\begin{pmatrix}1,0\\2,0\end{pmatrix}$ state by an energy Γ_1, i.e., the energy difference between states $\begin{pmatrix}1,0\\2,0\end{pmatrix}$ and $\begin{pmatrix}0,0\\1,0\end{pmatrix}$ is larger than the energy difference between $\begin{pmatrix}1,0\\1,1\end{pmatrix}$ and $\begin{pmatrix}0,0\\0,1\end{pmatrix}$ by an amount Γ_1. This additional energy shift moves the degeneracy point between $\begin{pmatrix}1,0\\2,0\end{pmatrix}$ and $\begin{pmatrix}1,0\\1,1\end{pmatrix}$ (indicated by number 1) to a slightly lower value of the applied electric field. Similarly, $\begin{pmatrix}1,0\\0,2\end{pmatrix}$ has an additional energy shift Γ_2, which arises from the Coulomb repulsion between the two holes and the weaker interdot electron–hole Coulomb interactions. Consequently, the degeneracy of $\begin{pmatrix}1,0\\1,1\end{pmatrix}$ and $\begin{pmatrix}1,0\\0,2\end{pmatrix}$ (indicated by number 2) shifts to higher electric fields.

In Fig. 10.9b we show a schematic depiction of the transition spectrum, which is determined by taking the energy difference between all optical ground and excited states. The two horizontal lines arise from the two direct recombination lines $\begin{pmatrix}1,0\\2,0\end{pmatrix}$ and $\begin{pmatrix}1,0\\1,1\end{pmatrix}$. These lines are separated by the energy offset Γ_1. The diagonal line with negative slope originates from the indirect recombination $\begin{pmatrix}1,0\\1,1\end{pmatrix}$. The diagonal line with positive slope (dashed) originates from the recombination $\begin{pmatrix}1,0\\2,0\end{pmatrix} \rightarrow \begin{pmatrix}0,0\\0,1\end{pmatrix}$, which is optically allowed only in the vicinity of anticrossings that mix it with other states. The combination of these four lines creates the distinctive x-shaped pattern. The fifth line in the transition spectrum originates from the recombination $\begin{pmatrix}1,0\\0,2\end{pmatrix}$.

When we include tunnel coupling, anticrossings will arise at the energy degeneracy points in both the optical ground and excited states (numbers 1–4). The anticrossing at point 3 will be substantially smaller because it requires a second order tunneling process. Because of the energy shifts (Γ_1 and Γ_2), these degeneracy points happen at different values of the applied electric field for the optical ground and excited state. The six highlighted intersections in Fig. 10.9b mark where anticrossings will appear in the transition spectrum. The two anticrossings on the left (number 1) are due to anticrossing 1 in the optically excited state, while the two on the right (number 4) are due to anticrossing 4 of the optical ground state. At higher values of the electric field we observe two additional anticrossings (2 and 3) that arise from the optically excited state. The spin interactions in each of these states induce unique fine structure in the anticrossings, which will be the subject of the next section. In the case of the singly charged exciton, the optical ground state anticrossings come from the tunneling of a single hole or electron with no spin interactions. It is from these anticrossings that we measure the tunnel rate for bare electrons and holes shown in Fig. 10.7c.

10.4 Spin Fine Structure

To consider spin interactions we will slightly modify our notation; instead of listing only the number of electrons and holes in each dot, we will explicitly list the spin orientation for each particle. We continue to consider only the lowest energy confined states for electrons and holes. Electrons have spin $\pm 1/2$; we will denote these spin projections as $\uparrow$ and $\downarrow$. Strain and confinement in the dots push light-hole states to significantly higher energies, so we consider only the heavy-hole states with spin $\pm 3/2$ which we denote in the pseudospin basis $\Uparrow$ and $\Downarrow$. Strictly speaking, these are hole spinor projections where the heavy-hole state is the dominant spinor component. The spin–orbit interaction mixes in small light-hole components that can change the nature of the tunnel coupling, as we discuss in Sect. 10.6.

When two particles are in the same orbital energy level, the Pauli exclusion principle requires them to be in a spin singlet. When spins are located in separate dots, both singlet and triplet configurations are possible. We will use an additional subscript (S, T) to differentiate between singlet and triplet configurations. As an example of this notation, $\begin{pmatrix} \uparrow,0 \\ \Downarrow,\Uparrow \end{pmatrix}_S$ denotes a positive trion with one spin-up electron in the bottom dot and one hole in each of the bottom and top dots. The holes are in the antisymmetric spin singlet configuration, i.e., $\frac{1}{\sqrt{2}}(|\Downarrow_1 \Uparrow_2\rangle - |\Uparrow_1 \Downarrow_2\rangle)$. Just as in single QDs, optical selection rules require that recombining electrons and holes have opposite spin projections to conserve angular momentum.

We will begin by presenting the consequences of spin interactions at zero magnetic field [10, 14, 15], where all states are two-fold degenerate under a flip of all spins. For example, the optically bright neutral excitons $\begin{pmatrix} \uparrow,0 \\ \Downarrow,0 \end{pmatrix}$ and $\begin{pmatrix} \downarrow,0 \\ \Uparrow,0 \end{pmatrix}$ are degenerate at zero magnetic field (we neglect anisotropic exchange, which is largely below the resolution of our spectrometer). For simplicity, we label figures with only one of the spin-degenerate states.

In Fig. 10.10a we show the simple anticrossing that arises from the formation of molecular states for a single hole. There are two basis states, corresponding to the hole localized in either the top or bottom dot. When these states are degenerate in energy, the tunnel coupling introduces an anticrossing energy gap $\Delta = 2|t_h|$. This anticrossing is described with a simple matrix Hamiltonian in the basis $\begin{pmatrix} 0,0 \\ \Uparrow,0 \end{pmatrix}$ and $\begin{pmatrix} 0,0 \\ 0,\Uparrow \end{pmatrix}$. There can be no Coulomb interactions, so the Hamiltonian is

$$\widehat{H}^{h+} = \begin{pmatrix} 0 & -t_h \\ -t_h & -edF \end{pmatrix}. \tag{10.3}$$

In the next sections, we will describe the spin interactions that arise as more particles are added to the system. The tunneling matrix element depends on both Coulomb interactions and spin–orbit interactions. Coulomb

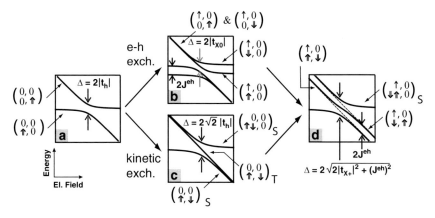

Fig. 10.10. Overview of spin interactions. (**a**) Anticrossing of a single hole. (**b**) Bright and dark exciton states split by the electron–hole exchange energy. (**c**) Kinetic exchange between two holes. (**d**) Competing electron–hole and kinetic exchange interactions for one electron and two holes. Figure adapted from [15]

interactions introduce small corrections to the tunneling rate when holes tunnel in the presence of additional particles. The contributions of the spin–orbit interaction will be discussed in Sect. 10.6.

10.4.1 Electron–Hole Exchange

Just as in single QDs, an electron and hole confined in the same dot experience an electron–hole exchange interaction J^{eh}. This interaction splits the bright and dark exciton states. J^{eh} depends on the overlap of electron and hole wavefunctions, and so is drastically suppressed when the electron and hole are in separate dots. Using the basis $\binom{\uparrow,0}{\Downarrow,0}$, $\binom{\uparrow,0}{\Uparrow,0}$, $\binom{\uparrow,0}{0,\Downarrow}$, and $\binom{\uparrow,0}{0,\Uparrow}$ the X^0 Hamiltonian is

$$\widehat{H}^{X^0} = E_0 + \begin{pmatrix} J^{\text{eh}} & 0 & -t_{X^0} & 0 \\ 0 & -J^{\text{eh}} & 0 & -t_{X^0} \\ -t_{X^0} & 0 & -edF & 0 \\ 0 & -t_{X^0} & 0 & -edF \end{pmatrix}. \quad (10.4)$$

The energies of each individual basis state appear on the diagonal. $\binom{\uparrow,0}{\Downarrow,0}$ has energy J^{eh} because of the antiparallel electron and hole spins. Similarly, $\binom{\uparrow,0}{\Uparrow,0}$ has energy $-J^{\text{eh}}$ because of the parallel electron and hole spins. $\binom{\uparrow,0}{0,\Downarrow}$ and $\binom{\uparrow,0}{0,\Uparrow}$ have an energy $(-edF)$ that depends on applied electric field because they have a single hole in the top dot. The energies of these states do not depend on the relative spins of the electron and hole because the particles are in separate dots. The off-diagonal matrix elements t_{X^0} give the tunnel coupling. Because tunneling is a spin-conserving process, t_{X^0} only couples states with the same hole spin orientation. t_{X^0} differs slightly from t_h, the tunneling matrix element for a single hole, because of the different Coulomb interactions.

In Fig. 10.10b we show calculated energy levels for the neutral exciton. The bright and dark direct exciton states appear as two horizontal lines split by the electron–hole exchange energy. As these lines pass through the anticrossing region and become indirect excitons, the energy splitting between these two lines collapses. The anticrossing formed by the coherent tunneling of the hole leads to an energy gap of $\Delta = 2|t_{X^0}|$. This gap is measured between two states with the same spin projection. The lines in Fig. 10.10b are the calculated energies of all neutral exciton states. In a PL experiment, the dark states cannot emit photons, and thus are not observed. However, application of a Voigt geometry magnetic field mixes the bright and dark states and allows for observation of the dark states [15]. The electron–hole exchange interaction also appears directly in PL spectra of the doubly charged exciton and the neutral biexciton at zero magnetic field [12, 15].

10.4.2 Hole–Hole (Electron–Electron) Exchange

We now consider the interactions between two holes, as shown in Fig. 10.10c. There are three possible spatial distributions: both holes in the bottom dot, one hole in each of the top and bottom dots, and both holes in the top dot. When the holes are located in separate dots, singlet and triplet configurations of the spins are possible. When both holes are in the same dot, the Pauli principle requires that the holes be in a spin singlet. There are six possible spin configurations. To understand the spin fine structure, however, we only need to consider three of these states: $\begin{pmatrix} 0,0 \\ \Downarrow\Uparrow,0 \end{pmatrix}_S$, $\begin{pmatrix} 0,0 \\ \Downarrow,\Uparrow \end{pmatrix}_S$ and $\begin{pmatrix} 0,0 \\ \Downarrow,\Uparrow \end{pmatrix}_T$. Here $\begin{pmatrix} 0,0 \\ \Downarrow,\Uparrow \end{pmatrix}_S$ is the antisymmetric hole spin singlet ($\frac{1}{\sqrt{2}}(|\Downarrow_1\Uparrow_2\rangle - |\Uparrow_1\Downarrow_2\rangle)$) and $\begin{pmatrix} 0,0 \\ \Downarrow,\Uparrow \end{pmatrix}_T$ the symmetric hole spin triplet with net angular momentum zero ($\frac{1}{\sqrt{2}}(|\Downarrow_1\Uparrow_2\rangle + |\Uparrow_1\Downarrow_2\rangle)$). In this basis, the h^{2+} state is described with the following Hamiltonian:

$$\widehat{H}^{h^{2+}} = \begin{pmatrix} 0 & -t_{h^{2+}} & 0 \\ -t_{h^{2+}} & -edF & 0 \\ 0 & 0 & -edF \end{pmatrix} \tag{10.5}$$

In our energy reference frame, the state with both holes in the bottom dot has no dependence in energy on applied electric field, while both states with a single hole in the top dot have an energy $-edF$ that depends on the applied field. Because tunneling is a spin-conserving process, only the $\begin{pmatrix} 0,0 \\ \Downarrow,\Uparrow \end{pmatrix}_S$ singlet state can tunnel couple with $\begin{pmatrix} 0,0 \\ \Downarrow\Uparrow,0 \end{pmatrix}_S$, which must be in a spin singlet because the holes are in the same dot. The triplet state passes through the anticrossing region with no coupling. This triplet state is actually triply degenerate: the additional triplet lines $\begin{pmatrix} 0,0 \\ \Downarrow,\Downarrow \end{pmatrix}_T$ and $\begin{pmatrix} 0,0 \\ \Uparrow,\Uparrow \end{pmatrix}_T$, which have parallel hole spins, behave exactly like the $\begin{pmatrix} 0,0 \\ \Downarrow,\Uparrow \end{pmatrix}_T$ triplet. The magnitude of the anticrossing energy gap is $\Delta = 2\sqrt{2}|t_h|$. The extra factor of $\sqrt{2}$ arises because there are two holes that can tunnel.

The energy separation between singlet and triplet states normally arises from exchange Coulomb interactions. In this case, it is a "kinetic exchange" interaction that arises primarily from the tunnel coupling of singlets, but not triplets [36]. This kinetic exchange interaction is different from the usual exchange Coulomb interaction. First, the singlet line has both upper and lower branches. The branches result from the formation of molecular orbitals with different symmetry: the bonding and antibonding combinations of the singlet basis states. Second, the degree of separation between singlet and triplet states depends on the applied electric field. The separation reaches its maximum at the electric field where the states would be degenerate in the absence of tunnel coupling.

In Fig. 10.10c we show the kinetic exchange for a single anticrossing. However, a second anticrossing with exactly the same spin interactions will occur between the states $\begin{pmatrix} 0,0 \\ \Downarrow\Uparrow,0 \end{pmatrix}_S$ and $\begin{pmatrix} 0,0 \\ 0,\Downarrow\Uparrow \end{pmatrix}_S$, where the two holes are in the top dot. We will return to this point in Sect. 10.4.3. The kinetic exchange interaction can be seen directly in PL spectra of the doubly positively charged exciton, which has only two holes in the optical ground state. Kinetic exchange is similarly present for the electron–electron exchange interaction in a QDM charged with two electrons.

10.4.3 Competing Processes

Electron–Hole Exchange and Hole–Hole Exchange

If both electron–hole and hole–hole exchange are present in a single state, they compete to determine the state's spin character [15]. This is the case for the positive trion, X^+, which has two holes and one electron in the optically excited state. In Fig. 10.10d we show calculated energy levels for the X^+ state. There are 6 basis states with electron spin up: $\begin{pmatrix} \uparrow,0 \\ \Uparrow\Downarrow,0 \end{pmatrix}_S$, $\begin{pmatrix} \uparrow,0 \\ \Downarrow,\Uparrow \end{pmatrix}_S$, $\begin{pmatrix} \uparrow,0 \\ \Downarrow,\Uparrow \end{pmatrix}_T$, $\begin{pmatrix} \uparrow,0 \\ \Downarrow,\Downarrow \end{pmatrix}_T$, $\begin{pmatrix} \uparrow,0 \\ \Uparrow,\Uparrow \end{pmatrix}_T$, $\begin{pmatrix} \uparrow,0 \\ 0,\Uparrow\Downarrow \end{pmatrix}_S$. In this basis, the Hamiltonian is

$$\hat{H}^{X^+} = E_0 + \begin{pmatrix} \Gamma_1 & -t_{X^+} & 0 & 0 & 0 & 0 \\ -t_{X^+} & -edF & J^{eh} & 0 & 0 & -t_{X^+} \\ 0 & J^{eh} & -edF & 0 & 0 & 0 \\ 0 & 0 & 0 & -edF+J^{eh} & 0 & 0 \\ 0 & 0 & 0 & 0 & -edF-J^{eh} & 0 \\ 0 & -t_{X^+} & 0 & 0 & 0 & \Gamma_2-2edF \end{pmatrix}.$$

(10.6)

Here Γ_1 and Γ_2 are energy corrections due to the Coulomb interactions when all particles are in the same dot (Γ_1) or when both holes are in the top dot while the electron is in the bottom dot (Γ_2). t_{X^+} is the Coulomb-corrected tunneling energy for a hole in the presence of an electron and additional hole. As before, only singlet states are coupled together by tunneling. The two anticrossing points for singlet states have similar fine structure, so for our explanation we only need to consider two of the singlet states: $\begin{pmatrix}\uparrow,0\\\Uparrow\Downarrow,0\end{pmatrix}_S$ and $\begin{pmatrix}\uparrow,0\\\Downarrow,\Uparrow\end{pmatrix}_S$. Like the case of only two holes, there are three triplet states. However, the degeneracy of the triplet states is lifted by the electron–hole exchange interaction. Consequently, $\begin{pmatrix}\uparrow,0\\\Downarrow,\Downarrow\end{pmatrix}_T$ and $\begin{pmatrix}\uparrow,0\\\Uparrow,\Uparrow\end{pmatrix}_T$ are separated by $2J^{\mathrm{eh}}$. These two states are the dashed lines in Fig. 10.10d, which are unlabeled in the figure. The anticrossing energy gap is determined by both the tunneling matrix element and the electron–hole exchange: $\Delta = 2\sqrt{2|t_{X^+}|^2 + (J^{\mathrm{eh}})^2}$.

The electron–hole exchange also lifts the energy degeneracy between $\begin{pmatrix}\uparrow,0\\\Downarrow,\Uparrow\end{pmatrix}$ and $\begin{pmatrix}\uparrow,0\\\Uparrow,\Downarrow\end{pmatrix}$, which are the two states that combine to create the $\begin{pmatrix}\uparrow,0\\\Downarrow,\Uparrow\end{pmatrix}_S$ singlet and $\begin{pmatrix}\uparrow,0\\\Downarrow,\Uparrow\end{pmatrix}_T$ triplet states. As a result of this energy splitting, the electron–hole exchange introduces a mixing between $\begin{pmatrix}\uparrow,0\\\Downarrow,\Uparrow\end{pmatrix}_S$ and $\begin{pmatrix}\uparrow,0\\\Downarrow,\Uparrow\end{pmatrix}_T$. The mixing is included in the Hamiltonian by the off-diagonal J^{eh} term. This mixing causes the triplet-like state to "wiggle" as it passes through the anticrossing region [27]. This "wiggling" can also be seen in the experimentally measured PL spectral map of Fig. 10.11a.

To compare this theoretical model to the experimentally measured spectra for the positive trion (Fig. 10.11a) we use (10.6) to calculate the energies of the X^+ states, which are plotted as functions of electric field in the upper part of Fig. 10.11b. Since the energy of PL lines is determined by the energy difference between the optically excited and optical ground states, we must also calculate the energy of the optical ground states. We do this using (10.3). At zero magnetic field, the same Hamiltonian applies to both hole spin projections. The energies of the optical ground state are plotted in the lower part of Fig. 10.11b. The calculated energy levels are a fit to the data in the sense that the values of individual parameters used in the calculation are extracted directly from the data. Individual parameters are varied to achieve a best fit only when incomplete spectral maps prevent direct extraction of the actual value from the data. The value of E_0 is determined from the overall photoluminescence energy. The value of t_{X^+} and t_{h} are determined by the anticrossing energy gaps. J^{eh} is determined by the splitting of the two asymptotic triplet lines. The specific values extracted from the data and used in the calculation are: $\Delta_{\mathrm{h}} = 2|t_{\mathrm{h}}| = 390\,\mu\mathrm{eV}$, $\Delta_{X^+} = 2\sqrt{2|t_{X^+}|^2 + (J^{\mathrm{eh}})^2} = 750\,\mu\mathrm{eV}$, $2J^{\mathrm{eh}} = 430\,\mu\mathrm{eV}$, $\Gamma_1 = 2.21\,\mathrm{meV}$, $ed = 0.99\,\mathrm{meV}\,\mathrm{kV}^{-1}\,\mathrm{cm}$).

By taking the difference between the calculated energies of the optically excited and optical ground states, we generate the calculated PL spectral map

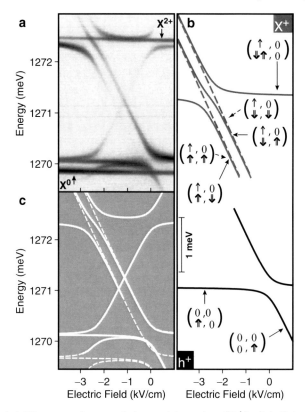

Fig. 10.11. (a) PL spectral map of the positive trion (X^+). (b) Calculated energy levels for optically excited (X^+) and optical ground (h^+) states. (c) Calculated PL spectra showing optically allowed (*solid*) and forbidden or weak (*dashed*) transitions. Experimental data is obtained for a QDM with a 4 nm high bottom dot separated from a 2.5 nm high top dot by a 4 nm barrier and embedded in an n-type diode. Calculation parameters are given in the text. Figure adapted from [15]

shown in Fig. 10.11c. Transitions that are optically dark or very weak are indicated by dashed lines. The excellent agreement between the calculated and experimentally measured spectra confirms that the matrix Hamiltonians presented here capture the essential spin interactions.

The eigenstates of (10.6) are superpositions of the basis states that describe the molecular states of the positive trion with all spin interactions included. The weighting of the basis states that make up any particular eigenstate is determined by the applied electric field, which controls the magnitude of the various exchange interactions. Consider the triplet-like state that "wiggles" as it passes through the anticrossing region. At one extreme (left side of Fig. 10.10d) this state asymptotically approaches the basis state $\left(\begin{smallmatrix}\uparrow,0\\\Uparrow,\Downarrow\end{smallmatrix}\right)$. At the other extreme (right side of Fig. 10.10d) the triplet-like state asymptotically approaches the $\begin{pmatrix}\uparrow,0\\\Downarrow,\Uparrow\end{pmatrix}$ state. Following this state continuously, we see that the mixing induced by the electron–hole exchange effectively swaps the spins of the two holes in separate dots. A similar mixing induced by the hyperfine interaction was recently used to induce a spin swap of electrons in lithographically defined quantum dots [39].

To understand the spin fine structure we considered only the tunnel coupling between $\begin{pmatrix}\uparrow,0\\\Uparrow\Downarrow,0\end{pmatrix}_S$ and $\begin{pmatrix}\uparrow,0\\\Downarrow,\Uparrow\end{pmatrix}_S$. In many cases this is the only tunnel coupling of the X^+ that we observe because the Coulomb interaction (Γ_2) shifts the $\begin{pmatrix}\uparrow,0\\0,\Uparrow\Downarrow\end{pmatrix}_S$ state to significantly higher energy. However, tunnel coupling of $\begin{pmatrix}\uparrow,0\\\Downarrow,\Uparrow\end{pmatrix}_S$ and $\begin{pmatrix}\uparrow,0\\0,\Uparrow\Downarrow\end{pmatrix}_S$ leads to the same kinetic exchange that separates singlet and triplet states. Away from the anticrossing, the singlet–triplet splitting asymptotically returns towards zero.

In some cases the relative contributions of Coulomb shifts (Γ_1 and Γ_2) create two anticrossings that are separated by a relatively small change in applied electric field. Schematically, this is represented by the difference in applied electric field between points 1 and 2 in Fig. 10.9. If the two anticrossings occur at sufficiently similar values of the applied electric field, the singlet–triplet splitting does not reach its asymptotic value between them. Consequently, the singlet and triplet states remain separated over the entire range of applied electric fields between these two anticrossing points. The proximity of the two anticrossings required to maintain the separation of singlet and triplet states depends on the magnitude of the tunnel coupling, which determines how rapidly the kinetic exchange asymptotes to zero.

10.5 QDMs in a Magnetic Field

10.5.1 Zeeman Splitting

At zero magnetic field, all states are at least two-fold degenerate. If we continue to neglect the anisotropic exchange interaction, states that differ by a total flip of all spins have the same energy. In a nonzero magnetic field along the optical axis (Faraday geometry), the opposite spin projections have a Zeeman splitting. The magnitude of this Zeeman splitting is given by the g factors for any unpaired electrons or holes.

To include magnetic fields, we must add a Zeeman term to the Hamiltonians. If we restrict ourselves to magnetic fields in the Faraday geometry, the Zeeman term has the form [40]

$$\widehat{H}_{\text{Zeeman}}(B) = \mu_B \left(g_e S_e - \frac{g_h}{3}\widehat{J}_h \right) B \tag{10.7}$$

with μ_B the Bohr magneton, g_e the longitudinal electron g factor, g_h the longitudinal heavy-hole g factor, S_e the electron spin projection ($\pm 1/2$), J_h the heavy-hole spin projection ($\pm 3/2$) and B the magnetic field. We assume that the heavy-hole g factor is the same in both dots. This assumption is well supported by experimental measurements of Zeeman splitting when holes are in each of the dots.

We consider first the states of the neutral exciton. In the same basis as (10.4) ($\begin{pmatrix}\uparrow,0\\\Downarrow,0\end{pmatrix}$, $\begin{pmatrix}\uparrow,0\\\Uparrow,0\end{pmatrix}$, $\begin{pmatrix}\uparrow,0\\0,\Downarrow\end{pmatrix}$ and $\begin{pmatrix}\uparrow,0\\0,\Uparrow\end{pmatrix}$), the Zeeman Hamiltonian for the neutral exciton states with electron spin up is

$$\widehat{H}_Z^{X^0} = \frac{\mu_B B}{2} \begin{pmatrix} (g_e + g_h) & 0 & g'_{X^0} & 0 \\ 0 & (g_e - g_h) & 0 & -g'_{X^0} \\ g'_{X^0} & 0 & (g_e + g_h) & 0 \\ 0 & -g'_{X^0} & 0 & (g_e - g_h) \end{pmatrix}. \quad (10.8)$$

The Zeeman Hamiltonian for the neutral exciton states with electron spin down are the same as (10.8) with a sign change for all terms in the matrix. The diagonal elements contain the g factor for each basis state, which depend only on the spin orientations and not on the spatial distribution of carriers. Electron spin up (down) corresponds to $+g_e$ ($-g_e$). Because of the minus sign in (10.7), hole spin up (down) corresponds to $-g_h$ ($+g_h$). g'_{X^0} is the contribution to the g factor from the amplitude of the wavefunction in the barrier, which will be discussed in Sect. 10.5.2.

For charged states, we must consider the Zeeman energy splitting in both the optically excited and optical ground states. There are no off-diagonal elements in the Zeeman Hamiltonian for the positively charged trion (X$^+$). The Zeeman terms for each basis state are

$$\begin{pmatrix}\uparrow,0\\\Uparrow\Downarrow,0\end{pmatrix}_S \qquad \frac{\mu_B B}{2} g_e,$$
$$\begin{pmatrix}\uparrow,0\\\Downarrow,\Uparrow\end{pmatrix}_S \qquad \frac{\mu_B B}{2} g_e,$$
$$\begin{pmatrix}\uparrow,0\\\Downarrow,\Uparrow\end{pmatrix}_T \qquad \frac{\mu_B B}{2} g_e,$$
$$\begin{pmatrix}\uparrow,0\\\Downarrow,\Downarrow\end{pmatrix}_T \qquad \frac{\mu_B B}{2}(g_e + 2g_h), \qquad (10.9)$$
$$\begin{pmatrix}\uparrow,0\\\Uparrow,\Uparrow\end{pmatrix}_T \qquad \frac{\mu_B B}{2}(g_e - 2g_h),$$
$$\begin{pmatrix}\uparrow, 0\\0,\Uparrow\Downarrow\end{pmatrix}_S \qquad \frac{\mu_B B}{2} g_e.$$

The Zeeman terms for the basis states with electron spin down are simply the negative of those in (10.9). The Zeeman Hamiltonian for the optical ground (h) state, in the $\begin{pmatrix}0,0\\\Uparrow,0\end{pmatrix}$, $\begin{pmatrix}0,0\\\Downarrow,0\end{pmatrix}$, $\begin{pmatrix}0,0\\0,\Uparrow\end{pmatrix}$, $\begin{pmatrix}0,0\\0,\Downarrow\end{pmatrix}$ basis, is

$$\widehat{H}_Z^{h^+} = \frac{\mu_B B}{2} \begin{pmatrix} -g_h & 0 & -g'_h & 0 \\ 0 & g_h & 0 & g'_h \\ -g'_h & 0 & -g_h & 0 \\ 0 & g'_h & 0 & g_h \end{pmatrix}. \quad (10.10)$$

Here we have explicitly written out the Zeeman Hamiltonian for all basis states. Using (10.9) and (10.10), we can calculate the energies of optically

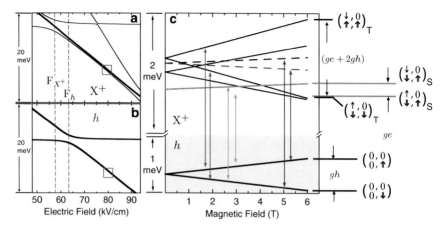

Fig. 10.12. Calculated energies of positive trion (X^+, panel **a**) and hole (h, panel **b**) states as a function of applied electric field in a QDM with 2 nm barrier at zero magnetic field. (**c**) Calculated energies of states indicated by the boxed regions in panels (**a**) and (**b**) for a fixed value of electric field and increasing magnetic field. The X^+ states are split into singlet and triplet states. Net Zeeman splittings are proportional to the net g factor, indicated on the right-hand side, which is determined by the sum of all spin projections. Optically allowed transitions are indicated by *arrows*. Values used in the calculation (in meV) are taken to match a real QDM: $\Gamma_1 = 3.2145$, $\Gamma_2 = 9.9$, $t_h = 0.855$, $t_{X^+} = 0.877$, $J^{eh} = 0.11$, $g_h = -1.4$, $g_e = -0.745$. $ed = 0.429\,\text{meV}\,\text{kV}^{-1}\,\text{cm}$. The singlet and triplet states of the positive trion are split in energy by the proximity of two anticrossings, as discussed in Sect. 10.4.3. The diamagnetic shift has been neglected. Figure adapted from [16]

excited and optical ground states as a function of applied magnetic field. In Fig. 10.12 we show such calculated energies for the optical ground state (one hole) and four of the optically excited states (positive trion).

As shown in the above discussion, there are many possible magnitudes of the Zeeman splitting in the optically excited and optical ground states. Fortunately, optical selection rules impose some order. The Zeeman splitting of optically active photoluminescence lines is given by the difference between the Zeeman splitting in the optically excited and optical ground states. Since the selection rules require that recombining electrons and holes have opposite spin orientations, the net g factor for an optically allowed transition is always given by the sum of the electron and hole g factor. In Fig. 10.13 we illustrate the Zeeman splittings for three optically allowed transitions. The constant factor $\mu_B B/2$ has been suppressed.

Because of the tunnel coupling, the optically excited and optical ground states in Fig. 10.13 are superpositions of multiple basis states. For simplicity, we have labeled each state in Fig. 10.13 with only one of the basis states. The g factors for electrons and holes in InAs dots are generally negative [40], so states with net electron spin down and hole spin up are at higher energies. In (10.11) we list the full superposition states for the optical transitions of

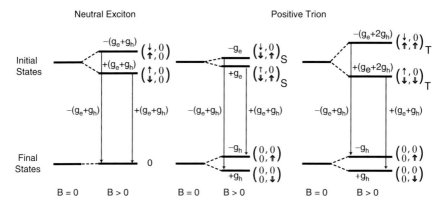

Fig. 10.13. Zeeman splitting of optically excited and optical ground states of the neutral exciton and positive trion. State energies are determined by the product of the listed g factors and the constant factor $\mu_B B/2$. Optically allowed transitions always split by $\mu_B B(g_e + g_h)$

Fig. 10.13. We also list the g factors for each state and the corresponding g factor of the optical transitions.

$$\begin{aligned}
&\alpha \begin{pmatrix} \uparrow,0 \\ \Downarrow,0 \end{pmatrix} + \beta \begin{pmatrix} \uparrow,0 \\ 0,\Downarrow \end{pmatrix} \quad &&\rightarrow \quad 0 \\
&\quad (g_e + g_h) &&\quad - \quad 0 \quad = (g_e + g_h), \\[4pt]
&\alpha \begin{pmatrix} \downarrow,0 \\ \Uparrow,0 \end{pmatrix} + \beta \begin{pmatrix} \downarrow,0 \\ 0,\Uparrow \end{pmatrix} \quad &&\rightarrow \quad 0 \\
&\quad -(g_e + g_h) &&\quad - \quad 0 \quad = -(g_e + g_h), \\[4pt]
&\alpha \begin{pmatrix} \uparrow,\,0 \\ \Uparrow\Downarrow,0 \end{pmatrix}_S + \beta \begin{pmatrix} \uparrow,0 \\ \Downarrow,\Uparrow \end{pmatrix}_S + \gamma \begin{pmatrix} \uparrow,\,0 \\ 0,\Uparrow\Downarrow \end{pmatrix}_S \rightarrow \alpha \begin{pmatrix} 0,0 \\ \Uparrow,0 \end{pmatrix} + \beta \begin{pmatrix} 0,0 \\ 0,\Uparrow \end{pmatrix} \\
&\quad g_e &&\quad - \quad -g_h \quad = (g_e + g_h), \\[4pt]
&\alpha \begin{pmatrix} \downarrow,\,0 \\ \Uparrow\Downarrow,0 \end{pmatrix}_S + \beta \begin{pmatrix} \downarrow,0 \\ \Downarrow,\Uparrow \end{pmatrix}_S + \gamma \begin{pmatrix} \downarrow,\,0 \\ 0,\Uparrow\Downarrow \end{pmatrix}_S \rightarrow \alpha \begin{pmatrix} 0,0 \\ \Downarrow,0 \end{pmatrix} + \beta \begin{pmatrix} 0,0 \\ 0,\Downarrow \end{pmatrix} \\
&\quad -g_e &&\quad - \quad g_h \quad = -(g_e + g_h), \\[4pt]
&\begin{pmatrix} \uparrow,0 \\ \Downarrow,\Downarrow \end{pmatrix}_T \quad &&\rightarrow \alpha \begin{pmatrix} 0,0 \\ \Downarrow,0 \end{pmatrix} + \beta \begin{pmatrix} 0,0 \\ 0,\Downarrow \end{pmatrix} \\
&\quad (g_e + 2g_h) &&\quad - \quad g_h \quad = (g_e + g_h), \\[4pt]
&\begin{pmatrix} \downarrow,0 \\ \Uparrow,\Uparrow \end{pmatrix}_T \quad &&\rightarrow \alpha \begin{pmatrix} 0,0 \\ \Uparrow,0 \end{pmatrix} + \beta \begin{pmatrix} 0,0 \\ 0,\Uparrow \end{pmatrix} \\
&\quad -(g_e + 2g_h) &&\quad - \quad -g_h \quad = -(g_e + g_h).
\end{aligned}$$

(10.11)

10.5.2 Resonant Changes in g Factor

Given the constraints of the optical selection rules, one would expect that the application of magnetic fields would result in a doubling of the spectral map observed at zero field. The two copies would be separated by a constant Zeeman splitting given simply by the total g factor ($g_e + g_h$). In Fig. 10.14 we show the energies of neutral exciton PL lines from a QDM with 2 nm barrier at a magnetic field of 6 T. It is clear that the Zeeman splitting of the molecular states is not constant, but rather a strong function of applied electric field. The splitting of the low energy molecular ground states is suppressed at about 83 kV cm^{-1}. The splitting of the high-energy molecular excited states is enhanced. This resonant change in Zeeman splitting comes from barrier contributions to the g factor.

Since the g factor depends on material parameters, it is sensitive to the amplitude of the wavefunction in regions comprised of different materials [41, 42]. By measuring far away from any anticrossing, we obtain the electron and hole g factors in the InAs dots of the QDM in Fig. 10.14. Obtaining independent values for g_e and g_h requires analysis of spectra taken at several angles of the magnetic field (not shown). The values obtained, $g_e = -0.745$ and $g_h = -1.4$, are consistent with other measurements of InAs dots [40]. The solid lines shown in Fig. 10.14 are calculated using these values in (10.4) and (10.8), along with a fit value for g'_{X^0} described below. Other calculational parameters include $t_{X^0} = 0.58$ meV and $ed = 0.429$ meV kV^{-1} cm.

Unlike in the InAs QDs, the g factor for holes in bulk GaAs is positive [43]. Resonant changes in the amplitude of the wavefunction in the GaAs barrier arise at anticrossings due to the formation of molecular states [16]. Bonding orbitals have a large amplitude in the barrier and thus add a large contribution from the GaAs to the overall hole g factor. This adds a positive component (g') to the otherwise negative heavy-hole g factor and suppresses the Zeeman splitting on resonance. In Fig. 10.14, the low-energy molecular ground states

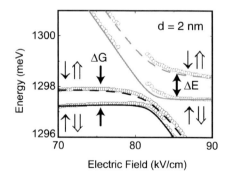

Fig. 10.14. Energies of X^0 PL lines for a QDM with 2 nm barrier measured at 6 T. The QDs are both nominally 2.5 nm high and embedded in an n-type diode. *Solid lines* are calculated as described in the text. Figure taken from [17]

have a suppressed splitting and must therefore have bonding orbital character. In contrast, antibonding states suppress the contributions from the tail of the wavefunction in the barrier, enhance the negative g factor, and result in increased Zeeman splitting on resonance. The increased splitting for the molecular excited states in Fig. 10.14 is a consequence of their antibonding character. The magnitude of g' is expected to differ from the value in bulk GaAs because of additional strain and boundary conditions in the barrier.

In Fig. 10.15 we plot the Zeeman splitting at $B = 6\,\text{T}$ for the three pairs of optical transitions described in (10.11) measured in a QDM with a 2 nm barrier. X^0_B and X^0_A are the splitting of the bonding and antibonding states shown in Fig. 10.14. The other two curves come from the Zeeman splitting of X^+ states that we will discuss shortly. At the left and right sides of Fig. 10.15, all four curves approach a Zeeman splitting of approximately 0.75 meV. Each curve reaches a resonant maximum or minimum in the Zeeman splitting at a particular value of the applied electric field. The bonding and antibonding states of the neutral exciton reach a resonant minimum and maximum, respectively, at the electric field indicated by F_{X^0}. F_{X^0}, which can be measured in data taken at zero magnetic field, is the electric field of the anticrossing where

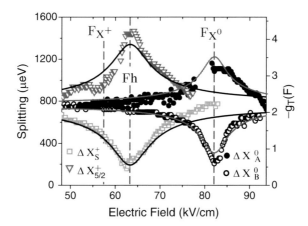

Fig. 10.15. Zeeman splitting of four PL lines as a function of applied electric field. Data is taken at $B = 6\,\text{T}$ for a QDM with nominally symmetric dots 2.5 nm high separated by a 2 nm barrier. ΔX^0_B and ΔX^0_A are the Zeeman splitting of PL lines from the bonding and antibonding molecular states of the neutral exciton, respectively. $\Delta X^+_{5/2}$ is the splitting between PL lines that originate in the $\begin{pmatrix}\uparrow,0\\\Downarrow,\Downarrow\end{pmatrix}_T$ and $\begin{pmatrix}\downarrow,0\\\Uparrow,\Uparrow\end{pmatrix}_T$ states. ΔX^+_S is the splitting between PL lines originating in the two molecular singlet states described in (10.11). The *solid curves* are calculated using (10.13) with different values of the tunneling coefficient for the neutral exciton (t_{X^0}: 0.58 meV) and single hole (t_h: 0.85 meV). Additional calculation parameters are given in the text. Figure adapted from [16]

the atomic-like basis states would be degenerate in energy. At F_{X^0} tunnel coupling is maximized and the states have fully molecular character.

The resonant contribution of the barrier to the g factors of molecular states is captured in the Hamiltonian (10.8) by the term g'_{X^0}. g'_{X^0} is a positive number since it originates in the g factor of a heavy hole in GaAs. The sign of g'_{X^0} at each matrix element in the Zeeman Hamiltonian is determined by the spin orientation of the tunneling hole: if the tunneling hole has spin up, the additional matrix element is $-g'_{X^0}$ because of the minus sign for holes in (10.7).

The full Hamiltonian for the neutral exciton is given by the sum of (10.4) and (10.8):

$$\hat{H}^{X^0} = E_0 + \begin{pmatrix} J^{\text{eh}} + \eta g_B & 0 & -t_{X^0} + \eta g'_{X^0} & 0 \\ 0 & -J^{\text{eh}} + \eta g_D & 0 & -t_{X^0} - \eta g'_{X^0} \\ -t_{X^0} + \eta g'_{X^0} & 0 & -edF + \eta g_B & 0 \\ 0 & -t_{X^0} - \eta g'_{X^0} & 0 & -edF + \eta g_D \end{pmatrix}.$$
(10.12)

Here $\eta = \frac{\mu_B B}{2}$, $g_B = (g_e + g_h)$, and $g_D = (g_e - g_h)$. Equation (10.12) is restricted to the basis with electron spin up ($\begin{pmatrix}\uparrow,0\\\downarrow,0\end{pmatrix}$, $\begin{pmatrix}\uparrow,0\\\uparrow,0\end{pmatrix}$, $\begin{pmatrix}\uparrow,0\\0,\downarrow\end{pmatrix}$, and $\begin{pmatrix}\uparrow,0\\0,\uparrow\end{pmatrix}$). The same Hamiltonian applies to the case of electron spin down if all spins are flipped and the sign of all Zeeman terms is reversed. Equation (10.12) is used to calculate the solid lines in Fig. 10.14.

Equation (10.12) allows us to calculate the energy of each state, including the resonant contributions of the barrier. To more clearly understand the resonant contribution to the g factor, it is helpful to take the difference in energy between Zeeman split molecular states calculated using (10.12). The Zeeman splitting for the molecular ground (excited) state as a function of applied electric field is given by

$$\Delta G(F) = \mu B \left(g_e + g_h + \frac{2t_{X^0} g'_{X^0}}{\sqrt{e^2 d^2 (F - F_{X^0})^2 + 4t^2_{X^0}}} \right), \quad (10.13)$$

$$\Delta E(F) = \mu B \left(g_e + g_h - \frac{2t_{X^0} g'_{X^0}}{\sqrt{e^2 d^2 (F - F_{X^0})^2 + 4t^2_{X^0}}} \right). \quad (10.14)$$

t_{X^0} and F_{X^0} are determined from measured spectral maps of the same QDM at zero magnetic field. g_e and g_h are determined from the Zeeman splitting far from resonance. Because the signs of g_e, g_h and g'_{X^0} are fixed, the sign of t_{X^0} determines whether the Zeeman splitting of the ground state is enhanced or suppressed at the resonance. In the 2 nm case, the Zeeman splitting of the molecular ground state (ΔG) decreases at the anticrossing point. t_{X^0} must therefore be positive, which is consistent with the identification that the molecular ground state has bonding character. In contrast, the Zeeman splitting of the upper level (ΔE) has a minus sign in (10.14); a positive t_{X^0} leads to an increase in the splitting at resonance, consistent with antibonding character. We will return to the sign of t_{X^0} in Sect. 10.6.

The magnitude of g'_{X^0} is determined by fits to the data using (10.13); in Fig. 10.15, $g'_{X^0} = 1.32$. The excellent agreement between the observed linewidth of the resonant change in g factor and the linewidth calculated using independently measured values of $|t_{X^0}|$ provides strong confirmation that the g factor resonance arises from the formation of molecular states with bonding and antibonding orbitals. Notice that the only difference between the fit to the Zeeman splitting of the molecular ground and excited states is the sign change between (10.13) and (10.14); both are fit with the same value of g'_{X^0}.

In Fig. 10.15 we also plot the Zeeman splitting for optical transitions from a singlet and triplet state of the positive trion, X^+. Both of the resonances peak at the electric field indicated by F_h, which is the electric field of the anticrossing for the single hole in the optical ground state. The resonant change in g factor does not peak at F_{X^+}, the electric field of the anticrossing in the optically excited state. There can be no resonant contribution from the barrier for the optically excited state (X^+) because there is no single-particle tunneling. Triplet states can not tunnel, and thus experience no resonant change in the amplitude of their wavefunction in the barrier. The spin singlet states that do tunnel contain two oppositely oriented spins. Any resonant change in g factor is canceled.

Both singlet and triplet states plotted in Fig. 10.15 come from optical recombination to the bonding orbital of the hole in the optical ground state. The singlet state shows a resonant decrease in Zeeman splitting exactly like the bonding orbital of the neutral exciton, X^0. The resonance is fit with an equation analogous to (10.13) using parameters (t_h, F_h) measured for X^+ at zero magnetic field. The resonant contribution from the barrier (g'_h) is found to be 1.65. g'_h likely differs from g'_{X^0} because the Coulomb interaction with the additional electron in the X^0 case shifts the magnitude of the wavefunction in the barrier. This same Coulomb interaction is believed to cause the difference in tunneling rates for the neutral exciton (t_{X^0}) and the single hole (t_h).

The splitting for the triplet state in Fig. 10.15 has a resonant increase in Zeeman splitting because of the relative g factor contributions of the optically excited and optical ground states. Returning to (10.11), we see that the optically excited state ($\left(\begin{smallmatrix}\uparrow,0\\\Downarrow,\Downarrow\end{smallmatrix}\right)_T$) has a g factor given by ($g_e + 2g_h$). There is no resonant change in this g factor for the triplet state. The resonant change in g factor only affects the optical ground state ($\left(\begin{smallmatrix}0,0\\0,\Downarrow\end{smallmatrix}\right)$) g factor, g_h. The total g factor for the optical transition is thus given by

$$(g_e+2g_h)-\left(g_h+\frac{2t_h g'_h}{\sqrt{e^2 d^2(F-F_h)^2+4t_h^2}}\right) = g_e+g_h-\frac{2t_h g'_h}{\sqrt{e^2 d^2(F-F_h)^2+4t_h^2}}. \tag{10.15}$$

The incomplete cancelation of the g_h term in the optically excited and optical ground states results in the resonant increase in Zeeman splitting. The experimental data are well fit with this model using the same values obtained for the recombination from the X^+ singlet state.

We have discussed only resonant changes in the g factor for holes. To date, no resonant change in g factor has been observed for electrons in QDMs. This comes as no surprise, because the electron g factor in bulk GaAs (-0.44) [44] is similar to the electron g factor in InAs quantum dots (~ -0.7). Any resonant contribution from the barrier therefore does not significantly change the total electron g factor.

Finally, we note that g' appears in the same matrix position as the tunneling matrix elements. The full Hamiltonians therefore have tunneling matrix elements given by $-t \pm \eta g'$. See, for example, (10.12). This illustrates an important consequence of the resonant barrier contribution to the molecular state g factor: the magnitude of the net tunneling matrix element is spin-dependent. Whether the tunneling rate is enhanced or suppressed depends on the relative signs of t and g'. In the next section we will discuss how the sign of t_{X^0} can be controlled by changing the thickness of the barrier [17].

10.6 Antibonding Molecular Ground States

In Fig. 10.16 we show the resonant increases and decreases in Zeeman splitting for neutral excitons in QDMs with $d = 2, 3, 4,$ and 6 nm. Figure 10.16a shows the resonant changes for a QDM with $d = 2$ nm. These are the same data and

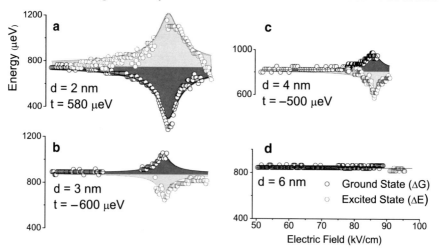

Fig. 10.16. Zeeman energy splitting as a function of applied electric field for $B = 6$ T. The QDMs have a barrier thickness of 2, 3, 4, and 6 nm. The *solid curves* are calculated with (10.13) using $(g_e + g_h, h) = (-2.12, 1.32), (-2.57, 0.47), (-2.32, 0.44)$ meV for 2, 3, and 4 nm, respectively. Figure adapted from [17].

theoretical fits as plotted above in Fig. 10.15, but here the theory curves have been filled in to emphasize that the molecular ground state (dark shading) has the resonant decrease in splitting indicative of a bonding orbital. The molecular excited state (light shading) has the resonant increase indicative of an antibonding orbital.

As the thickness of the barrier increases (Fig. 10.16b–d) the amplitude of the resonance decreases, and is below our noise level for $d = 6$ nm (Fig. 10.16d). This results from the reduction of the amplitude of the wavefunction in the barrier with increasing barrier thickness. In addition, there is a *qualitative* change in the nature of the resonance. When $d = 3$ and 4 nm the molecular ground state (dark shading) shows a resonant *increase* in Zeeman splitting. The resonant increase indicates that the molecular ground state now has antibonding character. This reversal of the energies of the bonding and antibonding states occurs at a barrier thickness of 3 nm [17].

Several recent atomistic calculations have predicted that the orbital character of the molecular ground state in QDMs would reverse as a function of increasing barrier thickness [22–25]. This reversal of the orbital character is a consequence of the spin–orbit (SO) interaction. For holes, the SO interaction couples spin and orbital degrees of freedom to create a total (Bloch) angular momentum $J = 3/2$, where $J_z = \pm 3/2$ projections correspond to heavy holes and $J_z = \pm 1/2$ to light holes. The SO interaction does not affect electrons because they have *s*-type (instead of *p*-type) atomic orbitals [45].

In QDs, the light-hole states are shifted up in energy by confinement and strain. As a result it is often useful and sufficient to consider the two low-energy (heavy-hole) states as pseudospin 1/2 particles ($\Uparrow, \Downarrow$). However, in vertically stacked QDMs, the mixing between heavy and light holes can become significant. When this mixing is included, low-energy hole states can be described within the Luttinger–Kohn $k.p$ Hamiltonian formalism [45] as four component Luttinger spinors [46], which are an admixture of all four projections of J_z.

The envelope function of the light-hole spinor component has approximate parity along z opposite to that of the heavy-hole spinor component. Because of the opposite parity, the tunnel-coupled light-hole component has antibonding character whereas the dominant heavy-hole component has bonding character. For thin barriers, the light-hole correction is small compared to the dominant heavy-hole component and the hole molecular ground state is bonding-like. As barrier thickness increases, the light-hole component does not decrease as fast as the heavy-hole component because of their relative effective masses. When the barrier is sufficiently thick (Fig. 10.16b,c) the light-hole correction is sufficiently large to reverse the orbital character of the molecular ground state.

Theoretical calculations predict that the reversal of orbital character should occur at a barrier thickness of about 2 nm. Experimentally, we find that all examples with $d = 2$ nm show that the molecular ground state has bonding orbital character. All examples with $d = 4$ nm show that the ground state has antibonding character. The intermediate case ($d = 3\,\text{nm}$) displays both types of behavior. The coexistence of both behaviors at $d = 3\,\text{nm}$ most likely arises from the inhomogeneous distribution of parameters that is also responsible for the fluctuation in anticrossing energies. The discrepancy between the predicted and observed barrier thickness at which the orbital character switches is likely due to specific details of dot structure and alignment that are not accounted for in the theoretical models.

10.7 Future Prospects

10.7.1 Summary

In this chapter we have reviewed the current understanding of the properties and interactions of spins confined in QDMs. The QDMs are constructed of two InAs QDs separated by a GaAs barrier and are grown using molecular beam epitaxy. The lowest confined energy levels in each dot are generally not at equal energy, but can be tuned into resonance by an applied electric field. The asymmetric energy levels and applied electric field generally trap one type of charge carrier in the bottom dot. Carriers of the other type can be distributed between the two dots. For simplicity, we have confined the discussion to QDMs where the electron is trapped in the bottom dot, holes tunnel between the dots, and the QDMs can be optically charged with excess holes. We have demonstrated above that the analysis applies equally well to QDMs containing excess or tunneling electrons.

Spin interactions between multiple charge carriers result in fine structure in the PL spectra of QDMs. When a single electron and single hole are in the same dot, the electron–hole exchange splits bright (electron and hole spin antiparallel) and dark (electron and hole spin parallel) configurations. Unlike single QDs, dark spin configurations can be observed when they are either a final state of optical recombination or there is an additional hole located in the second dot. When a state has two holes, both singlet and triplet spin configurations are possible. Because tunneling is a spin-conserving process, only the spin singlet configuration of one hole in each dot can tunnel couple with the states that have two holes in the same orbital energy level of a single dot. Triplet configurations can not participate in these tunnel couplings. As a result, singlet and triplet states are separated by a kinetic exchange energy. The proximity of multiple singlet anticrossings can induce a kinetic exchange energy separating singlet and triplet states over a wide range of electric fields. Both electron–hole and kinetic exchange can be present simultaneously; the combination can induce additional spin mixing.

Resonant tunneling of holes between the two dots results in the formation of molecular orbitals. These orbitals can be understood by constructing the symmetric and antisymmetric combinations of basis states with a single hole localized in one dot or the other. The symmetric and antisymmetric molecular orbitals have bonding and antibonding character. Due to SO interactions, the character of the lowest energy molecular orbital depends on the thickness of the barrier separating the two dots. For small barrier thicknesses, the bonding orbital has lowest energy.

The molecular orbitals have wavefunctions that span the entire QDM, including the GaAs barrier. The orbital character determines the resonant contribution of the GaAs barrier to the molecular state g factor. This results in a resonant decrease or increase in the heavy-hole g factor. In turn, this creates a spin-dependent tunneling barrier that manifests as a resonant increase or decrease of the Zeeman splitting in photoluminescence lines.

10.7.2 Materials Issues

Like single QDs, QDMs suffer from a substantial inhomogeneous distribution in energy levels. This remains one of the most significant challenges to the development of QDMs for device applications. In some ways, QDMs make it easier to deal with this distribution of energy levels. Indirect PL lines have a very strong Stark shift, which provides a mechanism for tuning the transition energies to desired values while preserving large optical matrix elements. QDMs also make it possible to measure the excited states of the valence and conduction bands by observing the anticrossings that arise each time the ground state in one dot is tuned into resonance with the excited state in the other dot [13]. The energies of excited valence and conduction band states are a crucial input to accurate theoretical modeling of QDs and to the development of optical spin manipulation protocols.

The inhomogeneous variation in energy levels poses an additional challenge to the design of QDM structures with predictable tunnel coupling. The discovery that the tunneling of electrons vs. holes can be controllably selected by introducing intentional dot asymmetry larger than the inhomogeneous fluctuation eliminates one major limitation. Nonetheless, the exact alloying of the barrier, and consequently the exact tunneling rate, remain incompletely controlled. Within the range of energy levels that normally occur in the separate dots, the applied electric field can generally tune the energy levels into resonance. However, in currently available structures, this applied electric field also controls the total charge state by tuning the energy of the QDM relative to the Fermi level. Although many charge states can be observed as a result of optical charging, useful devices will require QDMs that are electrically charged with a definitive number of carriers. Additional engineering is necessary to develop structures in which the electric field tunes the energy levels into resonance while simultaneously keeping the dots charged with the desired number of carriers.

The development of QDMs for device applications will benefit from the continued research into improving the homogeneity of single QDs. In addition, the development of growth techniques using predefined nucleation sites and/or the preparation of ordered arrays of QDs will facilitate the integration of

QDMs into more complex devices. Prospects for these improvements in MBE growth techniques are discussed in Chap. 2.

10.7.3 Doubly Charged QDMs and Quantum Information Processing

One of the earliest proposals for the implementation of quantum information processing in the solid state proposed using a single electron confined to a single quantum dot as the qubit [47]. Two-qubit operations were to be executed by lowering the potential barrier between neighboring dots to allow exchange operations to take place. Doubly charged QDMs provide an excellent test bed for such a scheme. The ground state contains two holes that can be located in separate dots. The kinetic exchange interaction results in a separation in energy between singlet and triplet states. The magnitude of this separation can be varied by tuning the applied electric field to control the proximity to anticrossings of the singlet states. In samples with a thin barrier, the singlet and triplet states can remain separated over a wide range of applied electric field.

In an alternative configuration, the singlet and triplet states of a doubly charged QDM could be used as the basis states for a qubit. Recent experiments in lithographically defined quantum dots have demonstrated Rabi oscillations in the singlet–triplet basis [39] and measured coherence times up to $10\,\mu s$ [39,48]. In self-assembled QDMs, asymmetric nuclear fields can make electric dipole transitions between singlet and triplet states allowed [49]. This, in turn, allows for superconducting-cavity-mediated gate operations between qubits separated by large distances. The formation and controllable separation of singlet and triplet states through the kinetic exchange interaction detailed here will be a crucial element in any implementation of such a scheme. Another crucial element will be the mechanism for coupling singlet and triplet configurations. One possibility is the combined influence of kinetic exchange and electron–hole exchange discussed in Sect. 10.4.3.

10.7.4 Optical Control

Optical manipulation of qubits can be substantially faster than electrical gating. For this reason there has been extensive research into experimental and theoretical methods for optical spin orientation and coherent optical control in both single and coupled quantum dots [50–58]. One possible spin-manipulation mechanism is the optical gating of exchange interactions. Because Coulomb interactions cause optical ground and excited states to have anticrossings at different values of the applied electric field, it is possible to bias a QDM so that tunnel coupling is only substantial in the optically excited state. The addition of an electron–hole pair would promote the QDM to a state where exchange interactions are turned on. After a defined period of time, stimulated emission can be used to drive the QDM back down to the optical ground state and turn off exchange interactions.

Recently, Emary and Sham proposed a mechanism to optically control the interaction between two spins confined in the two dots of a QDM [59]. The proposal uses the optically excited state of a QDM charged with two electrons, but with the excess electron tunneling between excited energy levels of the two dots. The tunneling electron mediates interactions between electrons localized in the ground states of each dot. Determining the proper QDM configuration for implementation of such a scheme requires considering the spin interactions described here.

10.7.5 Molecular Orbitals and g Factors

Electrically tunable g factors have been previously observed in quantum wells [41, 42]. However, the magnitude of the resonant change in g factor for molecular states of QDMs is substantially larger and may provide a new method for the manipulation of single spins. Moreover, unlike quantum wells, QDMs can easily be charged with a single spin. Spin manipulations could be implemented by tuning the spin splitting in and out of resonance with a constant AC magnetic field or by using g tensor modulation [60].

As discussed in Sect. 10.5.2, the resonant contribution of the barrier creates a spin-dependent tunnel barrier. This provides another powerful tool for the design of spin manipulation protocols in QDMs. The specific properties of the spin-dependent barrier could be engineered by varying the alloy composition of the dots and the barrier to control the individual g factors. The SO contributions to tunneling also allow for the design of ground state molecular orbitals that have a particular symmetry and thus provide a mechanism to control which spin orientation has a higher tunnel barrier.

10.7.6 Nuclear Spins

QDMs provide additional opportunities for controllable interactions with nuclear spins. In QDMs, the spatial locations of holes and electrons can be controlled. As described in Chap. 5, electrons couple to nuclear spins via the hyperfine interaction while hole coupling to nuclear spins is drastically weaker [61]. By controlling the location of electrons, QDMs could be used to generate local variations in the nuclear spin orientation, which are important to some schemes for executing spin manipulations [49].

10.7.7 Holes

Throughout this chapter we have used holes as an example case to discuss spin interactions. All of the spin interactions discussed at zero magnetic field apply equally well to electrons, which have long been considered the most likely candidate for use in spin-based devices. To date, no resonant changes in g factor have been observed in QDMs where electrons tunnel.

Recent research suggests that there are good reasons to consider holes for spin-based devices. Hole spin states can be optically initialized in a manner similar to electrons [61] and have surprisingly long spin relaxation times [61–63]. Unlike electrons, holes have an extremely small hyperfine interaction with nuclei, which suppresses a primary decoherence mechanism. The use of the SO interaction to design molecular orbitals with tailored resonant changes in g factor provides another spin-manipulation opportunity that is unique to holes.

References

1. G. Schedelbeck, W. Wegscheider, M. Bichler, G. Abstreiter, Science **278**(5344), 1792 (1997)
2. T. Hatano, M. Stopa, S. Tarucha, Science **309**(5732), 268 (2005)
3. L. Goldstein, F. Glas, J.Y. Marzin, M.N. Charasse, G. Leroux, Appl. Phys. Lett. **47**(10), 1099 (1985)
4. Q. Xie, A. Madhukar, P. Chen, N.P. Kobayashi, Phys. Rev. Lett. **75**(13), 2542 (1995)
5. G.S. Solomon, J.A. Trezza, A.F. Marshall, J.S. Harris, Phys. Rev. Lett. **76**(6), 952 (1996)
6. J. Stangl, V. Hol, G. Bauer, Rev. Mod. Phys. **76**(3), 725 (2004)
7. M. Bayer, P. Hawrylak, K. Hinzer, S. Fafard, M. Korkusinski, Z.R. Wasilewski, O. Stern, A. Forchel, Science **291**(5503), 451 (2001)
8. H.J. Krenner, M. Sabathil, E.C. Clark, A. Kress, D. Schuh, M. Bichler, G. Abstreiter, J.J. Finley, Phys. Rev. Lett. **94**, 057402 (2005)
9. G. Ortner, M. Bayer, Y.L. Geller, T.L. Reinecke, A. Kress, J.P. Reithmaier, A. Forchel, Phys. Rev. Lett. **94**, 157401 (2005)
10. E.A. Stinaff, M. Scheibner, A.S. Bracker, I.V. Ponomarev, V.L. Korenev, M.E. Ware, M.F. Doty, T.L. Reinecke, D. Gammon, Science **311**, 636 (2006)
11. T. Nakaoka, E.C. Clark, H.J. Krenner, M. Sabathil, M. Bichler, Y. Arakawa, G. Abstreiter, J.J. Finley, Phys. Rev. B **74**(12), 4 (2006)
12. M. Scheibner, I.V. Ponomarev, E.A. Stinaff, M.F. Doty, A.S. Bracker, C.S. Hellberg, T.L. Reinecke, D. Gammon, Phys. Rev. Lett. **99**(19), 197402 (2007)
13. M. Scheibner, M. Yakes, A.S. Bracker, I.V. Ponomarev, M.F. Doty, C.S. Hellberg, L.J. Whitman, T.L. Reinecke, D. Gammon, Nat Phys **4**, 291 (2008)
14. H.J. Krenner, E.C. Clark, T. Nakaoka, M. Bichler, C. Scheurer, G. Abstreiter, J.J. Finley, Phys. Rev. Lett. **97**(7), 076403 (2006)
15. M. Scheibner, M.F. Doty, I.V. Ponomarev, A.S. Bracker, E.A. Stinaff, V.L. Korenev, T.L. Reinecke, D. Gammon, Phys. Rev. B **75**(24), 245318 (2007)
16. M.F. Doty, M. Scheibner, I.V. Ponomarev, E.A. Stinaff, A.S. Bracker, V.L. Korenev, T.L. Reinecke, D. Gammon, Phys. Rev. Lett. **97**(19), 197202 (2006)
17. M.F. Doty, J.I. Climente, M. Korkusinski, M. Scheibner, A.S. Bracker, P. Hawrylak, D. Gammon, Phys. Rev. Lett. **102**, 047401 (2009)

18. Y.B. Lyanda-Geller, T.L. Reinecke, M. Bayer, Phys. Rev. B **69**(16), 161308 (2004)
19. J.M. Villas-Bas, A.O. Govorov, S.E. Ulloa, Phys. Rev. B **69**(12), 125342 (2004)
20. D. Bellucci, F. Troiani, G. Goldoni, E. Molinari, Phys. Rev. B **70**(20), 205332 (2004)
21. B. Szafran, T. Chwiej, F.M. Peeters, S. Bednarek, J. Adamowski, B. Partoens, Phys. Rev. B **71**(20), 205316 (2005)
22. G. Bester, J. Shumway, A. Zunger, Phys. Rev. Lett. **93**(4), 047401 (2004)
23. G. Bester, A. Zunger, J. Shumway, Phys. Rev. B **71**(7), 075325 (2005)
24. W. Jaskolski, M. Zielinski, G.W. Bryant, Acta Phys. Pol. A **106**(2), 193 (2004)
25. W. Jaskolski, M. Zielinski, G.W. Bryant, J. Aizpurua, Phys. Rev. B **74**(19), 195339 (2006)
26. I.V. Ponomarev, M. Scheibner, E.A. Stinaff, A.S. Bracker, M.F. Doty, S.C. Badescu, M.E. Ware, V.L. Korenev, T.L. Reinecke, D. Gammon, Phys. Stat. Sol. (b) **243**(15), 3869 (2006)
27. M.Z. Maialle, M.H. Degani, Phys. Rev. B **76**(11), 7 (2007)
28. B. Szafran, F.M. Peeters, Phys. Rev. B **76**(19), 9 (2007)
29. N.N. Ledentsov, V.A. Shchukin, M. Grundmann, N. Kirstaedter, J. Bohrer, O.G. Schmidt, D. Bimberg, V.M. Ustinov, A.Y. Egorov, A.E. Zhukov, P.S. Kopev, S.V. Zaitsev, N.Y. Gordeev, Z.I. Alferov, A.I. Borovkov, A.O. Kosogov, S.S. Ruvimov, P. Werner, U. Gosele, J. Heydenreich, Phys. Rev. B **54**(12), 8743 (1996)
30. J.M. Garcia, G. Medeiros-Ribeiro, K. Schmidt, T. Ngo, J.L. Feng, A. Lorke, J. Kotthaus, P.M. Petroff, Appl. Phys. Lett. **71**(14), 2014 (1997)
31. Z.R. Wasilewski, S. Fafard, J.P. McCaffrey, J. Cryst. Growth **202**, 1131 (1999)
32. Q. Gong, P. Offermans, R. Notzel, P.M. Koenraad, J.H. Wolter, Appl. Phys. Lett. **85**(23), 5697 (2004)
33. P. Offermans, P.M. Koenraad, J.H. Wolter, D. Granados, J.M. Garcia, V.M. Fomin, V.N. Gladilin, J.T. Devreese, Appl. Phys. Lett. **87**(13), 131902 (2005)
34. G. Costantini, A. Rastelli, C. Manzano, P. Acosta-Diaz, R. Songmuang, G. Katsaros, O.G. Schmidt, K. Kern, Phys. Rev. Lett. **96**(22), 226106 (2006)
35. A.S. Bracker, M. Scheibner, M.F. Doty, E.A. Stinaff, I.V. Ponomarev, J.C. Kim, L.J. Whitman, T.L. Reinecke, D. Gammon, Appl. Phys. Lett. **89**(23), 233110 (2006)
36. P. Fazekas, *Lecture Notes on Electron Correlation and Magnetism* (World Scientific, Singapore, 1999)
37. G. Burkard, D. Loss, D.P. DiVincenzo, Phys. Rev. B **59**(3), 2070 (1999)
38. M.F. Doty, M. Scheibner, A.S. Bracker, D. Gammon, Phys. Rev. B **78**, 115316 (2008)
39. J.R. Petta, A.C. Johnson, J.M. Taylor, E.A. Laird, A. Yacoby, M.D. Lukin, C.M. Marcus, M.P. Hanson, A.C. Gossard, Science **309**, 2180 (2005)
40. M. Bayer, O. Stern, A. Kuther, A. Forchel, Phys. Rev. B **61**(11), 7273 (2000)
41. G. Salis, Y. Kato, K. Ensslin, D.C. Driscoll, A.C. Gossard, D.D. Awschalom, Nature **414**, 619 (2001)
42. M. Poggio, G.M. Steeves, R.C. Myers, N.P. Stern, A.C. Gossard, D.D. Awschalom, Phys. Rev. B **70**, 121305(R) (2004)
43. M.J. Snelling, E. Blackwood, C.J. McDonagh, R.T. Harley, C.T.B. Foxon, Phys. Rev. B **45**, 3922(R) (1992)
44. H. Kosaka, A.A. Kiselev, F.A. Baron, K.W. Kim, E. Yablonovitch, Electron. Lett. **37**, 464 (2001)

45. J.M. Luttinger, W. Kohn, Phys. Rev. **97**(4), 869 (1955)
46. L.G.C. Rego, P. Hawrylak, J.A. Brum, A. Wojs, Phys. Rev. B **55**(23), 15694 (1997)
47. D. Loss, D.P. DiVincenzo, Phys. Rev. A **57**, 120 (1998)
48. F.H.L. Koppens, J.A. Folk, J.M. Elzerman, R. Hanson, L.H.W.v. Beveren, I.T. Vink, H.P. Tranitz, W. Wegscheider, L.P. Kouwenhoven, L.M.K. Vandersypen, Science **309**, 1346 (2005)
49. G. Burkard, A. Imamoglu, Phys. Rev. B **74**, 041307(R) (2006)
50. S. Cortez, O. Krebs, S. Laurent, M. Senes, X. Marie, P. Voisin, R. Ferreira, G. Bastard, J.M. Grard, T. Amand, Phys. Rev. Lett. **89**(20), 207401 (2002)
51. M. Kroutvar, Y. Ducommun, D. Heiss, M. Bichler, D. Schuh, G. Abstreiter, J.J. Finley, Nature **432**(7013), 81 (2004)
52. A.S. Bracker, E.A. Stinaff, D. Gammon, M.E. Ware, J.G. Tischler, A. Shabaev, A.L. Efros, D. Park, D. Gershoni, V.L. Korenev, I.A. Merkulov, Phys. Rev. Lett. **94**(4), 047402 (2005)
53. A. Zrenner, E. Beham, S. Stufler, F. Findeis, M. Bichler, G. Abstreiter, Nature **418**(6898), 612 (2002)
54. X.Q. Li, Y.W. Wu, D. Steel, D. Gammon, T.H. Stievater, D.S. Katzer, D. Park, C. Piermarocchi, L.J. Sham, Science **301**(5634), 809 (2003)
55. M. Atature, J. Dreiser, A. Badolato, A. Hogele, K. Karrai, A. Imamoglu, Science **312**(5773), 551 (2006)
56. M. Atature, J. Dreiser, A. Badolato, A. Imamoglu, Nat. Phys. **3**(2), 101 (2007)
57. E. Biolatti, R.C. Iotti, P. Zanardi, F. Rossi, Phys. Rev. Lett. **85**(26), 5647 (2000)
58. P. Chen, C. Piermarocchi, L.J. Sham, Phys. Rev. Lett. **87**(6), 067401 (2001)
59. C. Emary, L.J. Sham, Phys. Rev. B **75**(12), 125317 (2007)
60. Y. Kato, R.C. Myers, D.C. Driscoll, A.C. Gossard, J. Levy, D.D. Awschalom, Science **299**(5610), 1201 (2003)
61. B.D. Gerardot, D. Brunner, P.A. Dalgarno, P. Ohberg, S. Seidl, M. Kroner, K. Karrai, N.G. Stoltz, P.M. Petroff, R.J. Warburton, Nature **451**(7177), 441 (2008)
62. L.M. Woods, T.L. Reinecke, R. Kotlyar, Phys. Rev. B **69**(12), 125330 (2004)
63. D. Heiss, S. Schaeck, H. Huebl, M. Bichler, G. Abstreiter, J.J. Finley, D.V. Bulaev, D. Loss, Phys. Rev. B **76**, 241306 (2007)

11

Quantum Information with Quantum Dot Light Sources

M. Scholz, T. Aichele, and O. Benson

Summary. In this chapter we report on applications of single photon sources based on semiconductor quantum dots to quantum information processing. After a brief review of the quantum optical properties of quantum dots we introduce experiments concerning free-space and fiber-based quantum cryptography. Then, we describe first demonstrations along linear optics quantum computation using single photon emitters.

11.1 Introduction

Photons are ideal carriers to transmit quantum information over large distances due to their low interaction with the environment. In 1984, Bennett and Brassard proposed a secret key-distribution protocol [1] that uses the single-particle character of a photon to avoid the possibility of eavesdropping on an encoded message (for a review see [2]). Also, the implementation of efficient quantum gates based on photons and linear optics was proposed [3,4] and demonstrated [5,6]. Besides, photons have been suggested as information carriers in larger networks [7] between processing nodes of stationary qubits, like ions [8–11], atoms [12,13], quantum dots (QDs) [14,15], and Josephson qubits [16,17].

Applications of linear optics in quantum information processing require the reliable generation of single- or few-photon states on demand. However, due to their bosonic character, photons tend to appear in bunches. This characteristics hinders the implementation of classical sources particularly in quantum cryptographic systems since an eavesdropper may gain partial information by a beam splitter attack. Similar obstacles occur for linear optics quantum computation where photonic quantum gates [3], quantum repeaters [18], and quantum teleportation [19] require the preparation of single- or few-photon states *on demand* in order to obtain reliability and high efficiency.

A promising process for single-photon generation is the spontaneous emission from a single quantum emitter. Numerous emitters have been used to demonstrate single-photon emission [20]. Single atoms or ions are the most fundamental systems [21, 22]. Other systems capable of single-photon generation are single molecules and single nanocrystals [23–25]. However, their drawback is a susceptibility for photo-bleaching and blinking [26]. Stable alternatives are nitrogen-vacancy defect centers in diamond [27, 28], but they show a broad optical spectra and a rather long lifetimes (≈ 12 ns).

In this article, we focus on single-photon generation from self-assembled single QDs. Most experiments with QDs have to be conducted at cryogenic temperature in order to reduce electron–phonon interaction and thermal ionization. High count rates can be obtained due to their short transition lifetimes of a few nanoseconds, and their spectral lines are nearly lifetime-limited with material systems covering the ultraviolet, visible, and infrared spectrum. QDs also gain attractiveness by the possibility for electric excitation [29–32] and the implementation in integrated photonic structures [33].

This chapter is organized as follows. Section 11.2 provides an introduction to the general properties and optical characterization of single QDs. Sections 11.3 and 11.4, respectively, describe two applications in quantum information processing, i.e., in quantum cryptography and toward quantum computation. Section 11.5 concludes with a short summary.

11.2 Single- and Multiphoton Generation from Quantum Dots

Single photon sources based on spontaneous emission in single QDs provide several advantages compared to other sources: they are photostable, compatible with chip technology, have a short lifetime together with a lifetime limited emission line at low temperature, are available over a broad spectral range from the near-infrared to UV, and allow to produce complex nonclassical states of light via cascaded decay of higher excitations.

11.2.1 Microphotoluminescence from Single Quantum Dots

All experiments in this chapter were performed with self-assembled QDs fabricated by Stranski–Krastanov growth [34]. Overgrowth of QDs is required to obtain high-efficient and optically stable emission. As it is impossible to characterize the exact shape or material composition of an overgrown QD a priori, certain simplified models for the electronic structure are often used. Figure 11.1 shows a spherical potential which traps a single electron–hole pair (exciton, Fig. 11.1a) or two electron–hole pairs (biexciton, Fig. 11.1b). In Fig. 11.1c, the possible decay paths of a biexciton via the so-called bright excitons are depicted. A level splitting is shown for clarity. The recombination of a carrier pair leads to the emission of a single photon, generally at different wavelengths for exciton and biexciton due to Coulomb interaction.

The image of an ensemble of InP QDs is displayed in Fig. 11.2a. Figure 11.2b shows the spectrum of a single InP QD in a GaInP matrix with two dominant

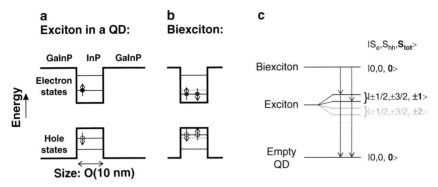

Fig. 11.1. Excitations in a QD: (**a**) An electron–hole pair forms an exciton. (**b**) Two electron–hole pairs form a biexciton, generally at an energy different from the exciton. (**c**) Schematic term scheme for the exciton and biexciton decay cascade

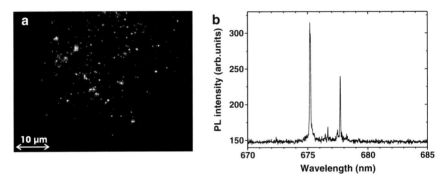

Fig. 11.2. (**a**) Microphotoluminescence image of InP QDs in GaInP. (**b**) Spectrum of a single InP/GaInP QD with spectral lines of exciton and biexciton decay

spectral lines originating from exciton and biexciton decay. This material system can generate single photons in the 620–750 nm region which perfectly fits the maximum detection efficiency of commercial silicon avalanche photo diodes (APD) with over 70% at wavelengths around 700 nm.

11.2.2 Single-Photon Generation

The single-photon character of the photoluminescence (PL) from a single QD can be tested by measuring the normalized second-order correlation function $g^{(2)}(t_1, t_2)$ via detecting the light intensity $\langle \hat{I}(t) \rangle$ at two points in time. For stationary fields, it reads

$$g^{(2)}(\tau = t_1 - t_2) = \frac{\langle : \hat{I}(0)\hat{I}(\tau) : \rangle}{\langle \hat{I}(0) \rangle^2},$$

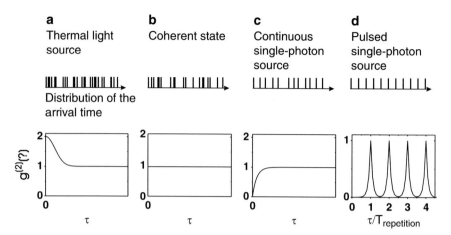

Fig. 11.3. $g^{(2)}$ function of (**a**) a thermal light source, (**b**) coherent light, (**c**) a continuously driven single-photon source, and (**d**) a pulsed single-photon source

where :: denotes normal operator ordering. For classical fields, $g^{(2)}(0) \geq 1$ and $g^{(2)}(0) \geq g^{(2)}(\tau)$ hold [35] which prohibits values smaller than unity. For thermal light sources, there is an increased probability to detect a photon shortly after another. This bunching phenomenon leads to $g^{(2)}(0) \geq 1$ (Fig. 11.3a). Coherent states show $g^{(2)}(\tau) = 1$ for all τ according to a Poisson photon number distribution (Fig. 11.3b). A single-photon state shows the antibunching effect of a sub-Poisson distribution with $g^{(2)}(0) \leq 1$ (Fig. 11.3c). For a Fock state $|n\rangle$, one has $g^{(2)}(0) = 1 - 1/n$ which yields $g^{(2)}(0) = 0$ for the special case of $|n = 1\rangle$. The pulsed excitation of a single-photon source leads to a peaked structure in $g^{(2)}(\tau)$ with a missing peak at zero delay indicating single-photon emission (Fig. 11.3d).

In order to circumvent the detector dead time ($\approx$50 ns for APDs [36] used in this experiment), the second-order correlation function is measured in a Hanbury Brown–Twiss arrangement [37] as depicted in Fig. 11.4, consisting of two photo detectors and a 50:50 beam splitter. A large number of time intervals between detection events is measured and binned together in a histogram.

11.2.3 Microphotoluminescence

Optical investigation of QDs is usually performed in a microphotoluminescence setup which combines excellent spatial resolution with high detection efficiency. Samples with low densities of 10^8–10^{11} dots cm^{-2} are required to isolate a single dot. In our setup, the sample is stored at 4 K inside a continuous-flow liquid Helium cryostat (Fig. 11.4). The dots are excited either pulsed (Ti:Sa, pulse width 400 fs, repetition rate 76 MHz, frequency-doubled to 400 nm) or continuously (Nd:YVO$_4$, 532 nm). The microscope

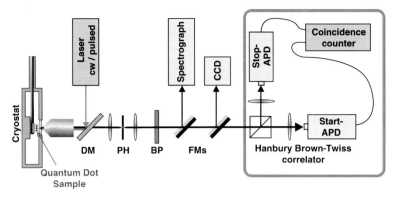

Fig. 11.4. Micro-PL setup (FM: mirrors on flip mounts, DM: dichroic mirror, PH: pinhole, BP: narrow bandpass filter, APD: avalanche photo diode)

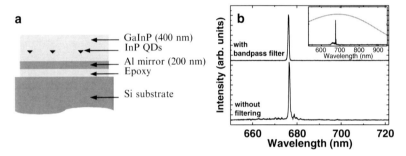

Fig. 11.5. (**a**) Structure of the InP/GaInP sample. (**b**) PL spectra from a single InP QD. An offset was added to separate the graphs. The *inset* shows a wider spectrum together with the APD detection efficiency

system (NA = 0.75) reaches a lateral resolution of 500 nm, and further spatial and spectral filtering selects a single transition of a single QD. The PL can be imaged on a CCD camera (see also Fig. 11.2a), on a spectrograph (Fig. 11.2b), or can be sent to a Hanbury Brown–Twiss setup with a time resolution of about 800 ps to prove single-photon emission.

11.2.4 Visible Single Photons from InP Quantum Dots

The sample used in our experiments was grown by metal-organic vapor phase epitaxy. An aluminum mirror was evaporated on the sample which was then thinned down to 400 nm and glued on a Si-substrate (see Fig. 11.5a). Figure 11.5b depicts the spectrum of a single InP QD on the sample under weak excitation showing only one dominant spectral line. Its linear dependance on the excitation power identifies an exciton emission. A 1-nm bandpass filter was used to further reduce the background. Figure 11.6 shows the corresponding cw correlation measurement at a count rate of 1.1×10^5 and

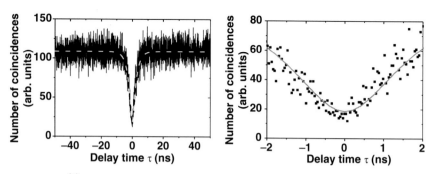

Fig. 11.6. $g^{(2)}$ function at continuous excitation. The fit function corresponds to an ideal single-photon source with limited time resolution. The *right image* shows a zoom into the central dip region

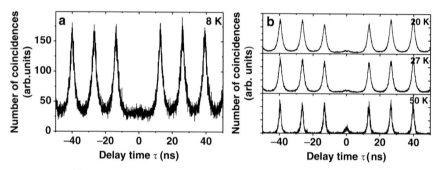

Fig. 11.7. $g^{(2)}$ function of a single InP QD at pulsed excitation: **(a)** at 8 K and **(b)** between 20 and 50 K

10 K. The function is modeled as a convolution of the expected shape of the ideal correlation function $g^{(2)}(\tau) = 1 - \exp(-\gamma\tau)$ and a Gaussian distribution according to a 800 ps time resolution of the detectors. The width of the antibunching dip (2.3 ns) depends on both the transition lifetime and the excitation timescale. Equivalent measurements at pulsed excitation show a vanishing peak at zero delay (Fig. 11.7a) which again proves single-photon generation. At rising temperature, phonon interactions lead to an increased incoherent background with other spectral lines overlapping the filter transmission window (Fig. 11.7b). However, characteristic antibunching was observed up to 50 K [38].

11.2.5 A Multicolor Photon Source

A great advantage of QDs as photon emitters is their potential to create more complex states of light consisting of few photons. For example, by trapping two or more electron–hole pairs (compare Fig. 11.1), photon cascades can be produced. These cascades can be used to enhance the transmission rate of cryptographic systems usually limited by spontaneous lifetime. Previous experiments have studied cascaded decays in InAs [39–41] and II–VI QDs [42,43]. Recently, also entangled photon pair generation has been demonstrated [44–46] following earlier proposals [47].

In order to study correlations between photons emitted in a cascade, the *cross-correlation function*

$$g^{(2)}_{\alpha\beta}(\tau) = \frac{\langle : \hat{I}_\alpha(t)\hat{I}_\beta(t+\tau) : \rangle}{\langle \hat{I}_\alpha(t)\rangle\langle \hat{I}_\beta(t)\rangle}$$

has to be measured with a Hanbury Brown–Twiss configuration modified by an interference filter in each arm. The filters can be tuned, e.g., to the exciton and biexciton transition lines in a two-photon cascade indicated by the index α and β in $g_{\alpha\beta}$, respectively. Cross-correlation measurements confirm assignments of spectral lines to exciton, biexciton, and triexciton transitions by the strong asymmetric shape of the correlation function (Fig. 11.8). If a biexciton decay starts the measurement and an exciton stops it, photon bunching occurs since an exciton photon is predominantly emitted shortly after a biexciton photon in a cascade. The opposite holds for switched start and stop channels. In this case, the detection of the exciton photon projects the QD to its empty (ground) state. The required reexcitation process results in an antibunching dip. Biexciton–triexciton correlations can be explained equivalently, but on different timescales. Since cross-correlation functions of two independent transitions show no (anti)bunching, the results in Fig. 11.8 confirm a

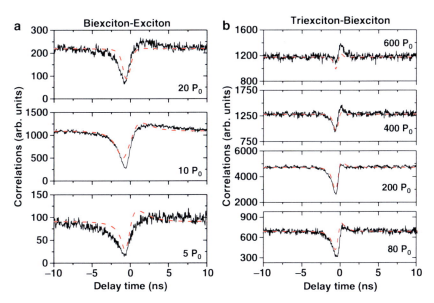

Fig. 11.8. Cross-correlation functions between the (**a**) exciton and biexciton line and (**b**) biexciton and triexciton line. *Dashed lines* are fits to a rate equation model [48,49]

three-photon cascade from the triexciton to the ground state of a single QD. The dashed lines in Fig. 11.8 are a fit to the rate equation model described in [48, 49].

11.3 Quantum Cryptography with Single Photon Sources

The most advanced quantum technology is quantum cryptography [2]. It is motivated by the possibility to transmit a secret key between a sender, Alice, and a receiver, Bob, without the possibility of eavesdropping. More precisely, any attack with the aim of extracting information will be detected by Alice and Bob. In this case their key will be discarded. The security is based on the laws of quantum physics. Whereas implementations of, e.g., the BB84 protocol [1] require true single photon sources, it has been shown that novel protocols, e.g., the protocol by Wang [50] using so-called *Decoy states*, provide unconditional security even with attenuated light sources. Still, these novel protocols require average photon numbers below one [51], so that optimized true on-demand single photon sources provide a much better efficiency.

11.3.1 Free-Space Quantum Cryptography

The first demonstration of quantum cryptography [52] was realized as a free-space transfer of qubits. Qubits where encoded in the polarization state of single photons using electrooptic modulators. Single photon pulses were mimicked by attenuated light pulses. Later, quantum cryptography has been realized with true single-photon states, e.g., from diamond defect centers [53] and single QDs [54] or with entangled photon pairs from parametric down conversion.

General requirements of a single-photon source for quantum communication are a high repetition rate and a high efficiency. Besides, passive elements like mirrors and solid immersion lenses, resonant techniques exploiting the Purcell effect can greatly enhance the emission rate [55, 56] if the source relies on the spontaneous decay of an excited state. A well-established technique from classical communication to enhance the information transfer rate is *multiplexing*. Each signal is marked with a physical label, like its wavelength, and identified at the receiver using filters tuned to the carrier frequencies. Here, we utilize multiplexing on a fundamental limit of low intensity, i.e., with single photons. For single photons, their wavelength can be used as their distinguishing label and polarization to encode quantum information [57]. If the two photons from a biexciton–exciton cascade in a single QD pass a Michelson interferometer, constructive and destructive interference will be observed at the two output ports by proper choice of the relative path difference (Fig. 11.9). These photons are fed into an optical fiber each, and one of them is delayed by half the repetition rate of the excitation laser. Recombination at a beam splitter then leads to a photon stream with a doubled count

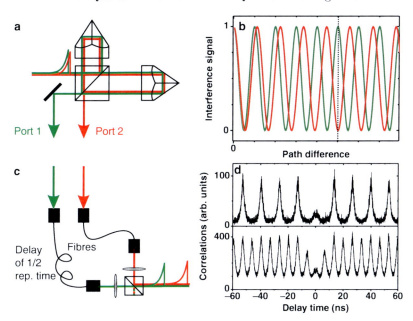

Fig. 11.9. (a) Michelson interferometer with two output ports as a single-photon add/drop filter. (b) Interference pattern for two distinct wavelengths vs. the relative arm length. (c) Merging the separated photons with one path delayed by half the excitation repetition rate. (d) Intensity correlation of the exciton spectral line and the multiplexed signal

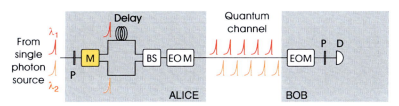

Fig. 11.10. Possible implementation of the multiplexer using the BB84 protocol. For a detailed description of the setup see text

rate. This is also reflected by autocorrelation measurements with a spacing of only 6.6 ns in the peak structure (see Fig. 11.9d) [57]. In our experiment, we implemented the BB84 protocol with such a multiplexed single photon source. Figure 11.10 shows an implementation of the interferometric multiplexing in the BB84 protocol using a cascaded photon source.

Behind an exciton–biexciton add/drop filter, Alice randomly prepares the photons' polarization. Bob's detection consists of a second EOM, an analyzing polarizer, and an APD. A visualization of a successful secret transmission of data is depicted in Fig. 11.11. After exchange of the key, Alice encrypts her data by applying an XOR operation between every image bit and the sequence

Fig. 11.11. Visualization of the quantum key distribution. Alice's original data (**a**), a photography taken out of our lab window in Berlin, is encrypted and sent to Bob (**b**). After decryption with his key, Bob obtains image (**c**)

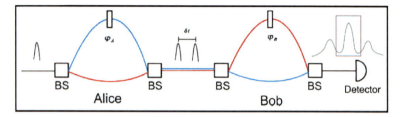

Fig. 11.12. Principle of time-bin encoding: A single photon pulse enters an unbalanced interferometer and is split into two wavepackets (separated in time by δt). After the second interferometer the configuration long-short path and short-long path (middle of the three peaks) interfere

of random bits (the secret key) which she has previously shared with Bob by performing the BB84 protocol. The encoded image (shown in Fig. 11.11b) is transmitted to Bob over a distance of 1 m with 30 bits s^{-1}. Another XOR operation by Bob reveals the original image with an error rate of 5.5%.

11.3.2 Fiber-Based Quantum Cryptography

Practical ground-based quantum cryptography requires transmission of photons via optical fibers. Then, the wavelength has to lie within the transparency window of standard fibers, i.e., at 1.3 or 1.55 μm. Due to the degradation of polarization in fibers, polarization encoding is impossible. *Time-bin-encoding* is used instead [2]. Figure 11.12 explains the principle.

A single photon pulse enters an unbalanced interferometer (here a Mach–Zehnder) where it can travel along a short path or a long path with a probability determined by the beamsplitter. The difference in armlength ΔL has to be larger than the coherence length l_c of the photons. After a second interferometer the configurations "short-long" path and "long-short" path are indistinguishable, and lead to constructive or destructive interference behind the last beam splitter. An additional ϕ is introduced to encode the four possible qubit states for the BB84 protocol. Time-bin encoding is robust against fluctuations of polarization and has been used in many demonstrations of fiber-based quantum cryptography [58, 59].

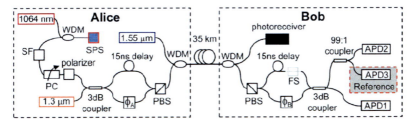

Fig. 11.13. Experimental setup for fiber-based quantum cryptography using time-bin encoding and a single photon source. SPS: single photon source at 1.3 µm; WDM: wavelength division multiplexer; SF: spectral filter; PC: polarization controller; PBS: polarizing beam splitter/combiner; and FS: fiber stretcher. From [63]

Recently, on-demand single photon generation at the telecommunication wavelengths 1.3 and 1.55 µm has been demonstrated using single QDs [60–62]. As an example Fig. 11.13 shows a typical experimental setup from [63], where time-bin encoded single photons at telecom wavelength were used to obtain secret key distribution over 35 km. These examples demonstrate that QD-based single photon sources, in particular with electrical excitation [30–32] have the potential as reliable nonclassical light emitters for applications in optical quantum information processing.

11.4 Toward Optical Quantum Computation with Single Photon Sources

Linear Optical Quantum Computation (LOQC) [64] is an attractive approach as it relies only on the use of linear optical elements and single photon detection. Knill et al. [3] propose the realization of a universal set of quantum gates for quantum computation, using linear optical elements and single-photon detectors. This scheme and related ones require massive resources of *on-demand* indistinguishable photons. Also entangled photon pairs are required to implement teleportation techniques for higher efficiency an error correction of LOQC gates. This motivates the development of *plug-and-play* single photon sources based on single QDs.

11.4.1 Generation and Application of Entanglement

Two breakthrough results were the demonstration of indistinguishable photons from a single photon source in 2002 by the Stanford group [65]. Meanwhile, the characteristic two-photon (Hong–Ou–Mandel) interference was successfully demonstrated by other groups as well [46, 66, 67]. Another important achievement was the generation of entangled photon pairs from a cascaded decay in single QDs [44, 45].

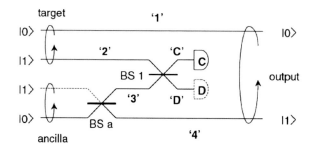

Fig. 11.14. Schematic of single mode teleportation. Target and ancilla qubits are each defined by a single photon occupying two optical modes (1–2 and 3–4). When detector C records a single photon, the state in modes 1–4 reproduces the initial state of the target. In particular, the coherence between modes 1 and 2 of the target can be transferred to a coherence between modes 1 and 4. From [68]

A particularly fascinating application of entanglement is the possibility to transfer an unknown quantum state from one place to another, the so-called teleportation, which plays a crucial role within quantum computer architectures. Usually, an entangled photon pair and a single photon carrying the state to be teleported are required in a teleportation scheme. However, a single mode teleportation can be mimicked using only two indistinguishable photons and postselection. This is apparent when a coherent superposition of two logical states $|0\rangle_L$ and $|1\rangle_L$ in a photon is expressed in the dual rail representation with two optical modes 3 and 4:

$$|\psi\rangle = \frac{1}{\sqrt{2}}(|0\rangle_L + |1\rangle_L) = \frac{1}{\sqrt{2}}(|1\rangle_3|0\rangle_4 + |0\rangle_3|1\rangle_4). \tag{11.1}$$

This state mimics an entangled state. A teleportation protocol can then be implemented as shown schematically in Fig. 11.14.

A fidelity of 80% was achieved with indistinguishable photons from a QD single photon source [68].

11.4.2 Implementation of the Deutsch–Jozsa Algorithm

So far, two-photon quantum gates (c-NOT gates) and basic quantum algorithms have been realized using photon pairs created by parametric down conversion [69, 70]. Here, we report the on demand operation of a two-qubit Deutsch–Jozsa algorithm [71] using a deterministic single-photon source. Previous all-optical experiments were restricted to emulate the Deutsch–Jozsa algorithm with attenuated classical laser pulses [72].

An intuitive and often cited interpretation compares this algorithm with a "coin tossing" game: While a classical observer has to check both sides of a coin in order to tell if it is fair, i.e., if its two sides are different (head and tail), or unfair with two equal sides (head–head or tail–tail), quantum computation

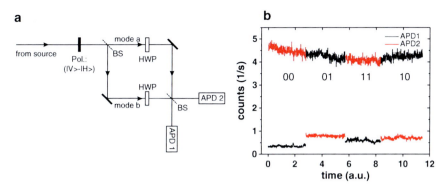

Fig. 11.15. (a) Setup of the Deutsch–Jozsa experiment. Pol: Polarizer, BS: beam splitter, HWP: half-wave plate. (b) Detector count rates of APD 1 and 2 for the four different combinations of the HWPs inside the interferometer: 00 no HWP, 10: HWP in mode a, 01: HWP in b, 11: HWP in a and b

can perform this task with a single shot. In the simplest case, two qubits are needed to store query and answer, respectively.

Here, we apply the proposal of [73] in order to implement two qubits using different degrees of freedom of a single photon (see also [74]). The first qubit corresponds to the which-way information about the photon path inside an Mach–Zehnder interferometer, namely the spatial modes a and b (Fig. 11.15a). The second qubit is implemented via the photon's polarization state. With these definitions of our two qubits, we are able to construct the proper states for a Deutsch–Jozsa algorithm. The initial state is realized by sending a photon into a linear superposition of the two spatial modes using a nonpolarizing 50:50 beam splitter after preparing it in a superposition of horizontal and vertical polarization.

In a next step, a unitary transformation is realized by selectively adding a half-wave plate (HWP) to each mode. A HWP in only one of the two paths corresponds to a balanced situation representing fair coins in the example above. In contrast, no HWP or a HWP in either path correspond to the constant situations, illustrated by the two possible false coins. Physically, the difference between these two situations is an additional local phase difference between the two paths, 0 for the constant and π for the balanced case. Recombining the two interferometer arms on a second 50:50 beam splitter implements a final Hadamard gate on the first qubit. The two interferometer outputs are monitored with one APD each. Constructive interference at the top (bottom) output indicates the balanced (constant) situation, which can thus be distinguished by only one detection event of one single photon. Figure 11.15b shows the outcome of many detection events (albeit only one would be enough) for the four possible combinations. In these measurements an overall success probability of 79% was achieved. This was mainly limited by a slight temporal and spatial mode mismatch in the interferometer.

11.4.3 Concepts of Noise-Resistant Encoding

A fundamental problem of quantum computation is dephasing, i.e., the tendency of a qubit to loose or change its quantum information due to coupling to the environment. A way to solve this might be active quantum error correction [75] or the use of decoherence-free subspaces of the Hilbert space spanned by several qubits [76]. In our experiment, we use the second approach in a modified setup insensitive to phase noise by extending the setup to the scheme in Fig. 11.16a. Horizontal and vertical polarization are separated and merged at polarizing beam splitters (PBS) so that modes, that interfere at the final beam splitter, travel along the same path. Thus, phase noise in the central interferometer does not cause perturbations of the algorithm, but only lead to a global phase change [74].

We simulate phase noise by modulating piezo-electric mirror mounts. The left graph in Fig. 11.16b displays the detection signals for two combinations (one constant and one balanced) of the HWPs in the first interferometer, without any *artificial* noise. Modulating piezo 2 in the central interferometer does also not affect the count rates as expected (middle graph). For comparison, phase noise induced by piezo 3 in the final interferometer changes the APD signals substantially (right graph). The same happens at modulation of piezo 1 (not shown). Thus, using this extended experimental setup, we showed the robustness of the algorithm against phase noise when information is properly encoded in an unaffected superposition of polarization states.

These experiments demonstrate that single photon sources can already be used as stable generators of nonclassical light states to test small-scale quantum computational tasks.

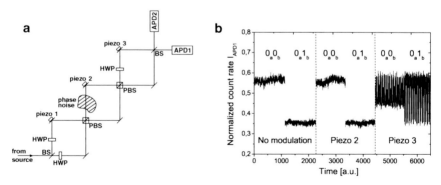

Fig. 11.16. (a) Extended setup. (P)BS: (polarizing) beam splitter. (b) Detector count rates for a constant (00) and a balanced (01) HWP combination. *Left graph:* No modulation of the piezos, *middle graph:* modulation on piezo 2, *right graph:* modulation on piezo 3

11.5 Summary

In this article, we have reviewed applications of QD single photon sources to quantum information processing. We described a proof-of-principle experiment that shows how multiphoton generation from single QDs may find applications in free-space quantum cryptography devices to enhance their maximum bandwidth. In this experiment, multiplexing on the single-photon level and its implementation in the BB84 quantum key distribution protocol has been accomplished. It demonstrates how classical and quantum mechanical aspects merge in optical information processing when the ultimate limit of intensity is reached. Concerning fiber-based quantum cryptography we briefly described the potential of single photon sources emitting at telecommunication wavelengths.

Another important application for single photon sources is quantum computation. First proofs of concepts which involve single photon emitters based on single QDs have been demonstrated. Quantum teleportation and an experimental realization of the Deutsch–Jozsa algorithm with linear optics was described. In the latter experiment, we proved the robustness of the setup against phase noise when encoding on adequate qubit bases.

The experiments show that today's single-photon sources based on QDs have reached the level of sophistication to be ready-to-use nonclassical light sources. The *on demand* character of the emission is extremely useful in all-solid state implementations of quantum information devices. Single QDs will therefore play an important role in the field of quantum cryptography and quantum computation, but also for realizing interfaces in larger quantum networks. Recent progress in electrical excitation of QD-based single photon sources are a promising step toward true quantum electrooptical devices for the generation of nonclassical light [30–32]. Also the generation of single photons at telecom wavelengths [32,60–63] paves the way toward up-scaling of single-photon quantum information and longer distance quantum cryptography by utilizing commercial fiber components. Key problems that still have to be solved are the generation of indistinguishable photons (ancilla states) from different QDs and the high-efficient detection of single photons at wavelengths $>1\,\mu m$ [77,78].

Acknowledgments

The authors want to thank V. Zwiller, G. Reinaudi, J. Persson, and W. Seifert for valuable assistance. This work was supported by Deutsche Forschungsgemeinschaft (SFB 296) and the European Union (EFRE). M. Scholz acknowledges financial support from Ev. Studienwerk Villigst and Deutsche Telekom Stiftung and T. Aichele by DAAD.

References

1. C.M. Bennett, G. Brassard, in *Proc. IEEE Conference on Computers, Systems, and Signal Processing*, Bangalore, India (IEEE, New York, 1984), p. 175
2. N. Gisin, G. Ribordy, W. Tittel, H. Zbinden, Rev. Mod. Phys. **74**, 145 (2002)
3. E. Knill, R. Laflamme, G.J. Milburn, Nature **409**, 46 (2001)
4. D. Gottesman, I.L. Chuang, Nature **402**, 390 (1999)
5. S. Gasparoni, J.-W. Pan, P. Walther, T. Rudolph, A. Zeilinger, Phys. Rev. Lett. **93**, 020504 (2004)
6. N. Kiesel, C. Schmid, U. Weber, R. Ursin, H. Weinfurter, Phys. Rev. Lett. **95**, 210505 (2005)
7. L. Tian, P. Rabl, R. Blatt, P. Zoller, Phys. Rev. Lett. **92**, 247902 (2004)
8. J.I. Cirac, P. Zoller, Phys. Rev. Lett. **74**, 4091 (1995)
9. J.I. Cirac, P. Zoller, Phys. Today **57**, 38 (2004)
10. F. Schmidt-Kaler, H. Häffner, M. Riebe, S. Gulde, G.P.T. Lancaster, T. Deuschle, C. Becher, C.F. Roos, J. Eschner, R. Blatt, Nature **422**, 408 (2003)
11. D. Leibfried, B. Demarco, V. Meyer, D. Lucas, M. Barrett, J. Britton, W.M. Itano, B. Jelenkovic, C. Langer, T. Rosenband et al., Science **422**, 412 (2003)
12. D. Kielpinski, C. Monroe, D.J. Wineland, Nature **417**, 709 (2002)
13. J.I. Cirac, P. Zoller, Nature **404**, 579 (2000)
14. D. Loss, D.P. DiVicenzo, Phys. Rev. A **57**, 120 (1998)
15. W.G. van der Wiel, S.D. Franceschi, J.M. Elzerman, T. Fujisawa, S. Tarucha, L.P. Kouwenhoven, Rev. Mod. Phys. **75**, 1 (2003)
16. J.E. Mooij, T.P. Orlando, L. Levitov, L. Tian, C.H. van der Wal, S. Lloyd, Science **285**, 1036 (1999)
17. Y.A. Pashkin, T. Yamamoto, O. Astafiev, Y. Nakamura, D.V. Averin, J.S. Tsai, Nature **421**, 823 (2003)
18. H.-J. Briegel, W. Dür, J.I. Cirac, P. Zoller, Phys. Rev. Lett **81**, 5932 (1998)
19. D. Bouwmeester, J.-W. Pan, K. Mattle, M. Eibl, H. Weinfurter, A. Zeilinger, Nature **390**, 575 (1997)
20. *New Journal of Physics* (special issue on single-photon sources) **6** (2004)
21. A. Kuhn, M. Hennrich, G. Rempe, Phys. Rev. Lett. **89**, 067901 (2002)
22. M. Keller, B. Lange, K. Hayasaka, W. Lange, H. Walther, Nature **431**, 1075 (2004)
23. C. Brunel, B. Lounis, P. Tamarat, M. Orrit, Phys. Rev. Lett **83**, 2722 (1999)
24. B. Lounis, W.E. Moerner, Nature **407**, 491 (2000)
25. P. Michler, A. Imamağlu, M.D. Mason, P.J. Carson, G.F. Strouse, S.K. Buratto, Nature **406**, 968 (2000)
26. M. Nirmal, B.O. Dabbousi, M.G. Bawendi, J.J. Macklin, J.K. Trautman, T.D. Harris, L.E. Brus, Nature **383**, 802 (1996)
27. C. Kurtsiefer, S. Mayer, P. Zarda, H. Weinfurter, Phys. Rev. Lett. **85**, 290 (2000)
28. A. Beveratos, S. Kühn, R. Brouri, T. Gacoin, J.-P. Poizat, P. Grangier, Eur. Phys. J. D **18**, 191 (2002)
29. Z.Yuan, B.E. Kardynal, R.M. Stevenson, A.J. Shields, C.J. Lobo, K. Copper, N.S. Beattie, D.A. Ritchie, M. Pepper, Science **295**, 102 (2002)
30. M. Scholz, S. Büttner, O. Benson, A.I. Toropov, A.K. Bakarov, A.K. Kalagin, A. Lochmann, E. Stock, O. Schulz, F. Hopfer, V.A. Haisler, D. Bimberg, Opt. Expr. **15**, 9107 (2007)
31. D.J.P. Ellis, A.J. Bennett, A.J. Shields, P. Atkinson, D.A. Ritchie, Appl. Phys. Lett. **90**, 233514 (2007)

32. C. Monat, B. Alloing, C. Zinoni, H.H. Li, A. Fiore, Nano Lett. **6**, 1464 (2006)
33. A. Badolato, K. Hennessy, M. Atatre, J. Dreiser, E. Hu, P.M. Petroff, A. Imamoğlu, Science **308**, 1158 (2005)
34. D. Bimberg, M. Grundmann, N. Ledentsov, *Quantum Dot Heterostructures* (Wiley, Chichester, UK, 1988)
35. L. Mandel, E. Wolf, *Optical Coherence and Quantum Optics* (Cambridge University Press, 1995)
36. *Single-Photon Counting Module – SPCM-AQR Series Specifications*, Laser Components GmbH, Germany (2004)
37. R. Hanbury Brown, R.Q. Twiss, Nature **178**, 1046 (1956)
38. V. Zwiller, T. Aichele, W. Seifert, J. Persson, O. Benson, Appl. Phys. Lett. **82**, 1509 (2003)
39. E. Moreau, I. Robert, L. Manin, V. Thierry-Mieg, J.M. Gérard, I. Abram, Phys. Rev. Lett. **87**, 183601 (2001)
40. D.V. Regelman, U. Mizrahi, D. Gershoni, E. Ehrenfreund, W.V. Schoenfeld, P.M. Petroff, Phys. Rev. Lett. **87**, 257401 (2001)
41. C. Santori, D. Fattal, M. Pelton, G.S. Salomon, Y. Yamamoto, Phys. Rev. B **66**, 045308 (2002)
42. S.M. Ulrich, S. Strauf, P. Michler, G. Bacher, A. Forchel, Appl. Phys. Lett. **83**, 1848 (2003)
43. C. Couteau, S. Moehl, F. Tinjod, J.M. Gérard, K. Kheng, H. Mariette, J.A. Gaj, R. Romestain, J.P. Poizat, Appl. Phys. Lett. **85**, 6251 (2004)
44. R.M. Stevenson, R.J. Young, P. Atkinson, K. Cooper, D.A. Ritchie, A.J. Shields, Nature **439**, 179 (2006)
45. N. Akopian, N.H. Lindner, E. Poem, Y. Berlatzky, J. Avron, D. Gershoni, B.D. Gerardot, P.M. Petroff, Phys. Rev. Lett **96**, 130501 (2006)
46. R. Hafenbrak, S.M. Ulrich, P. Michler, L. Wang, A. Rastelli, O.G. Schmidt, New J. Phys. **9**, 315 (2007)
47. O. Benson, C. Santori, M. Pelton, Y. Yamamoto, Phys. Rev. Lett. **84**, 2513 (2000)
48. J. Persson, T. Aichele, V. Zwiller, L. Samuelson, O. Benson, Phys. Rev. B **69**, 233314 (2004)
49. T. Aichele, V. Zwiller, M. Scholz, G. Renaudi, J. Persson, O. Benson, Proc. SPIE **5722**, 30 (2005)
50. X.-B. Wang, Phys. Rev. Lett. **94**, 230503 (2005)
51. D. Rosenberg, J.W. Harrington, P.R. Rice, P.A. Hiskett, C.G. Peterson, R.J. Hughes, A.E. Lita, S.W. Nam, J.E. Nordholt, Phys. Rev. Lett. **98**, 010503 (2007)
52. C.H. Bennett, F. Bessett, G. Brassard, L. Salvail, J. Smolin, J. Cryptol. **5**, 3–28 (1992)
53. A. Beveratos, R. Brouri, T. Gacoin, A. Villing, J.-P. Poizat, P. Grangier, Phys. Rev. Lett. **89**, 187901 (2002)
54. E. Waks, K. Inoue, C. Santori, D. Fattal, J. Vučković, G.S. Solomon, Y. Yamamoto, Nature **420**, 762 (2002)
55. G. Solomon, M. Pelton, Y. Yamamoto, Phys. Rev. Lett. **86**, 3903 (2001)
56. J.M. Gérard, B. Sermage, B. Gayral, B. Legrand, E. Costard, V. Thierry-Mieg, Phys. Rev. Lett. **81**, 1110 (1998)
57. T. Aichele, G. Reinaudi, O. Benson, Phys. Rev. B **70**, 235329 (2004)
58. P.A. Hiskett, D. Rosenberg, C.G. Peterson, R.J. Hughes, S. Nam, A.E. Lita, A.J. Miller, J.E. Nordholt, New J. Phys. **8**, 193 (2006)

59. D. Stucki, N. Brunner, N. Gisin, V. Scarani, H. Zbinden, Appl. Phys. Lett. **87**, 194108 (2005)
60. C. Zinoni, B. Alloing, C. Monat, V. Zwiller, L.H. Li, A. Fiore, L. Lunghi, A. Gerardino, H. de Riedmatten, H. Zbinden, N. Gisin, Appl. Phys. Lett. **88**, 131102 (2006)
61. M.B. Ward, O.Z. Karimov, D.C. Unitt, Z.L. Yuan, P. See, D.G. Gevaux, A.J. Shields, P. Atkinson, D.A. Ritchie, Appl. Phys. Lett. **86**, 201111 (2005)
62. T. Miyazawa, K. Takemoto, Y. Sakuma, S. Hirose, T. Usuki, N. Yokoyama, M. Takatsu, Y. Arakawa, JJAPL **44**, 620 (2005)
63. P.M. Intallura, M.B. Ward, O.Z. Karimov, Z.L. Yuan, P. See, A.J. Shields, P. Atkinson, D.A. Ritchie, Appl. Phys. Lett. **91**, 161103 (2007)
64. P. Kok, W.J. Munro, Kae Nemoto, T.C. Ralph, Jonathan P. Dowling, G.J. Milburn, Rev. Mod. Phys. **79**, 135 (2007)
65. C. Santori, D. Fattal, J. Vučković, G.S. Solomon, Y. Yamamoto, Nature **419**, 594 (2002)
66. S. Varoutsis, S. Laurent, P. Kramper, A. Lemaitre, I. Sagnes, I. Robert-Philip, I. Abram, Phys. Rev. B **72**, 041303 (2005)
67. A.J. Bennett, D.C. Unitt, A.J. Shields, P. Atkinson, D.A. Ritchie, Opt. Expr. **13**, 7772 (2005)
68. D. Fattal, E. Diamanti, K. Inoue, Y. Yamamoto, Phys. Rev. Lett. **92**, 037904 (2004)
69. J.L. O'Brien, G.J. Pryde, A.G. White, T.C. Ralph, D. Branning, Nature **426**, 264 (2003)
70. P. Walther, K.J. Resch, T. Rudolph, E. Schenck, H. Weinfurter, V. Vedral, M. Aspelmeyer, A. Zeilinger, Nature **434**, 264 (2005)
71. D. Deutsch, R. Jozsa, Proc. Roy. Soc. Lond. A **439**, 553 (1992)
72. S. Takeuchi, Phys. Rev. A **62**, 032301 (2000)
73. J. Cerf, C. Adami, P.G. Kwiat, Phys. Rev. A **57**, 1477 (1998)
74. M. Scholz, T. Aichele, S. Ramelow, O. Benson, Phys. Rev. Lett. **96**, 180501 (2006)
75. J. Chiaverini, D. Leibfried, T. Schaetz, M.D. Barrett, R.B. Blakestad, J. Britton, W. Itano, J.D. Jost, E. Knill, C. Langer et al., Nature **432**, 602 (2004)
76. D.A. Lidar, I.L. Chuang, K.B. Whaley, Phys. Rev. Lett. **81**, 2594 (1998)
77. Z.L. Yuan, B.E. Kardynal, A.W. Sharpe, A.J. Shields, Appl. Phys. Lett. **91**, 041114 (2007)
78. R. Leoni, A. Gaggero, F. Mattioli, M.G. Castellano, P. Carelli, F. Marsili, D. Bitauld, M. Benkahoul, F. Levy, A. Fiore, J. Low Temp. Phys. **151**, 580 (2008)
79. A. Soujaeff, T. Nishioka, T. Hasegawa, S. Takeuchi, T. Tsurumaru, K. Sasaki, M. Matsui, Opt. Expr. **15**, 726 (2007)

Index

active layer, 274, 279
antibunching, 3, 24, 90, 370
anticrossing, 109, 283
artificial atoms, 15, see quantum dots
atomic four-level system, 5

bad cavity regime, 22, 24
BB84 protocol, 375
Bell's inequality, 228
    parameters, 240
    violation, 250
β-factor, see spontaneous emission coupling factor
biexciton, 71, 373
biexciton cascade, 230, 234, 374
birth/death model, 7
Bohr radius, 32
Born–Markov approximation, 6
bunching, 3, 90

capillary model, 35
capping of SK QDs, 41
carrier–photon correlations, 21
carriers
    correlated, 13
    uncorrelated, 13, 18
cascade, 76
cavity QED with quantum dots, 307
cavity quantum electrodynamics, 268
CdSe, 71
CdTe, 71
chalcogenide glasses for resonator tuning, 305
charged exciton, 71

Cleaved edge overgrowth, 52
cluster expansion method, 16
coarsening of SK islands, 38
coherence
    first order, 24
coherence time, 24, 26
coherent, 93
coherent control, 71
coherent light, 3, 23
coherent optical probing, 318
coherent photonic coupling, 290, 291
communication, 72
concurrence, 248
confinement, 87
correlation, 89
    cross (polarised), 232
    measurement, 242
    polarised, 243
    second-order, 235
correlation function, 8
    first order, 2, 24, 25
        decay of, 27
    second order, 2, 10, 21
Coulomb interaction, 18, 71
coupled oscillator approach, 290
critical nucleus, 36
cross-correlation, 373
Curie law, 153

dark exciton, 75
decoherence
    Electron spin, 154
    Nuclear spin, 152

decoy states, 374
density matrix, 238, 246
density of SK QDs, 40
dephasing, 18, 85
dephasing time, 193
Deutsch–Jozsa algorithm, 378
device, 72
Distributed Bragg Reflectors, 234, 273
doped quantum dots, 15
dressed states, 270
droplet epitaxy, 49

eigenstates, 75
electric field, 72
electroluminescence, 71
electron beam lithography, 275
electron–electron exchange, 347
electron–hole exchange, 345
electronic properties, 71
enhancement and inhibition of spontaneous emission, 273, 280
entangled photon source, 227, 230
entanglement, 227, 243
    enhancement, 249
    evolution, 256
    limitations, 251
entanglement swapping, 229
epitaxy, 71
EPR paradox, 227
exchange interaction, 71
exciton, 5, 71
    bright exciton, 75
    charged, 235
    dark states, 231
    exchange interaction, 232
    intermediate bright states, 231, 232
    multiexcitons, 109
exciton magnetic polaron, 100
exciton–photon coupling strength, 270, 284, 293

Faraday, 75
fidelity, 240, 248
    Model, 253
fine structure, 194
fine-structure splitting, 232
    energy dependence, 237
    origin, 236
    phase evolution, 232, 251
    tuning, 237
first-order coherence, 189
frequency focusing of spin coherence, 136
FWM, 85

g-factor, 101
    quantum dot molecule, 350
    resonance, 351, 354
GaN, 78
geometric distribution, 187
giant optical nonlinearity, 321
Glauber, 187
Growth modes, 34

Hadamard gate, 379
Hamiltonian, 107
Hanbury Brown–Twiss setup, 280
Hanbury–Brown and Twiss (HBT), 17
Hanbury-Brown and Twiss setup, 189
Hartree–Fock factorization, 13, 16, 18
heavy hole, 74
hierarchy problem, 9, 16
high polarity, 71
hyperfine interaction, 96, 153
hyperfine interaction in quantum dots, 136

II–VI, 71
III-N, 71
in situ etching, 45
in situ laser processing, 61
InAs/GaAs(001) QDs - Facets, 39
InAs/GaAs(001) QDs - Morphology, 38, 39
indirect nuclear interactions
    measurement, 180
    theory, 156
indistinguishable photons, 191
information, 86
InGaN, 71
injection, 90
InP Quantum Dots, 371
input/output curve, 23
interband transition amplitude, 12, 18
    two-time, 25
intermixing in SK islands, 39
island nucleation, 36

Jaynes–Cummings Hamiltonian, 270
Jaynes–Cummings ladder, 270
Jaynes–Cummings model, 6

kinetic exchange, 347
Knight field, 97
    Measurement, 159
    Operator, 153

laser threshold, 9
lateral QD molecules, 48
lattice mismatch, 35
lifetime, 78
light-matter interaction, 16
lineshape, 103
linewidth, 85
Liouville equation, 6
lithographic QDs, 53
Lorentzian, 86
luminescence dynamics, 15

Mach–Zehnder interferometer, 379
magnetic field, 72
magnetic fluctuations, 102
magnetic ions, 103
magnetic moment, 101
magnetic properties, 71
magnetization, 71, 102
master equation, 7
material absorption coefficient, 278
Michelson interferometer, 189
micro photoluminescence spectroscopy, 279
microcavities, 2, 4, 19, 91
micropillars, 19, 26
migration enhanced growth, 275
mode locking of spin coherence, 127
Molecular Beam Epitaxy, 234, 273, 332
morphological transitions of SK islands, 36, 39
motional narrowing, 179
multiplexing, 374

nanoaperture, 72
nanomagnet, 93
nanomagnetism, 72
natural QDs, 44
nonresonant, 196
nuclear dipole interaction, 151
nuclear spin, 98

nuclear spin system
    bistability, 161
    decoherence, 152
    hysteresis, 166
    optical cooling, 158
    rate equation, 165
    relaxation due to electrons, 164
nuclear spin temperature, 152

one-atom laser, 6
operator average
    two-time, 24
optical cavity, 234
optical interconnects, 299
optical microcavities, *see* microcavities, micropillars
optical properties, 71
optical spectroscopy, 71
optical transition, 103
optical transmission measurement of quantum dot, 318
oscillator strength, 270, 271, 274, 284
Overhauser field, 97, 154
Overhauser shift
    buildup and decay times, 171
    buildup in high field, 177
    decay in high field, 178
    electron mediated decay, 173
    magnetic field dependance, 163
    zero field, 158
overlap integral, 193

parametric down conversion, 229
partial capping and annealing, 42
Pauli, 79
Pauli blocking, 24
phase noise, 380
phase synchronization condition, 129
phonon coupling, 71
photoluminescence, 15, 19, 72
    experimental, 235
    single dot, 235
photon
    antibunching, 190, 289
    anticrossing, 288
    autocorrelation, 288
    bunching, 190
    coupling, 71
    statistics, 8, 17, 21

photon-assisted polarization, *see* interband transition amplitude
photonic crystal cavity, 301
photonic crystal cavity optical switch, 323
photonic crystal cavity–waveguide coupling, 314
photonic crystal fabrication, 301
photonic crystal tuning - strain dependence, 305
photonic crystal tuning by index change - local, 305
photonic crystal waveguide, 313
photonic crystals, 299
photonic network, 313
plasma etching, 275
Poisson-distribution, 187
polarization, 84
polarization state, 379
pulsed excitation, 29
Purcell, 91
Purcell effect, 2, 273, 282, 308
Purcell factor, 199, 272, 282

Q-factor, 271
QD stacks, 43
QDs by local interdiffusion, 54
QDs in inverted pyramids, 55
QDs in nanowires, 51
quality (Q)-factor, 18
quantum algorithm, 378
quantum communication, 228
quantum computer, 229
quantum cryptography, 375
quantum dot
  charge tuning, 149
  photoluminescence, 149
quantum dot alignment, 302
quantum dot emission rate control, 309
quantum dot inhomogeneous linewidth, 302
quantum dot laser dynamics, 323
quantum dot molecule, 332
  Coulomb shift, 341
  g factor, 350
  Hamiltonian, 337, 345
quantum dot molecules
  molecular orbitals, 337
quantum dot saturation, 321
quantum dot temperature dependence, 304
quantum dot temperature tuning - local, 302
quantum dot–micropillar system, 269
quantum dots
  angular momentum, 230
  annealing, 237
  binding energy, 235
  coefficient, 238
  emission, 230, 234
  emission polarisation, 230, 236
  entangled state, 231
  excess charge, 255
  ideal, 230, 232
  in magnetic fields, 237, 252
quantum efficiency, 200
quantum error correction, 380
quantum imaging, 229
quantum information, 71
quantum information processing, 299, 367
quantum interferometry, 260
quantum key distribution, 228
quantum nature, 287
quantum regression theorem, 24
quantum repeater, 229
quantum rings, 47
quantum teleportation, 228
quantum-dot induced transparency, 319
quantum-dot laser model, 11
quantum-dot model, 4
quasiresonant, 196

Rabi oscillations, 268, 271
rapid thermal processing of QDs, 59
rate equations, 9
reabsorption processes, 9
recombination, 82
resonant, 196
resonant pumping, 287, 288
room temperature, 71

saturation effects, 10, 24
second-order coherence, 189
selection rules, 75
selective removal of cap above buried QDs, 46
self-assembled nanoholes, 46

self-assembled quantum dots, 273
self-organization, 71
single emitter regime, 286
single photon source indistinguishability, 316
single photon transfer, 317
single-photon emission, 313
single-photon generation, 368
single-photon source, 3, 22, 72
singly charged quantum dots, 125
site-controlled InAs/GaAs(001) QDs, 58
sp–d exchange, 95
spatial matching, 282
spectral jitter, 255
spectral map, 333
spectral matching, 282
spectroscopy, 85
spin, 71
    beats, 122
    coherence in quantum dots, 125
    coherence time, 122
    dephasing time, 122
    flip, 98
    orbit interaction, 96
    synchronization, 129
spontaneous emission, 9, 268, 271
    enhancement of, 2
    rate of, 12
    source term of, 18
spontaneous emission coupling factor, 2, 7, 20
spontaneous emission rate – modified, 308, 310
spontaneous emission rate coupling efficiency, 311
statistical properties of light, 2, 10
stimulated emission, 9, 18, 71
Strain-induced QDs, 51
Stranski–Krastanov growth, 35, 274, 368, 369
strong coupling, 269, 270, 280, 284
strong coupling-coherent probe of quantum dot, 319
strong-coupling regime, 219, 308, 311, 312
sub Poissonian, 190
submonolayer deposition, 274
Super Poissonian, 190
superradiance, 71

tangle, 248
teleportation, 378
temperature tuning, 281, 282
temporal gating, 248
thermal light, 3, 22, 23
three-particle effects, 16
threshold condition, 271
thresholdless laser, 2
time-bin encoding, 376
time-resolved, 72
time-resolved emission, 232
time-resolved Faraday rotation, 124
time-resolved Kerr rotation, 124
time-resolved photoluminescence, 15
    non-exponential decay of, 15
triexciton, 373
trion, 78
truncation scheme, 16
tunneling, 90
    electron, 340
    hole, 340
two-particle effects, 16
two-photon cascade, 373
two-photon quantum gates, 378

unstrained QDs, 47
unstrained QDs on patterned GaAs(001), 56

vacuum field, 268, 272
vacuum Rabi splitting, 269, 271, 283, 284, 291
vertical QD molecules, 43
visible, 71
Voigt, 75
von-Neumann equation, *see* Liouville equation

weak coupling, 270, 271, 280, 281
weak-coupling regime, 308
wetting layer, 4, 35
which-way information, 379
Wiener–Kintchine theorem, 203
wurtzide, 77

Zeeman, 75
    quantum dot molecule, 350
Zeeman Hamiltonian
    electron, 150
    nuclear, 151